OCR AS/A level
Biology A

Second Edition

1

Sue Hocking
Frank Sochacki
Mark Winterbottom

WAYS LEARNING

PEARSON

Published by Pearson Education Limited, 80 Strand, London, WC2R 0RL.

www.pearsonschoolsandfecolleges.co.uk

Text © Pearson Education Limited 2015
Edited by Priscilla Goldby and Caroline Needham
Designed by Elizabeth Arnoux for Pearson Education Limited
Typeset by Tech-Set Ltd, Gateshead
Original illustrations © Pearson Education Limited 2015
Illustrated by Tech-Set Ltd, Gateshead and Peter Bull Art Studio
Cover design by Juice Creative
Picture research by Alison Prior
Cover photo © **Alamy Images:** Scott Camazine

The rights of Sue Hocking, Frank Sochacki and Mark Winterbottom to be identified as authors of
this work have been asserted by them in accordance with the Copyright, Designs and Patents Act
1988.

First edition published 2008
This edition published 2015

20
10

British Library Cataloguing in Publication Data
A catalogue record for this book is available from the British Library

ISBN 978 1 447 99079 6

Websites
Pearson Education Limited is not responsible for the content of any external internet sites. It is
essential for tutors to preview each website before using it in class so as to ensure that the URL is still
accurate, relevant and appropriate. We suggest that tutors bookmark useful websites and consider
enabling students to access them through the school/college intranet.

Printed in Italy by Lego S.p.A

This resource is endorsed by OCR for use with specification OCR Level 3 Advanced Subsidiary GCE
in Biology A (H020) and OCR Level 3 Advanced GCE in Biology A (H420).
In order to gain OCR endorsement this resource has undergone an independent quality check. OCR
has not paid for the production of this resource, nor does OCR receive any royalties from its sale. For
more information about the endorsement process please visit the OCR website www.ocr.org.uk

Acknowledgements

Pearson would like to thank Peter Kennedy for his contribution to the previous edition.

The publisher would like to thank the following for their kind permission to reproduce their photographs:

(Key: b-bottom; c-centre; l-left; r-right; t-top)

Alamy Images: Chris Pearsall 114, Finn Dale Iversen 178, G P Bowater 262, Juniors Bildarchiv GmbH 293cr, Universal Images Group 293tr; **Ardea:** Jean Michel Labat 289bc; **College of Physicians of Philadelphia:** The image of Harry Eastlack is used by kind permission of The College of Physicians of Philadelphia. Photograph by Evi Numen. Copyright (2011) by The College of Physicians of Philadelphia 194; **Corbis:** Russel Glenister 289br; **DK Images:** Annabel Milne 88, Frank Greenaway 138, Spike Walker 218, Will Heap 258tr; **Fotolia.com:** artush 8-9, Eileen Kumpf 288, Sailorr 254tr, Stefan Simmeri 261bl; **Frank Sochacki:** 257, 258bl; **Getty Images:** Garry DeLong 210br, Javier Larrea 29tr, Kallista Images 29br, Martin Leigh 209tc, Visuals Unlimited / Dr. George Wilder 211tr, Visuals Unlimited, Inc. / Ken Wagner 210tr; **Martyn F. Chillmaid:** 209tr, 209cr; **Nature Picture Library:** Doug Allan 109l; **Pearson Education Ltd:** 70bl; **PhotoDisc:** 281 (Figure 7); **Press Association Images:** Sunti Tehpia / AP 270; **Science Photo Library Ltd:** A.B. Dowsett 281 (Figure 3), 158br, 228cr, Alex Rakosy, Custom Medical Stock Photo 280tr, Alfred Pasieka 101, 163, Anthony Cooper 222, 269bl, Astrid & Hanns-Frieder Michler 158cr, Biodisc, Visuals Unlimited 32, Biophoto Associates 34, 40, 47, 152, 156tl, 175tl, 175cl, 208, 238, Clause Nuridsany & Marie Perennou 73cl, CNRI 174, 228cl, 229, Courtesy of Crown Copyright Fera 233tr, D. Phillips 298, Darwin Dale 98-99, David M. Phillips 157, Don W. Fawcett 37cr, Dr Jeremy Burgess 56, 232, Dr Kari Lounatmaa 37br, Dr Keith Wheeler 26-27, 159tr, 209tl, 209bl, 219, Dr. Gladdon Wilis, Visuals Unlimited 157br, Dr. John Brackenbury 230, Dr. Richard Kessel & Dr. Gene Shih / Visuals Unlimited, Inc. 159cr, Eye of Science 280br, Garry Delong 158l, Gary Retherford 252-253, Georgette Douwma 233cr, Gunilla Elam 93, Hermann Schillers, Prof. Dr. H. Oberleithner, University Hospital of Muenster 126-127, J.C. Revy, ISM 148, J.W. Shuler 149bc, Jeff Rotman 123, Jerzy Gubernator 48-49, Leonard Lessin 150, Medimage 36, Microscape 158tr, Natural History Museum, London 286bc, NIAID / National Institute of Health 43cl, NOAA 265, Pan Xunbin 29l, Pascal Goetgheluck 269cr, Photo Insolite Realite 171bl, Power and Syred 156tr, 160, Pr. G. Gimenez-Martin 149c, Randy Moore, Visuals Unlimited 211bl, Robert and Jean Pollock 228bl, Saturn Stills 75, Science Picture Co 84-85, 228br, 249, Science Vu, Visuals Unlimited 209cl, SPL / Dr Alexey Khodjakov 149tc, St Mary's Hospital Medical School 246, Steve Gschmeissner 144-145, 171tl, 183, Thomas Deerinck, NCMIR 29cl, Thomas Derrinck, NCMIR 190tr, Tom McHugh 120, Valerie Giles 206-207; **Shutterstock.com:** Aleksey Stanmer 279cl, AlinaMD 254tc, bikeriderlondon 181, Doug Meek 264, FAUP 273, Galyna Andrushko 109r, hjschneider 73bl, Hung Chung Chih 268, Ian 2010 52, infinityyy 254c, Jezper 43tl, Jubal Harshaw 186, Kazakov Maksim 168-169, kurhan 73tl, Lincoln Rogers 261bc, Lodimup 192, maizhusein 266, Martynova Anna 240, Mayskyphoto 70cr, Meiqianbao 203, nito 281 (Figure 5), Nixx Photography 281 (Figure 4), optimarc 53, Pan Xunbin 189, RAJ CREATIONZS 226-227, Richard Griffin 281 (Figure 6), Rita Kochmar Jova 254cr, Sebastian Kaulitzki 45, Shilova Ekaterina 184-185, Steve Bower 286br, SuedeChen 279tl, Tony Wear 190bl, Tyler Olson 296

We are grateful to the following for permission to reproduce copyright material:

Article on p.22 adapted from 'Problems with scientific research, How science goes wrong', *The Economist*; Article on p.94 adapted from Gene editing comes of healthcare age, *Financial Times*, 16/02/2015 (Cookson,C), © The Financial Times Limited. All Rights Reserved; Article on p.122 adapted from 'The Bite That Heals', *National Geographic Magazine*, February 2013 (Holland JS); Article on p.140 adapted from 'Red blood cell membrane disorders', *British Journal of Haematology*, 104 Issue 1 (Published online on 8 February 2005), reproduced with permission of Blackwell Publishing Limited in the format Republish in a book via Copyright Clearance Center; Article on p.164 adapted from 'The Eastlack Skeleton', *The Biologist*, Vol. 61, no. 4, Aug/Sept, p46; Article on p.202 adapted from 'Mutation in key gene allows Tibetans to thrive at high altitude', *The Guardian*, 02/07/2010 (Cian O'Luanaigh), Guardian News and Media Limited 2010; Article on p.180 adapted from Asthma drugs stunt growth – but only by a centimetre © 2014 Reed Business Information – UK. All rights reserved. Distributed by Tribune Content Agency; Article on p.248 adapted from 'Male circumcision as a public health measure for the prevention of HIV transmission', *Southern African Journal of Infectious Diseases*, 2011;26(4)(Part II) (Titus MJ, Moodley J); Article on p.272 adapted from 'Gorilla "Paradise" Found; May Double World Numbers', *National Geographic News*, 05/08/2008 (Morrison,D),Dan Morrison/National Geographic Creative; Article on p.296 adapted from A malfunction that spawns Frankenstein bugs, *Financial Times*, 04/07/2014 (Ahuja, A), © The Financial Times Limited. All Rights Reserved.

The investigation on page 110 of this book was adapted from a resource developed through the Science and Plants for Schools (SAPS) programme. The original resource and others supporting biology education can be downloaded for free from the SAPS website: www.saps.org.uk

We would like to thank Richard Tateson for permission to redraw the image on page 242.

Contents

Module 3
Exchange and transport

Module 4
Biodiversity, evolution and disease

How to use this book

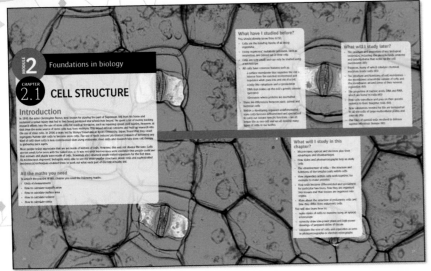

Welcome to your OCR AS/A Level Biology A student book. In this book you will find a number of features designed to support your learning.

Chapter openers

Each chapter starts by setting the context for that chapter's learning:

- Links to other areas of Biology are shown, including previous knowledge that is built on in the chapter and future learning that you will cover later in your course.
- The **All the maths I need** checklist helps you to know what maths skills will be required.

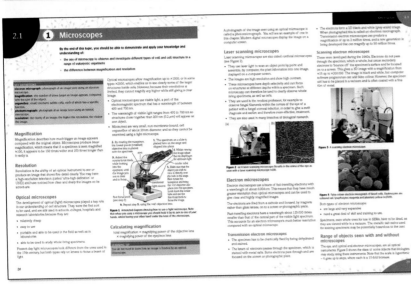

Main content

The main part of the chapter covers all of the points from the specification you need to learn. The text is supported by diagrams and photos that will help you understand the concepts.

Within each topic, you will find the following features:

- **Learning objectives** at the beginning of each topic highlight what you need to know and understand.
- **Key terms** are shown in bold and defined within the relevant topic for easy reference.
- **Worked examples** show you how to work through questions, and how your calculations should be set out.
- **Investigations** provide a summary of practical experiments that explore key concepts.
- **Learning tips** help you focus your learning and avoid common errors.
- **Did you know?** boxes feature interesting facts to help you remember the key concepts.

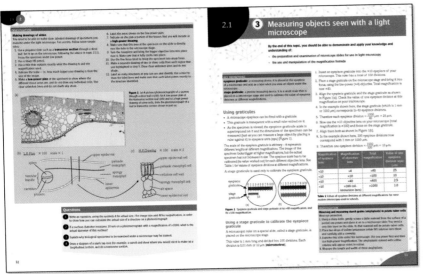

At the end of each topic, you will find **questions** that cover what you have just learned. You can use these questions to help you check whether you have understood what you have just read, and to identify anything that you need to look at again.

Thinking Bigger

At the end of each chapter there is an opportunity to read and work with real-life research and writing about science. These sections will help you to expand your knowledge and develop your own research and writing techniques. The questions and tasks will help you to apply your knowledge to new contexts and to bring together different aspects of your learning from across the whole course. The timeline at the bottom of the spread highlights which other chapters of your book the material relates to.

These spreads will give you opportunities to:

- read real-life material that's relevant to your course
- analyse how scientists write
- think critically and consider relevant issues
- develop your own writing
- understand how different aspects of your learning piece together.

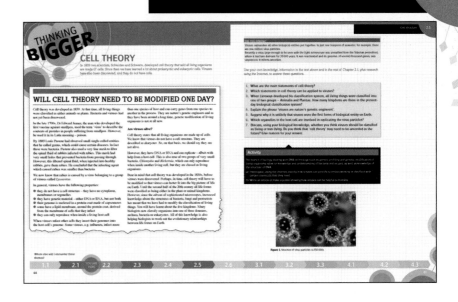

Practice questions

At the end of each chapter, there are **practice questions** to test how fully you have understood the learning.

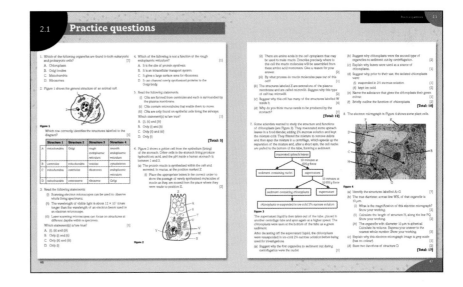

Getting the most from your ActiveBook

Your ActiveBook is the perfect way to personalise your learning as you progress through your OCR AS/A Level Biology A course. You can:

- access your content online, anytime, anywhere
- use the inbuilt highlighting and annotation tools to personalise the content and make it really relevant to you
- search the content quickly.

Highlight tool

Use this to pick out key terms or topics so you are ready and prepared for revision.

Annotations tool

Use this to add your own notes, for example, links to your wider reading, such as websites or other files. Or make a note to remind yourself about work that you need to do.

Development of practical skills in biology

PRACTICAL SKILLS ASSESSED IN A WRITTEN EXAMINATION

Introduction

In 2010 Ruth Brooks began an experiment. She was fed up with the snails that ate her flowers and lettuces but did not like killing them. Although scientists thought that snails were too simple to have a homing mechanism she wondered if they would return when moved, and over what distance they could find their way home. She marked the snails' shells with nail varnish and put them into her neighbour's garden. She asked her neighbour to mark snails and these were released into Ruth's garden. Both batches of snails returned to their original gardens, showing a strong homing instinct for distances up to 10 metres, with one snail travelling 100 metres. The conclusion was that if you want to move snails away from your garden, put them more than 100 metres away and where there is food for them. The retired special-needs teacher won the BBC Radio 4 amateur scientist of the year competition, and a senior lecturer in ecology at Exeter University, who was amazed by her findings, is conducting further research into this topic to find the mechanism that snails have.

Ruth had a question and used scientific method to investigate it. Her findings are likely to lead to a reassessment of snail behaviour. The development of practical skills is fundamental to the study and understanding of all aspects of Biology. All the theories you will learn, about how scientists think living organisms function and interact with each other and their environments, are based upon practical investigations carried out by many scientists over time. The theory and practical are intertwined.

All the maths you need

To unlock the puzzles of this chapter you need the following maths:

- Perform arithmetic and numerical calculations (e.g. finding the arithmetic mean of data set replicates)
- Be able to use an appropriate number of significant figures
- Construct and interpret tables and graphs
- Calculate the rate of change from a graph
- Understand the terms mean, median and mode
- Make order of magnitude calculations
- Understand the principles of sampling
- Select and use an appropriate statistical test, such as the chi squared test, Student's *t*-test or correlation coefficient, to analyse data
- Understand standard deviation and range
- Calculate percentage error

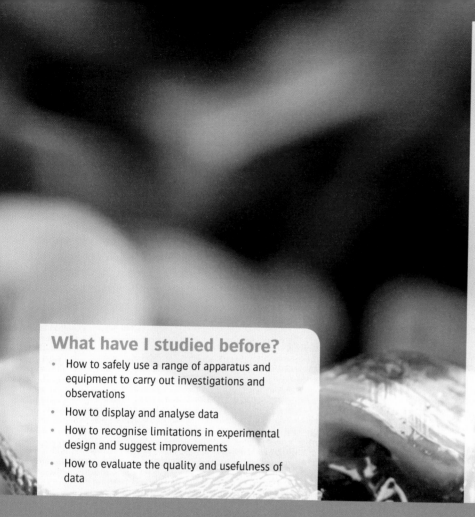

What will I study later?

Practical skills are embedded throughout the content of your course. You will carry out various practical activities for each module and may develop different skills with each one. Some skills will underlie and be used for many of the practical activities.

You will:

- use a microscope to observe, measure and make annotated and labelled drawings of specimens (AS and AL)
- carry out qualitative and quantitative assays for biological matter, including colorimetry (AS)
- use chromatography or electrophoresis (AS)
- investigate the factors affecting metabolic and physiological processes (including enzyme-catalysed reactions) in plants and animals (AS and AL)
- dissect animal and plant organs and observe whole specimens to see how they are adapted to their environments (AS and AL)
- use sampling to measure biodiversity and investigate the factors affecting it (AS)
- sample a variety of ecosystems and record distribution and abundance of organisms (AS and AL)
- investigate genetic inheritance patterns and use the chi-squared test to analyse results (AS and AL)
- use electrophoresis to separate DNA fragments (AL)

What have I studied before?

- How to safely use a range of apparatus and equipment to carry out investigations and observations
- How to display and analyse data
- How to recognise limitations in experimental design and suggest improvements
- How to evaluate the quality and usefulness of data

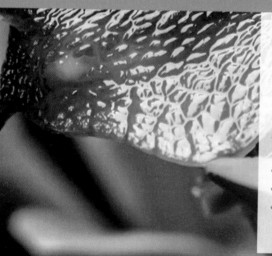

What will I study in this chapter?

- The principles of experimental design, including how to solve problems in a practical context,
- Identification of control variables and evaluating the methodology
- How to process, analyse and interpret qualitative and quantitative data
- How to interpret graphs
- How to evaluate results and draw conclusions
- How to recognise the limitations in experimental procedures and suggest improvements to the experimental design

Planning

By the end of this topic, you should be able to demonstrate and apply your knowledge and understanding of:

* experimental design, including how to solve problems set in a practical context
* identification of variables that must be controlled, where appropriate
* evaluation that an experimental method is appropriate to meet the expected outcomes

Solving problems in a practical context

In your written examination you may be asked, as part of a particular question, how you could test a prediction or investigate a hypothesis or question. This is an example of solving problems in a practical context.

Although this would be a written paper and you would not have to actually carry out the investigation, you should suggest a procedure that is possible.

You should:

* be able to state which apparatus, equipment and techniques would be needed for the proposed experiment.
* apply your scientific knowledge relating to that topic.
* identify and state the independent and dependent variables and the variables that need to be controlled.
* evaluate the proposed method to see if it would do the job and provide an answer to the question. It is quite likely that your proposed method would not provide a full answer and that is fine as long as you can recognise this and say so in your evaluation.

An example of a problem

Is the growth of the single-celled green alga *Pleurococcus* affected by its geographical position, e.g. north-facing or south-facing aspect? What factors might influence its distribution?

Applying some biological knowledge to the problem

Many living things are unevenly distributed both between and within ecosystems. Many factors affect their distribution. These may be temperature; habitat; availability of water, minerals, food, space and mates; light intensity; pollution and competition with other organisms for those limited resources.

Pleurococcus is a single-celled, photosynthetic green alga. It looks like green dust and you see it on vertical surfaces such as walls and tree trunks. You may notice that there is often more on the north-facing side of these surfaces or it may be more abundant in shaded and damp areas.

As it is photosynthetic you might expect it to grow more where light intensity is greater. However, it may be damaged by high light intensities or high temperatures, or be susceptible to desiccation, in which case it would grow more in shaded areas. It is living and so will need some water.

Observations have indicated that *Pleurococcus* may have greater abundance and distribution in cooler areas with lower light intensity, i.e. in areas with a north-facing aspect. However, you cannot draw any conclusions unless you carry out some systematic investigations.

Experimental design

Think about the type of data you will be collecting and whether you have a suitable statistical test for analysing that type of data. You would need to sample many trees in different locations. If you tied a piece of string, to form a **transect**, around the tree trunk and then used a compass to find North, you could sample around the trunk, by placing mini **quadrats** (of sides 10 cm) at intervals around the circumference where the string is, and give a score of 0–10 for density of *Pleurococcus* (see topic 4.2.2 for more on using transects and quadrats for sampling plants).

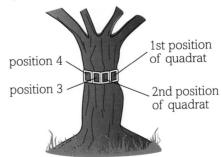

Figure 1 Using a transect and quadrat to sample the densities of *Pleurococcus* around a tree trunk.

Variables

* The independent variable (IV) is the aspect – whether north-, south-, east- or west-facing tree surface.
* The dependent variable (DV) is the density of *Pleurococcus* resulting from the different aspects.

Variables to be controlled

For example:

* species of tree
* ecosystem, whether a field or a wood
* sampling height above ground
* time of day/same day, so the weather and ambient temperature are the same
* the same person to assess density, as it is subjective.

What will you do with the data?

You could visually represent the data by constructing a bar chart for each tree, as shown in Figure 2.

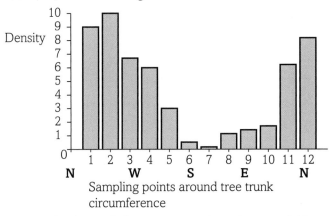

Figure 2 The distribution of *Pleurococcus* around an oak tree in a field, measured at noon during June.

Evaluation of the experimental method

There are limitations in this design:

- We have only sampled one tree, of one species, in one location.

- We have not sampled any other vertical surfaces such as walls.

- We have not used data loggers that can be left for a period of time to monitor the varying conditions.

- The data have not been analysed statistically to see if the difference between density on the north- and south-facing sides of the trees is significant.

- Even if we see a correlation between variables, for example light intensity and *Pleurococcus* distribution, correlation between two variables does not necessarily mean that one is causing the other.

Further investigations

Many experimental investigations lead to other questions that need investigating.

The data here show that the distribution of *Pleurococcus* is uneven but this does not solve the problem of what factors may *cause* this uneven distribution. We can make educated guesses, or hypotheses, as to the causes but we would need to investigate further. Those further investigations would also have to be evaluated.

Could it be light intensity? We could use a light meter to measure light intensity at the sampling areas and also look at the data on the bar chart to see if there is any pattern or correlation between light intensity and *Pleurococcus* distribution. Evaluation points: This would have to be done on the same day and at the same time of day, on a cloudy day and on a sunny day, and at the same sampling height.

Could it be temperature? Light heats surfaces so we might expect the temperature to be higher on the south-facing side of the tree trunk. Evaluation points: We could measure the temperature at each sampling area around the trunk, at the same sampling height, at the same time of day; this could be done for a cloudy day and a sunny day.

Could it be water availability? We could tape test tubes around the tree trunk and leave them to collect rain water that runs off the tree trunk. Evaluation points: Each tube would have to be left in place for the same length of time, at the same sampling height, and on the same days of the year. The tubes would have to be collected at the same time and covered to prevent evaporation, and then the water content measured by mass or volume.

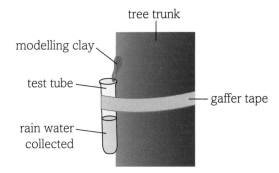

Figure 3 Collecting the water running off a tree trunk.

Could it be predation or infection? Does anything eat *Pleurococcus*? Do any microorganisms infect *Pleurococcus*? We might need to research to find this out and then examine the tree or its location to see if organisms might be infecting or eating *Pleurococcus*.

LEARNING TIPS

If you are stating that light is a possible factor that affects an organism's distribution, refer to the *intensity* of light.

Note that italics are used for proper names of living organisms. If you were writing *Pleurococcus*, for example in your field note book, by hand, you would underline it. As this is the generic name, it begins with an upper case letter.

DID YOU KNOW?

Pleurococcus is a genus of algae and has been said to be the most abundant organism on the planet.

If you use an artist's fine paint brush you can put a little of the green powdery *Pleurococcus* onto a microscope slide and examine it under low and high power. This is a eukaryotic organism; what features of its cell structure can you identify?

Questions

1. Suggest a more objective way of assessing the density of *Pleurococcus* on the bark of tree trunks.

2. Write a list of equipment you would need to carry out the investigation outlined above on *Pleurococcus* distribution.

3. Suggest improvements to this investigation, to reduce its limitations.

4. What are the possible sources of errors in this investigation?

② Implementing an investigation

By the end of this topic, you should be able to demonstrate and apply your knowledge and understanding of:

* how to use a wide range of practical apparatus and techniques correctly

* appropriate units for measurement

* presenting observations and data in an appropriate format

Using practical apparatus and techniques correctly

Throughout your course, you will carry out several practical investigations, some of which are outlined in this book. Bear in mind that our knowledge and ideas about biology stem from practical investigations that gather data to support hypotheses that then become theories or models.

You will already have carried out practical investigations for GCSE Science and will be familiar with a range of equipment and apparatus and be aware of how to use it safely.

In your written examination you may be asked about the use of apparatus in a practical investigation. Table 1 lists some of the apparatus and techniques that you should use during your course, as well as giving examples of suitable practical activities and areas of the specification that they cover.

Type of practical activity	Skills and techniques	Example(s) of suitable practical activities	Specification section
Light microscopy	• Prepare and stain material for slides • Use microscopes at a range of magnifications • Use a graticule and measure specimens • Produce annotated scientific drawings	• Study structure of plant, animal and prokaryotic cells • Study stages of mitosis • Observe plasmolysis and crenation • Observe a range of tissues	• Cells • Exchange and transport • Homeostasis (A Level only) • Respiration (A Level only) • Photosynthesis (A Level only)
Dissection	• Safely use dissecting instruments • Make annotated drawings	• Dissect mammalian heart • Dissect mammalian kidney • Dissect plant stems	• Homeostasis (A Level only) • Exchange and transport
Sampling techniques	• Sampling techniques used in fieldwork • Make annotated scientific drawings	• Calculate species diversity	• Biodiversity • Ecosystems (A Level only)
Rates of enzyme-controlled reactions	• Use a range of apparatus to record quantitative measurements • Use a range of glassware to make serial dilutions • Use data loggers to collect data or use computer software to process data	• Effects of temperature, pH, substrate and enzyme concentration on rate of enzyme-catalysed reactions	• Enzymes • Homeostasis (A Level only)
Colorimeter or potometer	• Use colorimeter to record quantitative data • Use potometer	• Effect of temperature on membrane permeability • Rate of enzyme-catalysed reaction • Investigate the factors affecting rate of transpiration	• Enzymes • Membranes • Exchange and transport

Table 1 Apparatus and techniques used in A Level Biology.

continued

Type of practical activity	Skills and techniques	Example(s) of suitable practical activities	Specification section
Chromatography or electrophoresis	• Thin layer or paper chromatography to separate biological compounds • Gel electrophoresis	• Analyse chlorophyll • Separate and identify a mixture of amino acids • Separate DNA fragments produced by treatment with restriction enzymes	• Biological molecules • Photosynthesis (A Level only) • Nucleic acids, genetic manipulation
Microbiological techniques	• Aseptic techniques • Use of solid and liquid culture media • Colorimetry • Serial dilutions	• The effect of antibiotics on microbial growth	• Cloning and biotechnology (A Level only) • Genetic manipulation (A Level only)
Transport into and out of cells	• Serial dilutions • Data logging	• Investigate water potential of plant tissue, such as potato tuber	• Cells • Membranes
Qualitative testing	• Use qualitative reagents to identify biological molecules	• Test for biological molecules, such as proteins, lipids, sugars and starch	• Biological molecules
Investigation using a data logger or computer modelling	• Use ICT	• Investigate DNA structure using RasMol	• Nucleic acids
Investigate plant and animal responses	• Safe and ethical use of organisms to measure plant and animal responses and physiological functions • Use spirometer	• Investigate tropism in plants • Investigate growth requirements of bacteria • Measure human pulse rate at rest and after exercise • Investigate breathing rate and oxygen uptake by human at rest and during exercise • Use *Drosophila* for genetic investigations	• Plant and animal responses (A Level only) • Exchange and transport
Research skills	• Use online sources and books to research topics • Correctly cite sources of information	• Investigate respiration in yeast, *Saccharomyces cerevisiae*	• All topic areas

Table 1 Apparatus and techniques used in A Level Biology (*continued*).
Note that the types of practical activity listed are organised according to the practical activity groups (PAGs) referred to in the specification.

Appropriate units for measurement

In many practical investigations you are likely to be measuring something. It is important that you use the correct units and the correct symbols or abbreviations.

Below are some of the units you may use, with their correct symbols, e.g. kilograms (kg), metres (m), seconds (s), joules (J) or kilojoules (kJ) for energy, kilopascals (kPa) for pressure or water potential. However, the actual unit used depends on what you are measuring. If you are measuring the diameter of a cell, micrometres (μm) would be appropriate, but if measuring the height of a tree, metres would be a more appropriate unit. For certain studies involving energy flow through ecosystems, the units might be gigajoules per hectare per year (GJ ha^{-1} yr^{-1}).

Prefix	Order of magnitude
nano-	10^{-9}
micro-	10^{-6}
milli-	10^{-3}
centi-	10^{-2}
kilo-	10^{3}
mega-	10^{6}
giga-	10^{9}
tera-	10^{12}
peta-	10^{15}

Table 2 Prefixes denoting orders of magnitude.

Unit	Abbreviation	Number of metres
kilometre	km	1000
metre	m	1
centimetre	cm	0.01
millimetre	mm	0.001
micrometre	μm	0.000 001
nanometre	nm	0.000 000 001

Table 3 Units for length: SI base unit = metre.

Unit	Abbreviation	Number of square metres
kilometres squared	km²	1 000 000
hectare	ha	10 000
centimetres squared	cm²	0.0001
millimetres squared	mm²	0.000 001

Table 4 Units for area.

Unit	Abbreviation	Number of centimetres cubed
cubic decimetres	dm³	1000
cubic centimetres – also called millilitres	cm³ or ml	1
cubic millimetres – also called microlitres	mm³ or μl	0.001

Table 5 Units for volume.

Unit	Abbreviation	Number of grams
metric tonne	t	1 000 000
kilogram	kg	1000
gram	g	1
milligram	mg	0.001
microgram	μg	0.000 001

Table 6 Units for mass.

Presenting your observations and data

If you have been observing a structure, such as an organ or organ system via dissection, a labelled drawing is the way to present this. When you study transport in animals (Chapter 3.2) you will have the opportunity to dissect a mammalian heart and make annotated drawings of your observations.

A labelled drawing is also the way to present observations of cells or tissues on a microscope slide. In Chapter 2.1 you will have several opportunities to make such annotated drawings from microscope slides, in the correct way.

Besides drawings, figures, graphs and diagrams are also visual representations of observations and results of investigations. Topics 1.1.3 and 1.1.4 deal with different types of graphs and diagrams.

Tables

Often the best way to present initial data from an investigation is in a table – see Table 7 for an example:

- The table must have a clear title to inform the reader.
- The table should be ruled off.
- The independent variable should be in the first column (to the left side of the table).
- Each column should have an informative heading and the units for the quantities shown should be in the column heading, not in the column itself.

- You can tabulate data that are not quantitative, such as colour of reagents used in tests and the inference (what it tells you).
- If the data are quantitative, the same number of decimal places should be used for all the values in one particular column.
- If replicates have been carried out there should be a column for each and a column for the calculated mean values.
- The mean values should be calculated to the same number of decimal places or to one more decimal place than those of the raw data values, but all the mean values in a column must be to the same number of decimal places.

Temperature (°C)	Rate of hydrolysis of starch (mg s⁻¹)			Mean rate of hydrolysis of starch (mg s⁻¹)
	1	2	3	
10	11.54	11.36	11.43	11.44
20	21.90	21.59	22.01	21.83
30	35.30	36.00	35.85	35.72
40	36.54	37.01	36.97	36.84

Table 7 Rates of digestion of starch by the enzyme amylase, obtained from goat saliva, at different temperatures.

Questions

A student investigated the digestion of triglyceride (fat) by the enzyme lipase. He wanted to investigate the effect of increasing temperature on the rate of reaction. The enzyme-catalysed reaction produces fatty acids and these lower the pH. This change in pH can be detected by an indicator, such as bromothymol blue, which is blue at pH 7.6, green at pH 7.0 and yellow at pH 6.0. The time taken for the indicator to change to yellow can be measured and so the rate of digestion can be determined. The student presented his data in a table as shown below.

Time taken for indicator to become yellow (secs)			Temperature
1	2	3	
454	476	468	10 °C
287	295	305	15 °C
210	208	212	20 °C
121	123	126	25 °C
105	110	109	30 °C
68	63.5	65.5	35 °C

1 State six ways in which this table can be improved.

2 Calculate the mean rates of reaction for these data. Calculate rate as 1000 divided by time taken for indicator to become yellow. (We use $1000/t$ rather than $1/t$ to calculate the rate, so that the numbers in the calculation are more user friendly. As long as all values are treated in this way, the relative rate of reaction is the same, in effect $1/t \times 10^3$.)

3 Present these data in a properly constructed table.

4 Comment on the range of temperatures used in this investigation.

5 What are the limitations of this investigation in terms of determining the end point of the indicator?

6 Suggest how this investigation could be improved and include suggestions for other ways of measuring the fall in pH.

③ Analysis of data 1: Qualitative and quantitative data

By the end of this topic, you should be able to demonstrate and apply your knowledge and understanding of:

* processing, analysing and interpreting qualitative and quantitative experimental results
* use of appropriate mathematical skills for analysis of quantitative data
* appropriate use of significant figures

KEY DEFINITIONS

qualitative data: data that does not involve quantity (numbers).
quantitative data: data that does involve quantity (numbers).
significant figures: the digits of a number that have a meaning and contribute to the number's precision.

Processing, analysing and interpreting results

When you carry out tests to indicate the presence of glucose, starch, lipids or proteins (see topic 2.2.12 for more about these food tests), you will obtain **qualitative data**. You can represent such findings in a table and indicate the colour observed and the inference – this tells us whether a substance is present or not.

Benedict's reagent is used to test for reducing sugar (if positive, reagent changes from blue to red when heated); iodine/KI solution tests for starch (if positive, a blue-black colour is seen); ethanol emulsion test indicates the presence of lipids if a white emulsion is seen; biuret reagent indicates the presence of protein by a purple/mauve colour.

To make your data **quantitative**, for example to see how much glucose is in a particular drink, you would need to make up a range of glucose solutions of known concentrations, using serial dilution. You would then carry out a Benedict's test, keeping certain variables constant, such as:

* volume of reagent
* volume of solution being tested
* temperature at which heated
* length of time for heating.

You would then see a range of colours showing the positive Benedict's test result, from brick red for a high concentration, through orange, yellow to green for a very low concentration, corresponding to specific concentrations of glucose in solution.

You could use these, or a photograph of them, as standards against which to compare the results of carrying out a Benedict's test on solutions of glucose of unknown concentration.

Using mathematical skills to analyse quantitative data

Think about the measurement of water uptake by a potometer, as described in Chapter 3.3. If measurements are taken at different ambient temperatures we can see the effect of temperature on the rate of water uptake and therefore on the rate of transpiration.

Ambient temperature (°C)	Distance travelled by air bubble in 10 min (mm)				Rate of uptake of water (μl s^{-1})
	1	2	3	mean	
10	12.5	13.0	13.5	13.0	0.11
20	28.0	27.5	27.3	27.6	0.23
30	45.0	47.0	46.0	46.0	0.38
40	55.5	56.5	55.7	55.9	0.46

Table 2 Mean rate of water uptake (μl s^{-1}) in a leafy sycamore maple, *Acer pseudoplatanus*, shoot in a potometer, at different ambient temperatures.

Calculating the volume of water taken up

If you are told, for example, that the diameter of the bore of the capillary tube is 2.5 mm and the air bubble travelled 24 mm in 10 minutes, you can calculate the rate of uptake of water in μl per minute or per second.

Food tested	Colour/observation at end of test				Inference
	Benedict's reagent at 80 °C for 10 min	Iodine/KI solution	Ethanol emulsion test	Biuret reagent	
bread	blue	black	colourless	mauve	contains starch and protein
potato	blue	black	colourless	mauve	contains starch and protein
apple	red	brown	colourless	blue	contains reducing sugar and protein
cheese	blue	brown	white	mauve	contains lipid and protein
chicken	blue	brown	white	mauve	contains lipid and protein

Table 1 Results of tests carried out on a variety of foods.

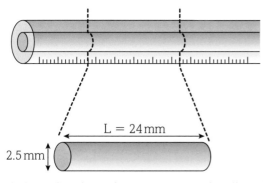

Figure 1 Calculating the volume of water in a section of capillary tube.

If the diameter is 2.5 mm then the radius is 1.25 mm.

L indicates the length moved by the air bubble, so the space in this cylinder is the same as the volume of water taken up by the shoot.

The formula for calculating the volume of a cylinder, V, is $V = \pi r^2 L$ So the volume of water taken up by the shoot in 10 minutes is $[3.142 \times (1.25)^2 \times 24]$ mm^3

$= 117.825$

$= 118 \, \mu l$

> **LEARNING TIP**
>
> Notice that the number in this calculated example has been rounded to a whole number. This is because you can only read this scale to one decimal place. You could express the answer as 117.8 μl, but to no more than one decimal place. mm^3 is not incorrect as a unit but μl is more often used.

Now to calculate rate of uptake, which is volume taken up per unit time.

If 118 μl is taken up in 10 minutes, then the rate of uptake is $118/10 = 11.8 \, \mu l$ min^{-1}.

You could also express this in terms of volume taken up per second, which would be $118/600 = 0.20 \, \mu l$ s^{-1}.

Calculating a median value

Suppose you measure the lengths of the leaves on a branch of a shrub. Their measurements in mm are:

62, 65, 75, 83, 55, 78, 77, 68, 57, 58, 54, 66, 72, 80, 48, 71, 72, 62, 49, 81.

The **arithmetic mean** is 66.7 mm.

The range is from 48 to 83 mm.

There are 10 numbers from 48 to 66 and 10 numbers from 68 to 83. The **median** is therefore 67 (between 66 and 68). This is correct even though there are no leaves of 67 mm in the sample.

Appropriate use of significant figures

In some cases we do not need a detailed answer or very precise number. When you work out an answer on your calculator you do not need to express it to 10 decimal places so you round it off to a certain number of decimal places.

Another method is to round it off using **significant** (meaningful) **figures**.

From the column in Table 2 showing the rate of transpiration, in the second row where the rate is 0.23, 2 is the most significant digit because it tells you that the rate is about 0.2 μl s^{-1}. The second number, 3, is the next significant figure. It tells us that the rate is faster than 0.2 μl s^{-1}. This therefore gives a more accurate and precise indication of the value of the rate calculated. Because this is a calculated value, it can be expressed to one more decimal place than the values in the other columns that were obtained by reading the apparatus and were therefore limited by the precision of the apparatus. The calculated values in this column in Table 2 are all to two significant figures.

As a general rule, the calculated values, in order to be significant, can be to one more decimal place than the values in the columns from which the calculation was made.

The following are not significant figures: leading zeros, trailing zeros and digits derived by calculation and giving several decimal places, which therefore give *far* greater precision than the original data or the instrument used for measurement.

> **Questions**
>
> **1** Express the following to two significant figures:
> (a) 5 374 641
> (b) 1.645 783 6
> (c) 0.985 342 1
> (d) 15.0
> (e) 0.678 000 0.
>
> **2** In an investigation using a potometer, the bubble of air moved 65 mm along the capillary tube in 15 minutes. The diameter of the bore of the capillary tube was 2 mm. Calculate the rate of water uptake by the plant in mm^3 s^{-1} (μl s^{-1}).
>
> **3** Suggest how you could adapt the use of the biuret test for protein to make it quantitative.

By the end of this topic, you should be able to demonstrate and apply your knowledge and understanding of:

* plotting and interpreting suitable graphs from experimental results

There is a variety of graphs and each type has specific uses, but each communicates information visually.

In a written examination you may be given a table of data and be asked to graph those data.

You may also be asked to:

* make deductions from graphical data
* draw conclusions from graphical data
* evaluate the data or its presentation (see next topic).

Line graphs

Line graphs are used to see if there is any correlation between two variables where the data are continuous.

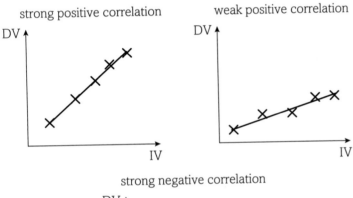

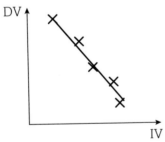

Figure 1 Examples of correlation between two variables.

* They involve a vertical *y*-axis and a horizontal *x*-axis forming a grid.
* Each axis should have a suitable linear scale and be labelled with quantities and units.
* The independent variable (IV) is usually plotted along the *x*-axis and the dependent variable (DV) along the *y*-axis.

* For biological data it is often best to join the plotting points with straight lines. If you do this then the line should go through the centre of each plot. Sometimes a smooth line of best fit can be drawn that goes through or very near to the points. Whichever type of line is drawn, it should *not* be extrapolated, that is, it should not be extended outside of the minimum and maximum value plot points.

LEARNING TIPS

A line on a graph is called a curve, even if it is a straight line. When you draw a graph, make it large and make sure you use a suitable scale and label each axis. Take care to plot the points accurately.

DID YOU KNOW?

Sometimes wrong conclusions have been drawn by extrapolating biological data. Data on high doses of ionising radiation were collected by physicists in the 1940s and 1950s and used to assess the risk to human health. When plotted and extrapolated, they suggest that low levels of radiation are harmful (Figure 2). However, more recent evidence suggests that lower levels of radiation are harmless (Figure 3).

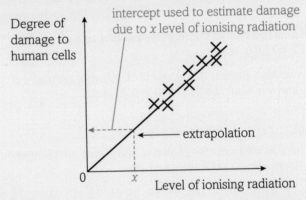

Figure 2 Predicted damage to human cells with increasing levels of ionising radiation.

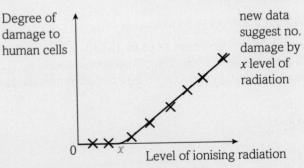

Figure 3 Actual damage to human cells with increasing levels of ionising radiation.

More than one curve can be drawn on the same set of axes, so comparisons can be made and a picture of what is happening during an investigation or observed phenomenon can be seen.

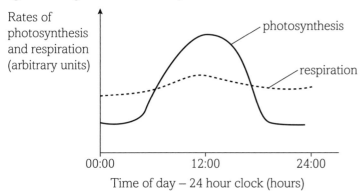

Figure 4 Graph showing the changes in rates of photosynthesis and respiration in a small pond over a 24 hour period during May.

- The rate of reaction can be calculated from the slope of a curve showing the progress of the reaction over time.

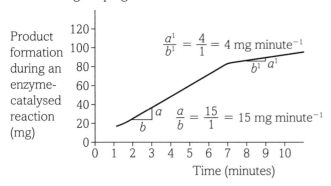

Figure 5 Calculating the rate of an enzyme-catalysed reaction from the slope of a graph

Scattergrams

Also called scatter diagrams or scatter plots, scattergrams are used when investigating the relationship between two naturally changing variables. For example, several plots can be made showing mean blood cholesterol level and death rates from heart disease and stroke in various countries. No line needs to be drawn, but the pattern of the plots can show if there is any correlation.

Bar graphs

Bar graphs are used to investigate relationships when the independent variable is categorical and the dependent variable is continuous, e.g. the concentration of Vitamin C (DV) in different fruit drinks (IV).

- The bars should be of the same width and equally spaced.

- If mean values are shown on the bars, the range bars can also be shown.

- If the data sets being compared have been analysed statistically, the error bars can be shown. If there is overlap it indicates that any apparent difference is not significant.

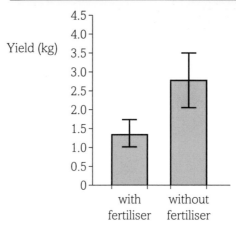

Figure 6 Comparison of yield of tomatoes grown with and without fertiliser. Error bars do not overlap, showing that the difference between these two data sets is significant.

Histograms

Histograms can be used for showing quantitative data organised into classes. For example, if we measured the height of a large number of human adults we may categorise the data, for example those between 140 and 149 cm and those between 150 and 159 cm. The number of people within each class shows the frequency. The class or category that contains the greatest frequency is the **mode.**

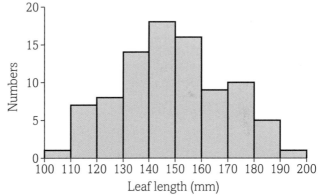

Figure 7 Histogram showing frequency of leaf length in sweet chestnut.

Questions

Which type of graph would you draw to display each of the following types of data?

1. Lengths of leaves on a tree branch.

2. Effect of changing pH on enzyme activity.

3. Sugar content of different types of biscuits.

4. Effect of light intensity on rate of photosynthesis.

5. Collagen content of skin and age in humans.

6. Amino acid content of beef and cheese.

(5) Evaluation

By the end of this topic, you should be able to demonstrate and apply your knowledge and understanding of:

* how to evaluate results and draw conclusions

* the identification of anomalies in experimental measurements

* limitations in experimental procedures

* the refining of experimental design by suggestion of improvements to the procedures and apparatus

* precision and accuracy of measurements and data, including margins of error, percentage errors and uncertainties in apparatus

---

KEY DEFINITIONS

accuracy: how close a measured or calculated value is to the true value.
anomaly: result that does not fit the expected trend or pattern.
precision: the closeness of agreement between measured values obtained by repeated measurements.

Evaluating results and drawing conclusions

You may be shown data and asked to evaluate them or to comment on a conclusion drawn from the data.

For example, in Table 1 are data about changes in blood cholesterol levels. One group of patients was given cholesterol-lowering drugs, called statins. Another group of patients within the same GP practice decided to try to lower their blood cholesterol levels by taking more exercise and altering their diet.

Patient group	Mean blood cholesterol level (mmol dm^{-3} [±SD])	
	Before treatment	6 months after treatment began
Group A – treated with statins ($n = 12$)	6.36 (±1.58)	4.21 (±0.19)
Group B – treated with lifestyle change ($n = 12$)	5.95 (±1.34)	4.87 (±1.60)

Table 1 Blood cholesterol levels of groups of patients in a GP practice.

Statins inhibit an enzyme in the liver from making cholesterol. The guidelines set by NICE (National Institute for Health and Care Excellence) in 2008 stated that people should have a blood cholesterol level of 5.2 mmol dm^{-3} or less. In 2014 the guidelines were changed to 4.0 mmol dm^{-3}.

What can we conclude from these data?

* It would appear that statins are more effective at lowering blood cholesterol level than a patient making lifestyle changes.

* Lifestyle changes appear to lower blood cholesterol, although the improvement was not as marked as in the group taking statins. However, the SD of this group was greater after treatment so some may not have shown any improvement.

* In group A the **standard deviation**, **SD** (indicating the variability of the data, or the size of its spread about the mean) is much larger before treatment than after, which shows that there was quite a high range of blood cholesterol values among these patients before treatment, but their values showed a much narrower range (they were all closer to the mean value) after treatment. This indicates that most of their blood cholesterol levels were probably brought close to the new guideline levels.

* However, we do not know if they suffered any side effects. Some people suffer muscular pains when taking statins and these prevent them from exercising, which is another way of helping to lower blood cholesterol.

* It is easy to monitor the dose of statins taken by each patient in group A, whereas lifestyle changes are harder to monitor as they depend on subjectivity of those making the changes.

* Group A could have also made some lifestyle changes; if they are concerned about their health and willing to take statins, they may also decide to eat more fruit and vegetables and take more exercise.

* There is no information here about the age, gender or family history of patients in each group.

* The two groups had different mean starting levels of blood cholesterol.

* The initial SD for group A is larger than for group B, so there may be more patients in group A with very high blood cholesterol levels.

* These are small groups and this is only one study. It would have to be replicated before any valid conclusions could be drawn.

Identifying anomalies in data

You have been trained to identify anomalies in data. These are results that do not fit the expected pattern. Seeing an anomaly can be an exciting moment, providing evidence that your expectation is wrong and a scientific breakthrough could be staring you in the face. On the other hand it could be due to a piece of grit in your detector or a leaky flask in your incubator. If you are certain that an anomalous piece of data was produced due to a failure in the experimental procedure, you might be justified in removing it before analysing the data. However, you must never discard data simply because they do not correspond with your expectation. By repeating the experiment and amassing more data one of two things could happen. If the anomaly was the result of an experimental error or was simply a very unusual result from naturally-occurring variation it will 'disappear' as the repeat measurements produce a mean in line with expectation. On the other hand, if the anomaly was in fact telling you something surprising about the system you are investigating it will be confirmed by repeat observations and your Nobel Prize is just around the corner.

Limitations in experimental procedures

- It is not always possible to control all extraneous variables.
- Some investigations would be unethical, such as deliberately damaging an area of children's brains to study the effects on their development.
- Results obtained from studying a small population cannot be generalised to the whole population.
- The resolution of the instruments and equipment used may impose limitations.
- The degree of accuracy of measurements may lead to limitations.
- Using a small sample size or having too few replicates is also a limitation, as it is difficult to see if the data are reliable; therefore a large enough sample or enough replicates should be used where possible.
- Not leaving a reaction for long enough to fully complete will give misleading data; therefore we should make sure that reactions are given long enough to complete.
- Not allowing reactants to reach the required temperature before adding them together will reduce validity; reactants should be placed, in their tubes, into a water bath to reach the required temperature before they are mixed.
- Some investigations that rely on questioning people or observing them in particular situations may be limited, because only certain types of people will volunteer to take part or people will behave differently when they think they are being observed.

- Lack of equipment to objectively measure something, such as a colour change, is a limitation as the observation is subjective and may change depending on the investigator.
- Limitations in equipment such as using a beaker of hot water for a waterbath; the investigation can be improved by using a thermostatically controlled water bath, with a thermometer to check the temperature, so as to maintain the desired temperature throughout the reaction.

Errors

Errors or experimental uncertainties arise because there are:
- inadequacies and imperfections in experimental procedures
- lapses of judgement by the experimenter
- limits to resolution, precision or accuracy of measuring apparatus.

Random errors due to judgement errors made by the experimenter are reduced when the procedure is repeated several times.

Systematic errors may be inherent in the equipment and are repeated at every replicate. However, if the percentage error is known, a calculation can be done to determine the margin of error.

LEARNING TIP

Be clear about the difference between accuracy and precision. A thermometer is inaccurate if it gives readings that are 5 °C above the true temperature but it could still be precise if it gives very consistent readings. By recalibrating an inaccurate instrument you can correct this and make accurate measurements. Still confused? An analogy might help: A precise archer will have her arrows tightly clustered somewhere on the target. By recalibrating her sight she can become accurate *and* precise and will have her arrows clustered on the bullseye!

Questions

1 A digital stopwatch can measure to the nearest 0.1 s. Explain why using this stopwatch to measure a reaction for 5 minutes is more accurate than using it to measure the reaction times of humans, which are around 0.3 s duration.

2 In school laboratories, thermometers filled with alcohol rather than mercury are used for safety reasons. They are precise and have an impressive resolution of 0.2 °C. However, the overall calibration could be up to 1 °C out. If you used one of these thermometers to measure the temperature of a water bath at 38 °C, within what range would the real temperature be?

3 Explain why using a gas syringe to collect oxygen given off from a well-illuminated aquatic plant, for 5 minutes, is better than counting the bubbles of oxygen produced during 5 minutes.

HOW SCIENCE CAN GO WRONG

Scientific research has changed how we perceive the world, and the scientific method has evolved to try and prevent flawed research that could misinform us. However, now it may itself need to change.

HOW SCIENCE GOES WRONG

A simple but powerful idea that underpins science: 'trust but verify', has generated a vast body of knowledge. Since its birth in the 17th century, modern science has changed the world beyond recognition, and overwhelmingly for the better. Results should always be subject to challenge from experiment.

However, success can breed complacency. There are many published academic studies that are the result of shoddy experiments or poor analysis. Less than half the published research on biotechnology can be replicated.

In the 1950s, following many successful applications of science during World War II, academic research was seen as important, and a few hundred thousand scientists were carrying out research. Since then, the numbers have grown to 6–7 million active researchers, and there is great pressure on them to 'publish or perish'. This overriding demand has led to a loss of self-policing and quality control. Many journals will not print studies that verify previous findings, so researchers see little value, in terms of advancing their careers, in replicating the studies of other scientists.

Failures to support a hypothesis are rarely offered for publication, and 'negative' results account for only 14% of published studies, down from 30% in 1990. However, in science, knowing what is false is as important as knowing what is true. The failure to report failures means that other researchers waste money and time exploring blind alleys that have already been explored. It may also cost lives. In March 2006, six healthy young men volunteered to take part in a clinical trial for a new drug TGN1412 that had not previously been given to humans. Within a day, all six were extremely unwell and in intensive care. With heroic efforts on the part of medical personnel over several weeks, the men all recovered, but lost fingers and toes. TGN1412 is an antibody molecule that attaches to a CD28 receptor on white blood cells of the immune system and interferes with the immune system in ways that are poorly understood. In 1996, a similar study using an antibody that attached to CD28 (as well as to CD3 and CD2) receptors, using one human subject, had similar results, but was not published. Had it been published then it may have prevented the ordeal of the six volunteers in 2006.

Even if flawed research does not always put people's lives at risk, it squanders money and effort.

The hallowed process of peer review may not be all it is cracked up to be. A prominent medical journal ran research past other experts in the field, and they failed to spot some deliberately inserted mistakes, even though they knew that they were being tested.

Ideally, research protocols should be registered in advance to prevent fiddling with experimental design midstream. Trial data should also be open to others to inspect and test. The most enlightened journals are becoming less averse to publishing less interesting papers, and some are encouraging replication studies. Younger scientists have a better understanding of statistics, and their use in analysing data needs to be extended. Peer review needs to be tightened so that science can correct its own mistakes and continue to command respect, rather than create barriers to understanding by shoddy research.

Sources

- Leader article: How science goes wrong, *The Economist*, 19 October 2013, p.11.
- Goldacre, B. (2012) *Bad Pharma*. Fourth Estate.
- http://www.newscientist.com/issue/2004.

DID YOU KNOW?

During the 19th century, there were many pseudoscientific claims, such as claims by Sylvester Graham, inventor of Graham crackers, that ketchup and mustard can cause insanity. Unfortunately, such pseudoscience still exists today. One 'celebrity' nutritionist has claimed that dark green leaves such as spinach are good for you, as they contain lots of chlorophyll, which is high in oxygen and so will give you more oxygen! She also says that certain foods are good sources of digestive enzymes. Another nutrition journalist has claimed that fructose is digested in the liver.

Where else will I encounter these themes?

1.1 YOU ARE HERE 2.1 2.2 2.3 2.4 2.5

Let's start by considering the nature of the writing in the article. This extract is from the magazine *The Economist*, which is aimed at the general public.

1. Discuss the style of writing – has the writer made too many assumptions about the knowledge of readers? Do you think that the 'scientific method' should have been explained? Do you think that all readers understand why replicating scientific investigations is important?

2. Is it clear what makes an experiment 'shoddy'?

Now let's look at the concepts about scientific methodology and the biology underlying the information in the article.

3. Explain why an investigation that fails to support a hypothesis is not a failed investigation.

4. Describe the process of peer review.

5. The results of many clinical trials, where the new treatment does not appear to be more effective than the present available treatment or a placebo, are not published. However, trials that show the new treatment in a good light are. What do you think are the disadvantages of such 'cherry-picking'?

6. The discovery of the structure of DNA was a piece of curiosity-driven research. At the time, it was not envisaged that it would have any practical applications. In today's economic climate, scientists have to show that their proposed research will have some economic benefits, in order to obtain funding. Briefly outline the disadvantages of this approach.

7. State two problems that are faced by humans today, that will need science to help find the solutions.

8. Discuss how the statements (a) made by a celebrity nutritionist – that eating foods rich in chlorophyll gives us more oxygen, and that some foods are sources of digestive enzymes; and (b) made by a journalist – that fructose is digested in the liver, are incorrect.

Activity

Complete ONE of the following two activities.

1 When scientific research challenges major dogma (accepted theory) it is said to be revolutionary and cause a paradigm shift. Examples of paradigm shifts are the Darwin–Wallace theory of evolution by natural selection; Pasteur's germ theory of disease; the theory of biogenesis; and Mendel's theory of inheritance.

Select one of the above paradigm shifts, and do research, using the Internet, to find out more about it. Prepare and deliver to others in your class a short (three-minute maximum) presentation outlining the essence of the displaced dogma and the now-accepted theory that replaced it.

2 There have been recent examples of flawed scientific investigations that have done considerable harm, such as the suggested link between the measles vaccine and autism, and the investigation that showed that rats fed on GM potatoes were harmed. Choose one of these studies, or another flawed study that you are aware of, and do research, using the Internet, to find out more about it. Prepare and deliver to others in your class a short (three-minute maximum) presentation detailing what the study purported to show; in what way it was found to be flawed; the harm that it caused; and what happened to the researcher.

3.1   3.2   3.3   4.1   4.2   4.3

At A level, assessment of your practical skills is carried out through the practical endorsement. However, you will also find questions in your written examinations that test your knowledge and understanding of practical procedures. These are likely to form part of a whole question rather than entire questions.

1. Which row shows the most appropriate units to use when measuring the size of different samples? [1]

	height of person	length of leaf	diameter of artery	length of cell
A	metres	millimetres	centimetres	micrometres
B	metres	centimetres	millimetres	micrometres
C	metres	centimetres	micrometres	millimetres
D	metres	centimetres	millimetres	nanometres

2. A student estimated the percentage cover of four different species in a field. What type of graph would be the best way to present these results? [1]

   A   a histogram

   B   a kite graph

   C   a line graph

   D   a pie chart

3. A student investigated the effect of temperature on the rate of enzyme action. Which row correctly identifies the dependent variable, independent variable and control variables? [1]

	dependent variable	independent variable	control variable	control variable
A	rate of enzyme action	temperature	volume of solutions	substrate concentration
B	temperature	rate of enzyme action	volume of solutions	substrate concentration
C	rate of enzyme action	temperature	volume of solutions	intensity of light
D	volume of solutions	rate of enzyme action	temperature	substrate concentration

4. Figure 1 shows the scale on a graticule. What is the level of precision in the measurement s that can be made using this scale? [1]

Figure 1 The scale on a graticule

   A   1 mm

   B   0.5 mm

   C   0.15 mm

   D   0.05 mm

5. A student investigated the rate of transpiration from a leafy shoot using a potometer. The student removed leaves from the shoot one at a time to investigate the effect of reduced surface area. As each leaf was removed, the student covered the damaged stem with petroleum jelly. The student identified a number of variables in this investigation:

   (i)   leaf size
   (ii) loss of water from damaged stem
   (iii) light intensity
   (iv) temperature
   (v)   species of plant used

   Which of these variables could be limiting factors for transpiration? [1]

   [Total: 5]

6. You are provided with seven beakers containing solutions of reducing sugar in beakers labelled 0.0, 0.2, 0.4, 0.6, 0.8, 1.0% and X. You are also provided with suitable reagents and glassware.

   (a) Outline how you would use the solutions provided to estimate the concentration of reducing sugar in solution X. [6]

   (b) State two limitations of your plan. [2]

   (c) Describe how you could modify your plan to make it more accurate. [5]

   [Total: 13]

7. Figure 2 shows a potometer.

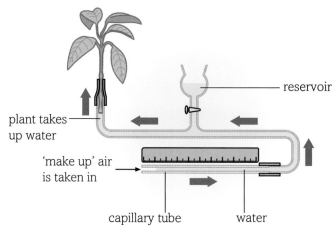

Figure 2

(a) Describe three precautions that should be taken to ensure the apparatus works properly. [3]

(b) You are asked to use the apparatus to investigate the effect of increasing air movement on the rate of transpiration. Draw a suitable blank results table, assuming you have a fan with three speed settings. [6]

(c) State the expected effect of increasing wind speed. [1]

(d) Suggest a suitable statistical test to assess whether your results match the expected results. [1]

[Total: 11]

8. You have been asked to measure the biodiversity of a field near your school. You decide to use a random sampling technique.

(a) Explain why you should use random samples. [2]

(b) Describe how you could generate random sample sites. [3]

(c) List the apparatus that you would need to take with you. [5]

(d) When you visit the site you realise that there are three distinct patterns of vegetation. Suggest how you should modify your plan. [3]

[Total: 13]

9. The table below shows the results of an investigation into the effect of reducing the number of leaves on the rate of transpiration.

number of leaves	rate of transpiration		
	1	2	3
8	65	62	64
7	53	27	56
6	42	45	44
5	45	31	32
4	25	26	27

(a) How could the table of results be improved? [1]

(b) One limitation of this experiment is that the leaves are not all the same size. How could you modify the experiment to overcome this limitation? [1]

(c) Identify one anomalous result and give a reason for your answer. [2]

(d) Do you think the results are reliable? Justify your answer by referring to the data. [9]

(e) Explain why the student collected three sets of data. [2]

[Total: 15]

10. Figure 3 shows an eyepiece graticule and a stage graticule.

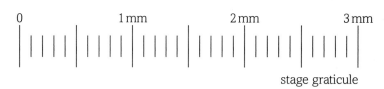

Figure 3

(a) What is the value of one eyepiece unit (epu)? [1]

(b) The diameter of a small artery was measured using the eyepiece graticule. It measured as 24 epu. What is the diameter of the artery? [2]

(c) Explain why it is important to use the same objective lens throughout the measuring process. [2]

[Total: 5]

Foundations in biology

CELL STRUCTURE

Introduction

In 1995 the actor Christopher Reeve, best known for playing the part of Superman, fell from his horse and sustained a spinal injury that led to him being paralysed and wheelchair-bound. He spent a lot of money backing research efforts into the use of stem cells for medical therapies, such as repairing spinal cord injuries. However, at that time the main source of stem cells was from embryos. This raised ethical concerns and held up research into the use of stem cells. In 2006 a team led by Shinya Yamanaka at Kyoto University, Japan, found that they could reprogram human skin cells to become stem cells. The use of such *induced pluripotent* (capable of becoming any kind of cell) stem cells is less controversial than using embryonic stem cells and research into stem cell therapy is gathering pace again.

Most people today appreciate that we are made of billions of cells. However, this was not always the case. Cells are too small to be seen with the naked eye, so it was not until microscopes were available that people could see that animals and plants were made of cells. Scientists also observed single-celled organisms for the first time. As microscopes improved, biologists were able to see the even smaller structures inside cells and sophisticated biochemical techniques enabled them to work out what each part of the cell actually did.

All the maths you need

To unlock the puzzles of this chapter you need the following maths:

- Units of measurement
- How to calculate magnification
- How to calculate surface area
- How to calculate volume
- How to calculate ratios

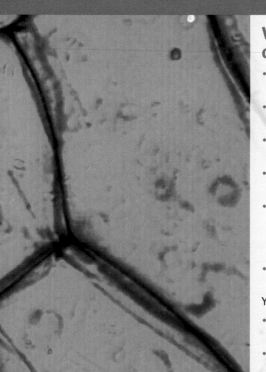
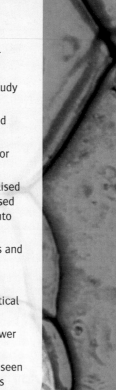

What have I studied before?

You should already know from GCSE:

- Cells are the building blocks of all living organisms
- Living organisms' metabolic processes, such as respiration, are carried out in their cells
- Cells are very small and can only be studied using a microscope
- All cells have common features such as:
    - a surface membrane that separates the cell's interior from the external environment and regulates what goes into and out of a cell
    - a jelly-like cytoplasm and a cytoskeleton
    - DNA that makes up the cell's genetic content (genome)
    - ribosomes where proteins are assembled
- There are differences between plant, animal and bacterial cells
- Within a developing organism undifferentiated stem cells become differentiated and specialised to carry out certain specific functions – we all began life as one cell and we all develop many types of cells in our bodies

What will I study later?

- The structure and properties of key biological molecules, including the phospholipids, proteins and carbohydrates that make up the cell membranes (AS)
- Enzymes, many of which catalyse chemical reactions inside cells (AS)
- The structure and functions of cell membranes – the membranes around the outside of cells and the membranes around some of their internal organelles (AS)
- The properties of nucleic acids, DNA and RNA, which are found in cells (AS)
- How cells reproduce and pass on their genetic material to their daughter cells (AS)
- How substances needed for life are transported to all the cells of large multicellular plants and animals (AS)
- The roles of special cells involved in defence against infectious disease (AS)

What will I study in this chapter?

- Microscopes, optical and electron, plus their advantages and disadvantages
- How slides and photomicrographs help us study cells
- The ultrastructure of cells – the structure and functions of the smaller parts within cells
- How organelles within cells work together, for example to make proteins
- How cells become differentiated and specialised for particular functions, how they are organised into tissues and that tissues are organised into organs
- More about the structure of prokaryotic cells and how they differ from eukaryotic cells

You will also learn how to:

- make slides of cells to examine using an optical microscope
- correctly draw low-power plans and high-power drawings of prepared slides of tissues
- calculate the size of cells and organelles as seen in photomicrographs or electron micrographs

(1) Microscopes

By the end of this topic, you should be able to demonstrate and apply your knowledge and understanding of:

* the use of microscopy to observe and investigate different types of cell and cell structure in a range of eukaryotic organisms

* the difference between magnification and resolution

Magnification

Magnification describes how much bigger an image appears compared with the original object. Microscopes produce *linear* magnification, which means that if a specimen is seen magnified ×100, it appears to be 100 times wider and 100 times longer than it really is.

Resolution

Resolution is the ability of an optical instrument to see or produce an image that shows fine detail clearly. You may have a high-resolution television (called 'ultra-high definition' or UHD) and have noticed how clear and sharp the images on its screen are.

Optical microscopes

The development of optical (light) microscopes played a key role in our understanding of cell structure. They were the first sort to be used, and are still used in schools, colleges, hospitals and research laboratories because they are:

* relatively cheap

* easy to use

* portable and able to be used in the field as well as in laboratories

* able to be used to study whole living specimens.

Present-day light microscopes look different from the ones used in the 17th century, but both types rely on lenses to focus a beam of light.

Optical microscopes allow magnification up to ×1500, or in some types ×2000, which enables us to see clearly some of the larger structures inside cells. However, because their **resolution** is limited, they cannot magnify any higher while still giving a clear image.

* Optical microscopes use visible light, a part of the electromagnetic spectrum that has a wavelength of between 400 and 700 nm.

* The wavelength of visible light ranges from 400 to 700 nm so structures closer together than 200 nm (0.2 μm) will appear as one object.

* Ribosomes are very small, non-membrane-bound, cell **organelles** of about 20 nm diameter, and so they cannot be examined using a light microscope.

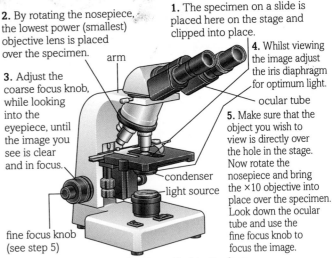

2. By rotating the nosepiece, the lowest power (smallest) objective lens is placed over the specimen.

arm

1. The specimen on a slide is placed here on the stage and clipped into place.

3. Adjust the coarse focus knob, while looking into the eyepiece, until the image you see is clear and in focus.

4. Whilst viewing the image adjust the iris diaphragm for optimum light.

ocular tube

5. Make sure that the object you wish to view is directly over the hole in the stage. Now rotate the nosepiece and bring the ×10 objective into place over the specimen. Look down the ocular tube and use the fine focus knob to focus the image.

condenser

light source

fine focus knob (see step 5)

6. Repeat step **5**. using the ×40 objective lens.

Figure 1 Annotated diagram showing how to use a light microscope. Note that when you carry a microscope you should hold it by its arm in one of your hands, whilst having your other hand under the base of the microscope.

Calculating magnification

total magnification = magnifying power of the objective lens × magnifying power of the eyepiece lens

A photograph of the image seen using an optical microscope is called a **photomicrograph**. You will see an example of one in this chapter. Modern digital microscopes display the image on a computer screen.

Laser scanning microscopes

Laser scanning microscopes are also called confocal microscopes (see Figure 2).

- They use laser light to scan an object point by point and assemble, by computer, the pixel information into one image, displayed on a computer screen.

- The images are high resolution and show high contrast.

- These microscopes have depth selectivity and can focus on structures at different depths within a specimen. Such microscopy can therefore be used to clearly observe whole living specimens, as well as cells.

- They are used in the medical profession, for example to observe fungal filaments within the cornea of the eye of a patient with a fungal corneal infection, in order to give a swift diagnosis and earlier, and therefore more effective, treatment.

- They are also used in many branches of biological research.

(a)
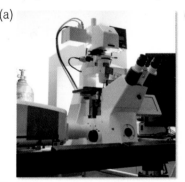

(b)
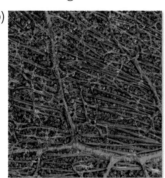

Figure 2 (a) A laser scanning microscope; (b) cells in the retina of the eye as seen with a laser scanning microscope (×64).

Electron microscopes

Electron microscopes use a beam of fast-travelling electrons with a wavelength of about 0.004 nm. This means that they have much greater resolution than optical microscopes and can be used to give clear and highly magnified images.

The electrons are fired from a cathode and focused, by magnets rather than glass lenses, on to a screen or photographic plate.

Fast-travelling electrons have a wavelength about 125 000 times smaller than that of the central part of the visible light spectrum. This accounts for an electron microscope's much better resolution compared with an optical microscope.

Transmission electron microscopes

- The specimen has to be chemically fixed by being dehydrated and stained.

- The beam of electrons passes through the specimen, which is stained with metal salts. Some electrons pass through and are focused on the screen or photographic plate.

- The electrons form a 2D black-and-white (grey-scale) image. When photographed this is called an **electron micrograph**. Transmission electron microscopes can produce a magnification of up to 2 million times, and a new generation is being developed that can magnify up to 50 million times.

Scanning electron microscopes

These were developed during the 1960s. Electrons do not pass through the specimen, which is whole, but cause secondary electrons to 'bounce off' the specimen's surface and be focused on to a screen. This gives a 3D image with a magnification from ×15 up to ×200 000. The image is black and white, but computer software programmes can add false colour. However, the specimen still has to be placed in a vacuum and is often coated with a fine film of metal.

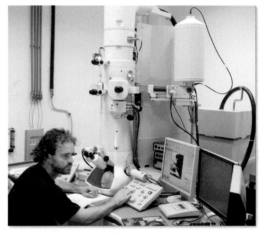

Figure 3 A scanning electron microscope.

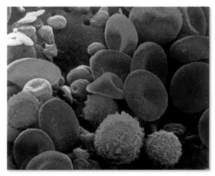

Figure 4 False-colour electron micrograph of blood cells. Erythrocytes are coloured red, lymphocytes magenta and platelets yellow (×1900).

Both types of electron microscope:

- are large and very expensive
- need a great deal of skill and training to use.

Specimens, even whole ones for use in SEMs, have to be dead, as they are viewed while in a vacuum. The metallic salt stains used for staining specimens may be potentially hazardous to the user.

Range of objects seen with and without microscopes

The eye, and optical and electron microscopes, are all optical instruments. Figure 5 shows the sizes of some objects that biologists may study, using these instruments. Note that the scale is *logarithmic* – it goes up in steps, where each is a 10-fold increase.

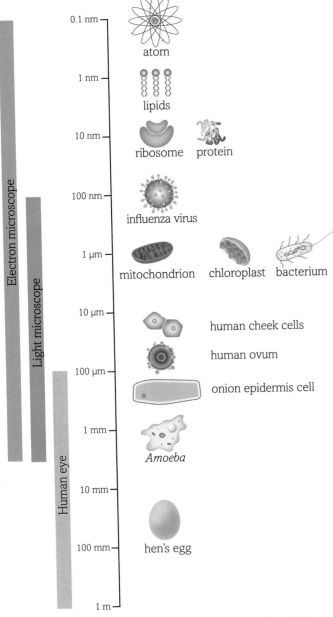

Figure 5 Relative sizes of some biological structures on a logarithmic scale, showing the scope of the electron microscope, light microscope and human eye for studying them.

Your eye can distinguish objects that are about 0.3–0.5 mm apart. This is the limit of its resolution, but it gives you quite good visual acuity for 'everyday' objects. In the retina, at the back of the eye, are photosensitive cells called cones that work in bright light and produce this visual acuity (sharpness). You have about 200 000 cones per mm². Eagles and hawks have many more cones in their retinas, around 1 million per mm², and therefore have greater resolution and visual acuity. When you see a hawk hovering 20 m high over a roadside grass verge, it can clearly see an insect scurrying amongst that vegetation. An eagle can spot a rabbit 2 miles away and, although it is much smaller than you, its eyes are about the same size as yours.

LEARNING TIP

You need to really get to grips with the units mm, μm and nm and be able to convert one to the other.
In the next topic you will carry out some maths exercises that require you to use and convert these units.

Questions

1 If you were to examine a slide of a protoctist, using a ×40 objective lens and a ×15 eyepiece lens, what would be the total magnification of the protoctist?

2 List or make a table to show the advantages and disadvantages of optical microscopes.

3 Suggest the most useful type of microscope to observe each of the following:

(a) living water-fleas in pond water during a biology field trip

(b) cells taken from a cervical smear to be examined for abnormalities that may indicate cancer

(c) virus particles

(d) the inner structure of a mitochondrion

(e) the ribosomes in a liver cell.

4 List or make a table to show the advantages and disadvantages of electron microscopes.

5 The wavelength of red light is 700 nm. How many times larger is this than the wavelength of electrons?

6 What is a logarithmic scale? Why do you think it is used for comparing sizes of biological structures?

Slides and photomicrographs

By the end of this topic, you should be able to demonstrate and apply your knowledge and understanding of:

* the preparation and examination of microscope slides for use in light microscopy

* the use of staining in light microscopy

* the use and manipulation of the magnification formula

* the representation of cell structure as seen under the light microscope, using drawings and annotated diagrams of whole cells or cells in sections of tissue

Making slides

You can use an optical microscope to view a wide range of specimens including:

* living organisms such as *Paramecium* and *Amoeba*

* smear preparations of human blood and cheek cells

* thin sections of animal, plant and fungal tissue, such as bone, muscle, leaf, root or fungal hyphae.

Observing unstained specimens

Many biological structures, including single-celled organisms such as *Paramecium*, are colourless and transparent. Some microscopes use light interference, rather than light absorption, in order to produce a clear image without staining. Some use a dark background against which the illuminated specimen shows up. These microscopes are particularly useful for studying living specimens. You can observe living specimens with a school light microscope by adjusting the iris diaphragm to reduce the illumination of the specimen.

Staining specimens

Stains are coloured chemicals that bind to molecules in or on the specimen, making the specimen easy to see. Methylene blue is an all-purpose stain. Some stains bind to specific cell structures, staining each structure differently so the structures can be easily identified within a single preparation. This is called **differential staining**. There are many such stains, including:

* acetic orcein binds to DNA and stains chromosomes dark red

* eosin stains cytoplasm; Sudan red stains lipids

* iodine in potassium iodide solution stains the cellulose in plant cell walls yellow, and starch granules blue/black (these will look violet under the microscope).

Observing prepared specimens

You will have access to many prepared and permanently fixed slides. They have been made by experts in a laboratory by:

* dehydrating the specimens

* embedding them in wax to prevent distortion during slicing

* using a special instrument to make very thin slices called sections – these are stained and mounted in a special chemical to preserve them.

Calculations involving magnification

On the photomicrograph in Figure 1 (showing a leaf in transverse section), because you know the magnification, you can then find the actual size of the structures.

* Measure the widest part of the leaf on the photomicrograph, in mm.

* Convert that measurement to μm by multiplying by 1000.

* Now divide this figure by the magnification. This tells you the actual thickness of the leaf at this point.

If you are told the actual size of a structure on a photomicrograph (A), and you measure its image size on the photomicrograph (I), in μm [mm × 1000], you can calculate the magnification factor (M) using the formula:

$$M = \frac{I}{A}$$

There are no units for magnification but if, for example, the magnification factor is 1000, then you must write it as ×1000.

LEARNING TIPS

When you observe structures in a microscope section, bear in mind the following:

* cells have a 3D structure and you are looking at a 2D section

* depending on whereabouts in the cell the section was cut, some structures may be absent from your slide section

* depending how certain structures were oriented in the cell, they will appear as different shapes. For example, mitochondria sliced lengthways (in **longitudinal section**) would appear sausage-shaped, but if sliced transversely (crossways) will appear round and if sliced obliquely (slanting) will appear elliptical.

When drawing your specimens, always draw what you see, not what you think the specimen should look like as remembered from a textbook diagram.

INVESTIGATION

Making drawings of slides

You need to be able to make clear, labelled drawings of specimens you examine under the light microscope. For success, follow some simple rules:

1. Use a prepared slide such as a **transverse section** through a dicot leaf. Set it up on the microscope, following the advice in topic 2.1.1. Focus the specimen under low power.
2. Use a sharp HB pencil.
3. Use a title that explains exactly what the drawing is and the magnification used.
4. Indicate the scale – i.e. how much bigger your drawing is than the size of the image.
5. Make a **low-power plan** of the specimen to show where the different tissue areas are, and do *not* draw any individual cells. Use clear unbroken lines and do *not* shade any areas.
6. Label the areas shown on the low-power plan.
7. Indicate on the plan a portion of the tissues that you will include in a **high-power drawing.**
8. Make sure that this area of the specimen on the slide is directly over the hole in the microscope stage.
9. Turn the nosepiece and bring the bigger objective lens into place over it. Make sure that it fully clicks into place.
10. Use the fine-focus knob to bring the specimen into sharp focus.
11. Make a separate drawing of two or three cells from each region that you highlighted in step 5. Draw clear unbroken lines and do not shade.
12. Label as many structures as you can see and identify. Use a ruler to draw the label lines and make sure that each label points exactly to the structure identified.

Figure 1 (a) A picture (photomicrograph) of a section through a dicot leaf (×100). (b) A low-power plan of the transverse section leaf slide and (c) a high-power drawing of some cells, from the photomicrograph of a leaf in transverse section shown in part (a).

(a)

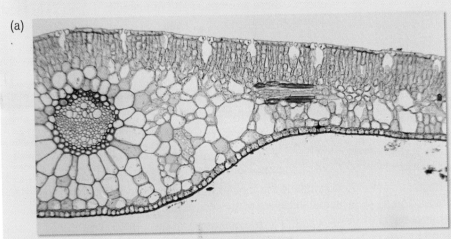

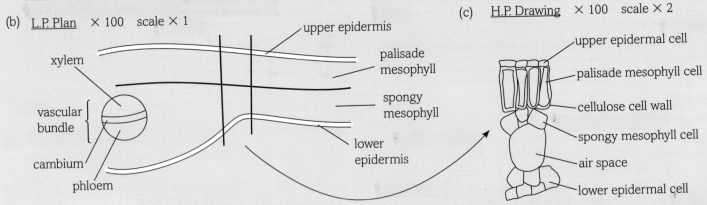

(b) L.P. Plan × 100 scale × 1

- upper epidermis
- xylem
- vascular bundle
- cambium
- phloem
- palisade mesophyll
- spongy mesophyll
- lower epidermis

(c) H.P. Drawing × 100 scale × 2

- upper epidermal cell
- palisade mesophyll cell
- cellulose cell wall
- spongy mesophyll cell
- air space
- lower epidermal cell

Questions

1. Write an equation, using the symbols A for actual size, I for image size and M for magnification, in order to show how you can calculate the actual size of a structure on a photomicrograph.

2. If a nucleus diameter measures 10 mm on a photomicrograph with a magnification of ×1000, what is the actual diameter of this nucleus?

3. Explain why biological specimens to be examined under a microscope may be stained.

4. Draw a diagram of a plant tap-root (for example, a carrot) and show where you would cut it to make (a) a longitudinal section, and (b) a transverse section.

3 Measuring objects seen with a light microscope

By the end of this topic, you should be able to demonstrate and apply your knowledge and understanding of:

* the preparation and examination of microscope slides for use in light microscopy

* the use and manipulation of the magnification formula

KEY DEFINITIONS

eyepiece graticule: a measuring device. It is placed in the eyepiece of a microscope and acts as a ruler when you view an object under the microscope.

stage graticule: a precise measuring device. It is a small scale that is placed on a microscope stage and used to calibrate the value of eyepiece divisions at different magnifications.

Using graticules

* A microscope eyepiece can be fitted with a graticule.
* This graticule is transparent with a small ruler etched on it.
* As the specimen is viewed, the **eyepiece graticule** scale is superimposed on it and the dimensions of the specimen can be measured (just as you can measure a large object by placing a ruler against it) in eyepiece units (epu) (Figure 1).

The scale of the eyepiece graticule is arbitrary – it represents different lengths at different magnifications. The image of the specimen looks bigger at higher magnifications, but the actual specimen has not increased in size. The eyepiece scale has to be calibrated (its value worked out) for each different objective lens. See Table 1 for values of eyepiece divisions at different magnifications.

A **stage graticule** is used only to calibrate the eyepiece graticule.

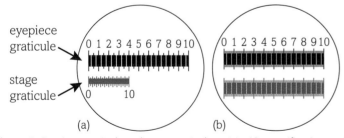

eyepiece graticule

stage graticule

(a)   (b)

Figure 1 Eyepiece graticule and stage graticule at (a) ×40 magnification, and (b) ×100 magnification.

Using a stage graticule to calibrate the eyepiece graticule

A microscopic ruler on a special slide, called a stage graticule, is placed on the microscope stage.

This ruler is 1 mm long and divided into 100 divisions. Each division is 0.01 mm or 10 μm (**micrometres**).

1. Insert an eyepiece graticule into the ×10 eyepiece of your microscope. This ruler has a total of 100 divisions.

2. Place a stage graticule on the microscope stage and bring it into focus, using the low-power (×4) objective. Total magnification is now ×40.

3. Align the eyepiece graticule and the stage graticule as shown in Figure 1(a). Check the value of one eyepiece division at this magnification on your microscope.

4. In the example shown here, the stage graticule (which is 1 mm or 1000 μm) corresponds to 40 eyepiece divisions.

5. Therefore each eyepiece division $= \dfrac{1000}{40} \mu m = 25 \ \mu m$.

6. Now use the ×10 objective lens on your microscope (total magnification is ×100) and focus on the stage graticule.

7. Align them both as shown in Figure 1(b).

8. In the example shown here, 100 eyepiece divisions now correspond with 1 mm or 1000 μm.

9. Therefore one eyepiece division $= \dfrac{1000}{100} \mu m = 10 \ \mu m$.

Magnification of eyepiece lens	Magnification of objective lens	Total magnification	Value of one eyepiece division (epu) (μm)
×10	×4	×40	25
×10	×10	×100	10
×10	×40	×400	2.5
×10	×100 (oil-immersion lens)	×1000	1.0

Table 1 Values of eyepiece divisions at different magnifications for most modern microscopes used in schools.

INVESTIGATION

Observing and measuring starch grains (amyloplasts) in potato tuber cells
Wear eye protection.

1. Using a sharp knife, gently scrape a little material from the surface of a peeled raw potato and place it on to a microscope slide. You need a very thin layer on the slide. In that material will be potato tuber cells.

2. Place two drops of iodine/potassium iodide (KI) solution onto them and carefully add a coverslip.

3. Examine this slide under the microscope. Use low power first and then use high-power magnification. The amyloplasts stained with iodine solution will appear violet in colour.

4. Measure the length and width of three amyloplasts.

WORKED EXAMPLE

If you are asked to measure a structure as it would appear with a light microscope with a graticule, then you will also be told the value of the divisions on the graticule. It is just like a ruler and allows you to measure the specimen.

Figure 2 shows a plant root in transverse section, seen under a light microscope. The smallest divisions on the graticule are 0.01 mm (10 μm) apart.

1. Use the graticule to measure the actual width of this root along the line AB.

   Answer
   The length of the line AB is 59 small divisions of the graticule × 10 μm
   = 590 μm = 0.59 mm.

2. Calculate the magnification of this image.

   Answer
   Measure the length of the image on the page with a ruler. It is approx. 59 mm = 59 000 μm. So the magnification = $\frac{59\,000}{590}$ = 100.

3. The eyepiece used in this microscope had a magnification factor of ×10. What was the magnification of the objective lens?

   Answer
   If the total magnification is ×100 and the eyepiece magnification is ×10, then the objective lens magnification is $\frac{100}{10}$ = 10.

LEARNING TIP

You have already seen the equation that **A**ctual size = **I**mage size/**M**agnification.
Therefore **M**agnification = **I**mage size/**A**ctual size
The IMA triangle shown in Figure 3 may help you to use and substitute in that equation.

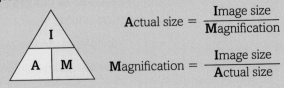

$$\text{Actual size} = \frac{\text{Image size}}{\text{Magnification}}$$

$$\text{Magnification} = \frac{\text{Image size}}{\text{Actual size}}$$

Figure 3

Questions

1. If a nucleus measures 100 mm on a diagram, with a magnification of ×10 000, what is the actual size of the nucleus?

2. Draw up a table to show each of the following measurements in metres (m), millimetres (mm) and micrometres (μm): 5 μm, 0.3 m, 23 mm, 75 μm.

3. Express the following measurements in micrometres: (a) 5 cm, (b) 25 mm, and (c) 100 nm.

4. Express the following measurements in **nanometres**: (a) 0.5 mm, (b) 0.4 μm, and (c) 0.1 cm.

5. You want to examine and measure the sizes of some living, single-celled eukaryotic organisms in pond water.
   (a) Describe how you would make the slide, and how you would illuminate the specimen under the microscope.
   (b) Explain why it may be difficult to focus this specimen under the highest magnification of your microscope (see topic 2.1.2 as well to help you answer this question).

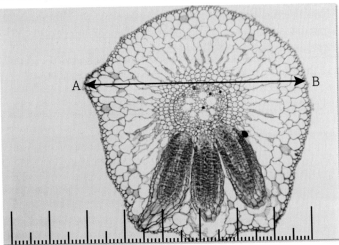

Figure 2 Transverse section of a plant root as seen under a light microscope and with a graticule in place.

(4) The ultrastructure of eukaryotic cells: membrane-bound organelles

By the end of this topic, you should be able to demonstrate and apply your knowledge and understanding of:

* the ultrastructure of eukaryotic cells and the functions of the different cellular components

* the importance of the cytoskeleton

* photomicrographs of cellular components in a range of eukaryotic cells

All animal, plant, fungal and protoctist cells are eukaryotic. This means that they have (see Figure 1):

* a nucleus surrounded by a nuclear envelope and containing DNA organised and wound into linear chromosomes

* an area inside the nucleus called the nucleolus, containing RNA, where chromosomes unwind; the nucleolus is also involved in making ribosomes

* jelly-like cytoplasm in which the organelles are suspended

* a cytoskeleton – a network of protein filaments (actin or microtubules) within the cytoplasm that move organelles from place to place within the cell; allow some cells (amoebae and lymphocytes) to move; and allow contraction of muscle cells

* a plasma membrane (also called cell surface membrane or cytoplasmic membrane)

* membrane-bound organelles, other than the nucleus, such as mitochondria, the Golgi apparatus and endoplasmic reticulum

* small vesicles

* ribosomes, which are organelles without membranes, where proteins are assembled.

Organelles

Cells are the fundamental units or building blocks of all living organisms. You will see in topics 2.6.4 and 2.6.5 that cells become specialised to do particular jobs. Within every cell there are various organelles, each having specific functions. This provides a division of labour, which means that every cell can carry out its many functions efficiently.

Membrane-bound organelles

Most of the organelles within eukaryotic cells are membrane bound, which means they are covered by a membrane (similar in structure to the plasma membrane or cell surface membrane); see topic 2.5.1. This keeps each organelle separate from the rest of the cell, so that it is a discrete compartment. Membrane-bound organelles are a feature of eukaryotic cells; prokaryotic cells do not have them.

Electron microscopy has enabled scientists to ascertain the *structure* of these organelles by making and examining several sections through an organelle in order to build up a 3D picture of it. Biochemistry research has enabled scientists to find the *functions* of each organelle.

(a)

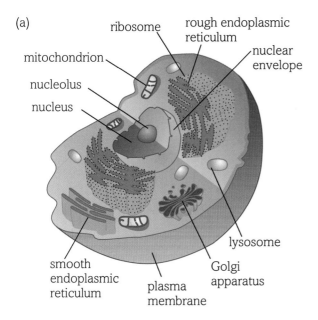

(b)

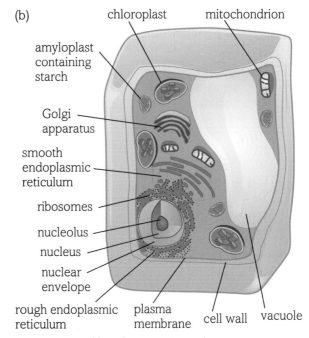

Figure 1 Structure of (a) generalised animal cell and (b) generalised plant cell, showing structures visible with a transmission electron microscope.

Nucleus, nuclear envelope and nucleolus

Structure	Function
• The nucleus is surrounded by a double membrane, called the nuclear envelope. There are pores in the nuclear envelope. • The nucleolus does not have a membrane around it. It contains RNA. • Chromatin is the genetic material, consisting of DNA wound around histone proteins. When the cell is not dividing, chromatin is spread out or extended. When the cell is about to divide, chromatin condenses and coils tightly into chromosomes (see topic 2.6.2). These make up nearly all the organism's genome (see Figure 2). 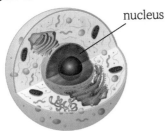 nucleus **Figure 2**	• The nuclear envelope separates the contents of the nucleus from the rest of the cell. • In some regions the outer and inner nuclear membranes fuse together. At these points some dissolved substances and ribosomes can pass through. • The pores enable larger substances, such as messenger RNA (mRNA) to leave the nucleus (see topic 2.3.3). Substances, such as some steroid hormones, may enter the nucleus, from the cytoplasm, via these pores. • The nucleolus is where ribosomes are made. • Chromosomes contain the organism's genes. In summary, the nucleus: • is the control centre of the cell • stores the organism's genome • transmits genetic information • provides the instructions for protein synthesis.

Rough endoplasmic reticulum (RER)

Structure	Function
• This is a system of membranes, containing fluid-filled cavities (cisternae) that are continuous with the nuclear membrane. • It is coated with ribosomes.	• RER is the intracellular transport system: the cisternae form channels for transporting substances from one area of a cell to another. • It provides a large surface area for ribosomes, which assemble amino acids into proteins (see topic 2.3.3). These proteins then actively pass through the membrane into the cisternae and are transported to the Golgi apparatus for modification and packaging.

Smooth endoplasmic reticulum (SER)

Structure	Function
• This is a system of membranes, containing fluid-filled cavities (cisternae) that are continuous with the nuclear membrane. • There are no ribosomes on its surface (see Figure 3). 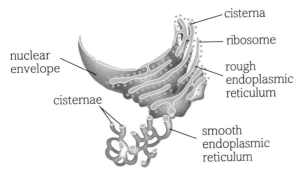 cisterna ribosome nuclear envelope rough endoplasmic reticulum cisternae smooth endoplasmic reticulum **Figure 3**	• SER contains enzymes that catalyse reactions involved with lipid metabolism, such as: 　○ synthesis of cholesterol 　○ synthesis of lipids/phospholipids needed by the cell 　○ synthesis of steroid hormones. • It is involved with absorption, synthesis and transport of lipids (from the gut). 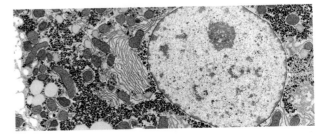 **Figure 4**

Golgi apparatus

Structure	Function

This consists of a stack of membrane-bound flattened sacs. Secretory vesicles bring materials to and from the Golgi apparatus (see Figure 5).
 	
vesicles bringing materials to and from the Golgi apparatus
 	
Figure 5
	• Proteins are modified for example by:
	  ○ adding sugar molecules to make glycoproteins
	  ○ adding lipid molecules to make lipoproteins
	  ○ being folded into their 3D shape.
	• The proteins are packaged into vesicles that are pinched off and then:
	  ○ stored in the cell or
	  ○ moved to the plasma membrane, either to be incorporated into the plasma membrane, or exported outside the cell.

Mitochondria (singular: mitochondrion)

Structure	Function
• These may be spherical, rod-shaped or branched, and are 2–5 μm long. • They are surrounded by two membranes with a fluid-filled space between them. The inner membrane is highly folded into cristae. • The inner part of the mitochondrion is a fluid-filled matrix. (see Figure 6). cristae matrix outer membrane inner membrane intermembrane space **Figure 6**	• Mitochondria are the site of ATP (energy currency) production during aerobic respiration. • They are self-replicating, so more can be made if the cell's energy needs increase. • They are abundant in cells where much metabolic activity takes place, for example in liver cells and at synapses between neurones where neurotransmitter is synthesised and released. **Figure 7**

Chloroplasts

Structure	Function
• These are large organelles, 4–10 μm long. • They are found only in plant cells and in some protoctists. • They are surrounded by a double membrane or envelope. The inner membrane is continuous with stacks of flattened membrane sacs called thylakoids (resembling piles of plates), which contain chlorophyll. Each stack or pile of thylakoids is called a granum (plural: grana). The fluid-filled matrix is called the stroma. • Chloroplasts contain loops of DNA and starch grains (see Figure 8). inner membrane outer membrane stroma granum intermembrane compartment intergranal lamallae thylakoids **Figure 8**	• Chloroplasts are the site of photosynthesis. • The first stage of photosynthesis, when light energy is trapped by chlorophyll and used to make ATP, occurs in the grana. Water is also split to supply hydrogen ions. • The second stage, when hydrogen reduces carbon dioxide, using energy from ATP, to make carbohydrates, occurs in the stroma. Chloroplasts are abundant in leaf cells, particularly the palisade mesophyll layer. **Figure 9**

Vacuole

Structure	Function
The vacuole is surrounded by a membrane called the tonoplast, and contains fluid.	• Only plant cells have a large permanent vacuole. • It is filled with water and solutes and maintains cell stability, because when full it pushes against the cell wall, making the cell turgid. • If all the plant cells are turgid then this helps to support the plant, especially in non-woody plants. There is a practical involving the pigments in beetroot cell vacuoles in topic 2.5.5.

Lysosomes

Structure	Function
• These are small bags, formed from the Golgi apparatus. Each is surrounded by a single membrane. • They contain powerful hydrolytic (digestive) enzymes. • They are abundant in phagocytic cells such as neutrophils and macrophages (types of white blood cell) that can ingest and digest invading pathogens such as bacteria (see topic 4.1.5).	• Lysosomes keep the powerful hydrolytic enzymes separate from the rest of the cell. • Lysosomes can engulf old cell organelles and foreign matter, digest them and return the digested components to the cell for reuse.

Cilia and undulipodia

Structure	Function
• These are protrusions from the cell and are surrounded by the cell surface membrane. • Each contains microtubules (see 'Cytoskeleton', in topic 2.1.5). • They are formed from centrioles (see 'Centrioles', in topic 2.1.5).	• The epithelial cells lining your airways each have many hundreds of cilia that beat and move the band of mucus. • Nearly all cell types in the body have one cilium that acts as an antenna. It contains receptors and allows the cell to detect signals about its immediate environment. • The only type of human cell to have an undulipodium (a longer cilium) is a spermatozoon. The undulipodium enables the spermatozoon to move.

LEARNING TIP

Sometimes the undulipodium of eukaryotic cells is called a flagellum (from the Latin word meaning 'whip'). However, some bacteria (prokaryotes) also have a flagellum or flagella (more than one flagellum), and as the internal structure of the prokaryotic flagellum is different from that of the eukaryotic 'flagellum', the eukaryotic structure should really be called an undulipodium.

DID YOU KNOW?

Scientists have recently discovered that nearly all of our cells have at least one cilium. Cells lining the kidney tubules each have one, and these cilia monitor the flow of urine. Brain cells also have one, and these are important for enabling learning. Individuals with genetic diseases that prevent the formation of these cilia have learning difficulties.

LEARNING TIP

Remember that there is a level of organisation within an organism. Organelles are small structures inside cells. Do not confuse them with organs, which are large structures made of many cells and tissues.

Questions

1. Name one substance that passes out of pores in the nuclear envelope, to the cell cytoplasm.

2. Name one substance that passes into the nucleus via the pores in the nuclear envelope.

3. Describe the role of the nucleolus.

4. State three functions of the nucleus.

5. Name one type of human cell that does not contain a nucleus.

6. Suggest why hydrolytic enzymes within cells need to be inside a vesicle.

7. If you carried out a physical training programme, how and why would you expect the number of mitochondria in your muscle cells to change?

(5) Other features of eukaryotic cells

By the end of this topic, you should be able to demonstrate and apply your knowledge and understanding of:

* the ultrastructure of eukaryotic cells and the function of the different cellular components
* the importance of the cytoskeleton

Organelles without membranes

Ribosomes and the cytoskeleton, including centrioles, are not covered by membranes.

The tables below explain how the structures of organelles without a membrane help them to carry out their functions.

Ribosomes

Structure	Function
• Small spherical organelles, about 20 nm in diameter. • Made of ribosomal RNA. • Made in the nucleolus, as two separate subunits, which pass through the nuclear envelope into the cell cytoplasm and then combine. • Some remain free in the cytoplasm and some attach to the endoplasmic reticulum.	• Ribosomes bound to the exterior of RER are mainly for **synthesising** proteins that will be exported outside the cell. • Ribosomes that are free in the cytoplasm, either singly or in clusters, are primarily the site of assembly of proteins that will be used inside the cell. Topic 2.3.3 outlines how proteins are made.

Centrioles

Structure	Function
• The centrioles consist of two bundles of microtubules at right angles to each other. The microtubules are made of tubulin protein subunits, and are arranged to form a cylinder (see Figure 1). centrosome — nucleus — two centrioles **Figure 1** Centrioles.	• Before a cell divides, the spindle, made of threads of tubulin, forms from the centrioles. • Chromosomes attach to the middle part of the spindle (see topic 2.6.2) and motor proteins walk along the tubulin threads, pulling the chromosomes to opposite ends of the cell. Centrioles are involved in the formation of cilia and undulipodia: • Before the cilia form, the centrioles multiply and line up beneath the cell surface membrane. • Microtubules then sprout outwards from each centriole, forming a cilium or undulipodium. Centrioles are usually absent from cells of (higher) plants but may be present in some unicellular green algae, such as *Chlamydomonas*.

Cytoskeleton

Structure	Function
A network of protein structures within the cytoplasm. It consists of: <ul><li>rod-like microfilaments made of subunits of the protein actin; they are polymers of actin and each microfilament is about 7 nm in diameter</li><li>intermediate filaments about 10 nm in diameter</li><li>straight, cylindrical microtubules, made of protein subunits called tubulin; about 18–30 nm in diameter.</li><li>The **cytoskeletal motor proteins**, myosins, kinesins and dyneins, are molecular motors. They are also enzymes and have a site that binds to and allows **hydrolysis** of ATP as their energy source.</li></ul>	The protein microfilaments within the cytoplasm give support and mechanical strength, keep the cell's shape stable and allow cell movement. Microtubules also provide shape and support to cells, and help substances and organelles to move through the cytoplasm within a cell. <ul><li>They form the track along which motor proteins (dynein and kinesin) walk and drag organelles from one part of the cell to another.</li><li>They form the spindle before a cell divides. These spindle threads enable chromosomes to be moved within the cell.</li><li>Microtubules also make up the cilia, undulipodia and centrioles.</li></ul> **Intermediate filaments** are made of a variety of proteins. They: <ul><li>anchor the nucleus within the cytoplasm</li><li>extend between cells in some tissues, between special junctions, enabling cell–cell signalling and allowing cells to adhere to a basement membrane, therefore stabilising tissues.</li></ul>

Cellulose cell wall

Structure	Function
The cell wall of plants is on the outside of the plasma membrane. It is made from bundles of cellulose fibres.  **Figure 2** There is more information on the structure of cellulose in Chapter 2.2.	Absent from animal cells, the cell wall is strong and can prevent plant cells from bursting when turgid (swollen). The cell walls of plant cells: <ul><li>provide strength and support</li><li>maintain the cell's shape</li><li>contribute to the strength and support of the whole plant</li><li>are permeable and allow solutions (solute and solvent) to pass through.</li></ul> Fungi have cell walls that contain chitin, not cellulose.

DID YOU KNOW?

There are many different types of kinesin. In the human genome there are 45 genes for kinesin proteins. Each type has a differently shaped tail domain so that it can attach to a different type of 'cargo' to move it within the cell.

Questions

1. Describe how the functions of ribosomes that are free in the cytoplasm differ from the functions of ribosomes that are attached to RER.

2. Name the monomeric units of microtubules that occur in the cytoskeleton.

3. Describe the functions of the cellulose cell wall found in plant cells.

4. On a dull day, the chloroplasts inside palisade leaf cells are moved up to near the surface, to absorb more light. By what mechanism do you think the chloroplasts are moved?

(6) How organelles work together in cells

By the end of this topic, you should be able to demonstrate and apply your knowledge and understanding of:

* the interrelationship between the organelles involved in the production and secretion of proteins

Making and secreting a protein

- The **gene** that has the coded instructions for a protein such as insulin, housed on chromatin in the nucleus, is **transcribed** into a length of RNA, called messenger RNA (mRNA).

- Many copies of this mRNA are made and they pass out of the pores in the nuclear envelope to the ribosomes.

- At the ribosomes, the instructions are **translated** and insulin molecules are assembled.

- The insulin molecules pass into the cisternae of the rough endoplasmic reticulum (RER) and along these hollow sacs.

- Vesicles with insulin inside are pinched off from the RER and pass, via microtubules and motor proteins, to the Golgi apparatus.

- The vesicles fuse with the Golgi apparatus, where the insulin protein molecules may be modified for release.

- Inside vesicles pinched off from the Golgi apparatus, these molecules pass to the plasma membrane.

- The vesicles and plasma membrane fuse, and the insulin is released to the outside of the cell.

This is a type of bulk transport called **exocytosis**, an active process for which energy is needed. (See topic 2.5.4 for more on exocytosis.)

> **LEARNING TIP**
>
> Proteins are made at ribosomes. They may pass into the RER to be taken to the Golgi apparatus for modification, but they are not actually made in the RER.

Questions

Hint: You may need to refer back to information in topics 2.1.4 and 2.1.5 in order to answer these questions.

1. Why do you think beta cells of the islets of Langerhans in the pancreas contain many mitochondria?

2. Explain how the vesicles containing insulin, that pinch off from the Golgi apparatus, are moved to the plasma membrane of the beta cells.

3. At which stage on the diagram in Figure 1 does exocytosis occur?

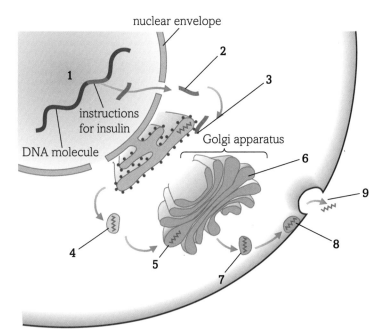

1. mRNA copy of the instructions (gene) for insulin is made in the nucleus.
2. mRNA leaves the nucleus through a nuclear pore.
3. mRNA attaches to a ribosome, in this case attached to rough endoplasmic reticulum. Ribosome reads the instructions to assemble the protein (insulin).
4. Insulin molecules are 'pinched off' in vesicles and travel towards Golgi apparatus.
5. Vesicle fuses with Golgi apparatus.
6. Golgi apparatus processes and packages insulin molecules ready for release.
7. Packaged insulin molecules are 'pinched off' in vesicles from Golgi apparatus and move towards plasma membrane.
8. Vesicle fuses with plasma membrane.
9. Plasma membrane opens to release insulin molecules outside.

Figure 1 How insulin is made in a beta cell in an islet of Langerhans in the pancreas.

(7) Prokaryotic cells

By the end of this topic, you should be able to demonstrate and apply your knowledge and understanding of:

* the similarities and differences in the structure and ultrastructure of prokaryotic and eukaryotic cells

Comparing prokaryotic and eukaryotic cells

Bacteria are microorganisms. They have prokaryotic cells (see Figure 1).

Their cells are *similar to* eukaryotic cells in that they have:

* a plasma membrane
* cytoplasm
* ribosomes for assembling amino acids into proteins
* DNA and RNA.

They are *different from* eukaryotic cells, as they:

* are much smaller
* have a much less well-developed cytoskeleton with no centrioles
* do not have a nucleus
* do not have membrane-bound organelles such as mitochondria, endoplasmic reticulum, chloroplasts or Golgi apparatus
* have a wall that is made of peptidoglycan and not cellulose

* have smaller ribosomes
* have naked DNA that is not wound around histone proteins but floats free in the cytoplasm, as a loop (not linear chromosomes).

Some prokaryotic cells also have:

* a protective waxy capsule surrounding their cell wall
* small loops of DNA, called plasmids, as well as the main large loop of DNA
* flagella – long whip-like projections that enable them to move. The structure of these flagella differs from that of eukaryotic undulipodia
* pili – smaller hair-like projections that enable the bacteria to adhere to host cells or to each other, and allow the passage of plasmid DNA from one cell to another.

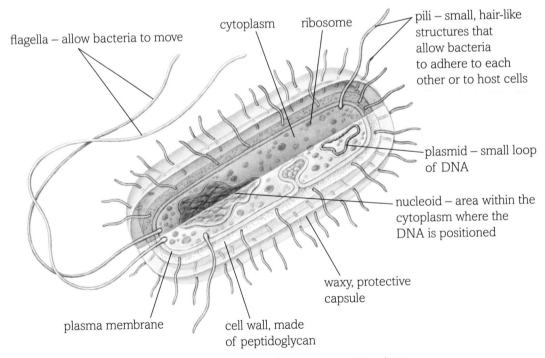

Figure 1 Many of the features of prokaryotic cells. Prokaryotic cells are usually between 1 and 5 μm long.

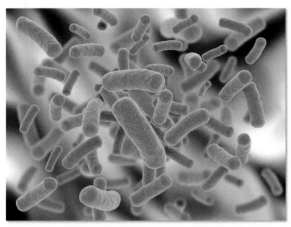

Figure 2 SEM of rod-shaped (bacillus) bacteria (×20 000).

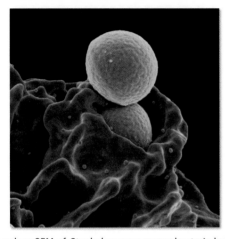

Figure 3 False-colour SEM of *Staphylococcus aureus* bacteria being engulfed by a neutrophil (×30 000).

Prokaryotic cells divide by **binary fission** and *not* by mitosis. They do not have linear chromosomes, so could not carry out mitosis. However, before they divide, their DNA is copied so that each new cell receives the large loop of DNA and any smaller plasmids.

Prokaryotes (bacteria) do not have any membrane-bound organelles, but they do have organelles that are not covered by a membrane, such as ribosomes.

Bacteria are microorganisms because they are very small. They are also prokaryotes because of their cell structure. However, not all microorganisms are prokaryotes. Yeast (which is a single-celled fungus) and amoebae have eukaryotic cells.

Viruses are microscopic but they do not have cells.

Questions

1. Make a list to show the similarities between prokaryotic and eukaryotic cells.

2. Make a table to show the differences between prokaryotic and eukaryotic cells.

3. Explain what is meant by the endosymbiont theory.

4. What features of chloroplast and mitochondrial structures support the endosymbiont theory of the origin of eukaryotic cells?

5. By what process do prokaryotic cells divide?

6. Name three diseases of humans caused by bacteria.

THINKING BIGGER

CELL THEORY

In 1839 two scientists, Schleiden and Schwann, developed cell theory that said all living organisms are made of cells. Since then we have learned a lot about prokaryotic and eukaryotic cells. Viruses have also been discovered, and they do not have cells.

WILL CELL THEORY NEED TO BE MODIFIED ONE DAY?

Cell theory was developed in 1839. At that time, all living things were classified as either animals or plants. Bacteria and viruses had not yet been discovered.

In the late 1700s, Dr Edward Jenner, the man who developed the first vaccine against smallpox, used the term 'virus' to describe the contents of pustules in people suffering from smallpox. However, he used it in its Latin meaning – poison.

By 1885 Louis Pasteur had observed small single-celled entities that he called germs, which could cause certain diseases. In fact these were bacteria. Pasteur also used a very fine mesh to filter the spinal fluid of rabbits infected with rabies. This mesh had very small holes that prevented bacteria from passing through. However, this filtered spinal fluid, when injected into healthy rabbits, gave them rabies. He concluded that the infecting agent which caused rabies was smaller than bacteria.

We now know that rabies is caused by a virus belonging to a group of viruses called *Lyssavirus*.

In general, viruses have the following properties:

● they do not have a cell structure – they have no cytoplasm, membranes or organelles
● they have genetic material – either DNA or RNA, but not both
● their genome is enclosed in a protein coat made of capsomeres
● some have a lipid membrane, around the protein coat, derived from the membrane of cells that they infect
● they can only reproduce when inside a living host cell

When viruses infect other cells they insert their genomes into the host cell's genome. Some viruses, e.g. influenza, infect more than one species of host and can carry genes from one species to another in the process. They are nature's genetic engineers and as they have been around a long time, genetic modification of living organisms is not at all new.

Are viruses alive?

Cell theory states that all living organisms are made up of cells. We know that viruses do not have a cell structure. They are described as akaryotes. So, on that basis, we should say they are not alive.

However, they have DNA or RNA and can replicate – albeit with help from a host cell. This is also true of two groups of very small bacteria, *Chlamydia* and *Rickettsia*, which can only reproduce when inside another living cell, but they are classed as living organisms.

Bear in mind that cell theory was developed in the 1830s, before viruses were discovered. Perhaps, in time, cell theory will have to be modified so that viruses can better fit into the big picture of life on Earth. Until the second half of the 20th century all life forms were classified as being either in the plant or animal kingdoms. However, since the advent of sophisticated microscopes, increased knowledge about the structures of bacteria, fungi and protoctists has meant that we have had to modify the classification of living things. You will have learnt about the five kingdoms. Many biologists now classify organisms into one of three domains, archaea, bacteria or eukaryotes. All of this knowledge is also helping biologists to work out the evolutionary relationships between life forms on Earth.

Where else will I encounter these themes?

1.1   2.1   YOU ARE HERE   2.2   2.3   2.4   2.5

Viruses outnumber all other biological entities put together. In just one teaspoon of seawater, for example, there are one million virus particles.

Recently a virus large enough to be seen with the light microscope was unearthed from the Siberian permafrost, where it had lain dormant for 30 000 years. It was reactivated and its genome, of several thousand genes, was sequenced. It infects amoebae.

Use your own knowledge, information in the text above and in the rest of Chapter 2.1, plus research using the Internet, to answer these questions.

1. What are the main statements of cell theory?
2. Which statements in cell theory can be applied to viruses?
3. When Linnaeus developed his classification system, all living things were classified into one of two groups – Animalia and Plantae. How many kingdoms are there in the present-day biological classification system?
4. Explain the phrase 'viruses are nature's genetic engineers'.
5. Suggest why it is unlikely that viruses were the first forms of biological entity on Earth.
6. Which organelles in the host cell are involved in replicating the virus particles?
7. Discuss, using your biological knowledge, whether you think viruses should be classified as living or non-living. Do you think that 'cell theory' may need to be amended in the future? Give reasons for your answer.

Activity

The branch of biology dealing with DNA technology such as genetic profiling and genetic modification of living organisms relies on knowledge and understanding of bacteria and viruses, as well as knowledge of the structure of DNA.

(a) Investigate, using the Internet, exactly how viruses can provide scientists working in this field with certain chemicals that they need.

(b) Write an article or make a poster showing how viruses can be helpful to humans.

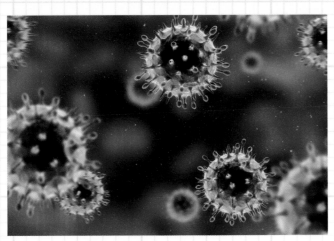

Figure 1 Structure of virus particles (×250 000).

Practice questions

1. Which of the following organelles are found in both eukaryotic and prokaryotic cells? [1]
   - A. Chloroplasts
   - B. Golgi bodies
   - C. Mitochondria
   - D. Ribosomes

2. Figure 1 shows the general structure of an animal cell.

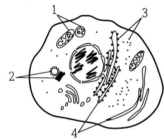

Figure 1

Which row correctly identifies the structures labelled in the diagram? [1]

	Structure 1	Structure 2	Structure 3	Structure 4
A	mitochondria	Golgi	rough endoplasmic reticulum	smooth endoplasmic reticulum
B	centrioles	mitochondria	vesicles	cytoskeleton
C	mitochondria	centrioles	ribosomes	endoplasmic reticulum
D	mitochondria	centrosome	ribosome	Golgi

3. Read the following statements:
   - (i) Scanning electron microscopes can be used to observe whole living specimens.
   - (ii) The wavelength of visible light is about 12×10^5 times longer than the wavelength of an electron beam used in an electron microscope.
   - (iii) Laser scanning microscopes can focus on structures at different depths within a specimen.

   Which statement(s) is/are true? [1]
   - A. (i), (ii) and (iii)
   - B. Only (i) and (ii)
   - C. Only (ii) and (iii)
   - D. Only (i)

4. Which of the following is not a function of the rough endoplasmic reticulum? [1]
   - A. It is the site of protein synthesis.
   - B. It is an intracellular transport system.
   - C. It gives a large surface area for ribosomes.
   - D. It can channel newly synthesised proteins to the Golgi body.

5. Read the following statements.
   - (i) Cilia are formed from centrioles and each is surrounded by the plasma membrane.
   - (ii) Cilia contain microtubules that enable them to move.
   - (iii) Cilia are only found on epithelial cells lining the airways.

   Which statement(s) is/are true? [1]
   - A. (i), (ii) and (iii)
   - B. Only (i) and (ii)
   - C. Only (ii) and (iii)
   - D. Only (i)

   [Total: 5]

6. Figure 2 shows a goblet cell from the epithelium (lining) of the stomach. Other cells in the stomach lining produce hydrochloric acid, and the pH inside a human stomach is between 1 and 2.

   (a) The protein mucin is synthesised within the cell and secreted, in mucus, at the position marked Z.
      - (i) Place the appropriate letters in the correct order to show the passage of newly synthesised molecules of mucin as they are moved from the place where they were made to position Z. [2]

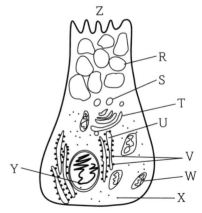

Figure 2

(ii) There are amino acids in the cell cytoplasm that may be used to make mucin. Describe precisely where in this cell the mucin molecules will be assembled from these amino acid monomers. Give a reason for your answer. [2]

(iii) By what process do mucin molecules pass out of this cell? [1]

(b) The structures labelled Z are extensions of the plasma membrane and are called microvilli. Suggest why this type of cell has microvilli. [2]

(c) Suggest why this cell has many of the structures labelled W inside it. [4]

(d) Why do you think mucus needs to be produced by the stomach? [1]

[Total: 12]

7. Some scientists wanted to study the structure and functions of chloroplasts (see Figure 3). They macerated some spinach leaves in a food blender, adding 2% sucrose solution and kept the mixture cold. They filtered the mixture to remove debris and then spun the mixture in a centrifuge, which speeds up the separation of the mixture and, after a short spin, the cell nuclei are pulled to the bottom of the tube, forming a sediment.

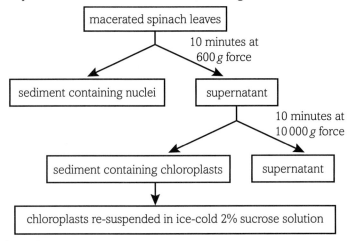

Figure 3

The supernatant liquid is then taken out of the tube, placed in another centrifuge tube and spun again at a higher speed. The chloroplasts were seen at the bottom of the tube as a green sediment.

After decanting off the supernatant liquid, the chloroplasts were resuspended in ice-cold 2% sucrose solution before being used for investigations.

(a) Suggest why the first organelles to sediment out during centrifugation were the nuclei. [1]

(b) Suggest why chloroplasts were the second type of organelles to sediment out by centrifugation. [2]

(c) Explain why leaves were used as a source of chloroplasts. [1]

(d) Suggest why, prior to their use, the isolated chloroplasts were:

(i) suspended in 2% sucrose solution [1]

(ii) kept ice cold. [2]

(e) Name the substance that gives the chloroplasts their green colour. [1]

(f) Briefly outline the function of chloroplasts. [2]

[Total: 10]

8. The electron micrograph in Figure 4 shows some plant cells.

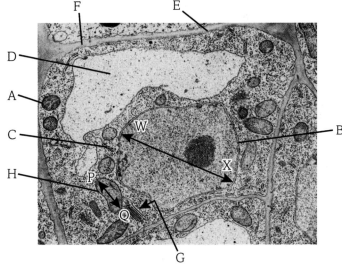

Figure 4

(a) Identify the structures labelled A–G. [7]

(b) The true diameter, across line WX, of that organelle is $10 \mu m$.

(i) What is the magnification of this electron micrograph? Show your working. [2]

(ii) Calculate the length of structure H, along the line PQ. Show your working. [2]

(iii) The organelle with diameter $10 \mu m$ is spherical. Calculate its volume. Express your answer to the nearest whole number. Show your working. [3]

(c) Explain why this electron-micrograph image is grey-scale (has no colour). [1]

(d) State two functions of structure D. [2]

[Total: 17]

Foundations in biology

BIOLOGICAL MOLECULES

Introduction

Life on Earth survives because of water. Water is unlike any other liquid. Its special properties allow trees to grow, blood to flow, and animals and plants to survive freezing temperatures. However, water is just one of a large number of biological molecules and ions from which living things are made. These are built from relatively few chemical elements, but these elements can form a vast array of biologically important molecules, some of which have millions of atoms bonded together. Without them, we would look very different. We would have no sources or stores of energy. We would have no bones or tendons. We would have no enzymes or hormones. Even our skin would lose its stretchiness.

In this module, you will learn about the spectrum of molecules upon which the survival of living things depends. You will learn about the special properties of water, which enable life to survive. You will learn how atoms and molecules are bonded together, and about three classes of biological molecules: carbohydrates, lipids and proteins, understanding their structure and function in living things. You probably remember learning about food tests at GCSE. You will revisit these, understand how they work, and learn how they and other techniques can be used to investigate the biological molecules in our food.

All the maths you need

To unlock the puzzles of this chapter you need the following maths

- Power (e.g. 2 to the power of 100)
- Units of measurement
- Ratios and division

What have I studied before?

- Humans and other animals need a balanced diet
- Foods have different nutrients in them, including carbohydrates, fats and proteins
- Plants make their own food in the form of glucose, which they convert to starch
- Animals and plants respire glucose to release energy

What will I study later?

- Nucleic acids are important biological molecules (AS)
- The structure and function of enzymes (AS)
- Membranes are made from phospholipids, cholesterol and proteins (AS)
- Antibodies are proteins involved in our immune response (AS)
- Cell signalling depends on biological molecules (AL)
- The role of the liver in storing glycogen (AL)
- The structure and function of hormones (AL)
- Photosynthesis and respiration (AL)

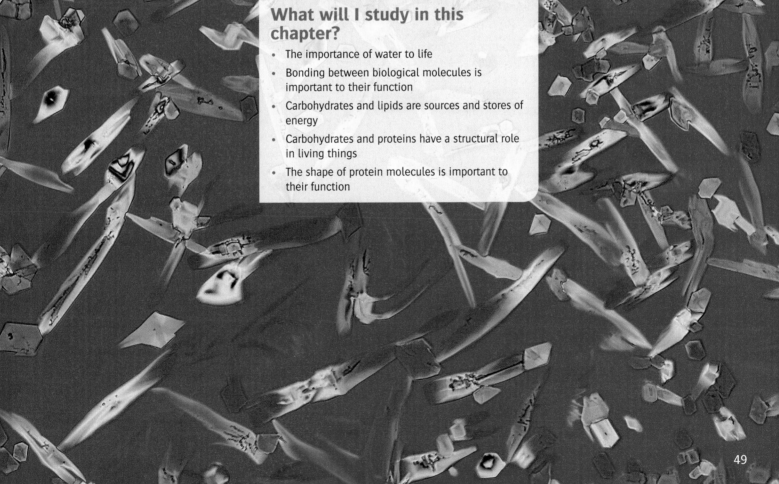

What will I study in this chapter?

- The importance of water to life
- Bonding between biological molecules is important to their function
- Carbohydrates and lipids are sources and stores of energy
- Carbohydrates and proteins have a structural role in living things
- The shape of protein molecules is important to their function

① Molecular bonding

By the end of this topic, you should be able to demonstrate and apply your knowledge and understanding of:

* the concept of monomers and polymers and the importance of condensation and hydrolysis reactions in a range of biological molecules

* how hydrogen bonding occurs between water molecules, and relate this, and other properties of water, to the roles of water for living organisms

* the chemical elements that make up biological molecules

KEY DEFINITIONS

condensation reaction: reaction that occurs when two molecules are joined together with the removal of water.

hydrogen bond: a weak interaction that can occur wherever molecules contain a slightly negatively charged atom bonded to a slightly positively charged hydrogen atom.

hydrolysis reaction: reaction that occurs when a molecule is split into two smaller molecules with the addition of water.

monomer: a small molecule which binds to many other identical molecules to form a polymer.

polymer: a large molecule made from many smaller molecules called monomers.

Understanding the structure of molecules is essential to understanding their function. In this topic, you will learn about the bonding within and between biological molecules.

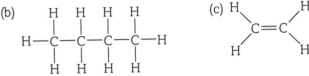

Figure 1 (a) The carbon–hydrogen bond is one of the most important in organic molecules. Hydrogen has one electron. Carbon can therefore bind with four hydrogen atoms. (b) The carbon–carbon bond is also very important in hydrocarbon chains. (c) Sometimes a double bond can form (the equivalent of sharing two electrons).

LEARNING TIP

When drawing the molecular structure of biological molecules, make sure that hydrogen atoms only have a single bond, that oxygen atoms only have two bonds, that nitrogen atoms only have three bonds, and carbon atoms have four bonds. A double bond counts as two bonds.

Covalent bonds

Atoms consist of a nucleus (of protons and neutrons) surrounded by shells of electrons. Most atoms tend to be stable when their outermost shell contains eight electrons. Atoms of different elements have different numbers of electrons in their outermost shells. For example, carbon has four electrons. By sharing electrons with other atoms, the atom's outermost shell can be 'filled' and it becomes strongly bonded with the other atom. This is called a **covalent bond**, and is drawn by a single line (the equivalent of sharing one electron). Because carbon's outer shell is filled by four electrons, carbon forms four covalent bonds. Hydrogen forms a single covalent bond, oxygen two covalent bonds and nitrogen three covalent bonds.

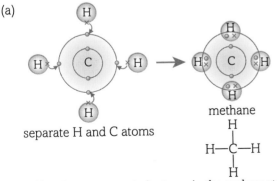

Blue dots represent electrons in the carbon atom.
Red crosses represent electrons in the hydrogen atoms.

Condensation, hydrolysis and polymerisation

When condensation happens on your windows on a cold day, water vapour in the air has changed state and settled on the window as a liquid. The water vapour has been removed from the air.

A **condensation reaction** occurs when two molecules are joined together with the removal of water. In the same way, two molecules can be split apart with the addition of water. This kind of reaction is called **hydrolysis** (literally, water-splitting). Almost all condensation reactions happen in the same way, when two –OH groups react together. The reaction involves the breaking and formation of covalent bonds.

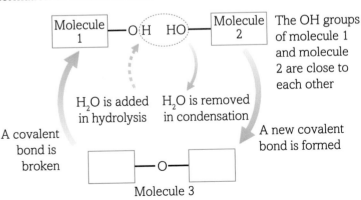

Figure 2 Hydrolysis and condensation reactions.

Monomers and polymers

Condensation and hydrolysis reactions are responsible for linking and splitting apart biological molecules in living things. The units which are joined together are called monomers. When two **monomers** join together they form a dimer. When lots of monomers join together they form a **polymer**.

We will look in more detail at the hydrolysis and condensation reactions involved in the structure of carbohydrates, lipids and proteins in the subsequent topics in this chapter (nucleic acids are covered in Chapter 2.3). All of these molecules contain carbon, hydrogen and oxygen atoms. Proteins also contain nitrogen and sulfur, and nucleic acids also contain nitrogen and phosphorus. Lipids do not form polymers like the others, but they do still contain C, H and O atoms.

Type of molecule	Monomer	Polymer
Carbohydrates (C, H and O)	Monosaccharides (e.g. glucose)	Polysaccharides (e.g. starch)
Proteins (C, H, O, N and S)	Amino acids	Polypeptides and proteins
Nucleic acids (C, H, O, N and P)	Nucleotides	DNA and RNA

Table 1 Monomers and polymers of biological molecules.

Hydrogen bonds

Water consists of two hydrogen atoms, each covalently bonded to one oxygen atom. However, because the oxygen atom has a greater number of positive protons in its nucleus, this exerts a stronger attraction for the shared electrons. This means the oxygen atom becomes slightly negative, and the hydrogen atoms become slightly positive. When this happens, we say the molecule is **polar**.

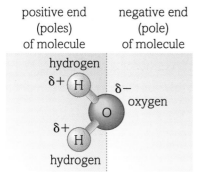

Figure 3 Distribution of charge on a polar water molecule. The Greek δ symbols ($z-$ or $\delta+$) mean 'slight' negative or positive charge.

A **hydrogen bond** is a weak interaction which happens wherever molecules contain a slightly negatively charged atom bonded to a slightly positively charged hydrogen atom. The bond is weaker than a covalent bond. However, in some polymers, thousands and thousands of hydrogen bonds form between chains of monomers. Having many bonds like this helps stabilise the structure of some biological molecules.

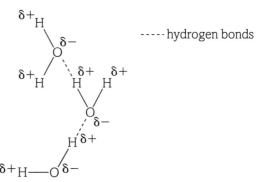

Figure 4 Hydrogen bond formation in water.

Questions

1. Explain why a water molecule is polar.

2. Explain why a hydrogen bond is weaker than a covalent bond.

3. What kind of molecule can catalyse hydrolysis reactions in living things?

4. Two nitrogen molecules form a triple covalent bond between them to form a N_2 molecule. Explain how a triple bond is formed.

5. Atoms form covalent bonds according to how many spaces they have in their outer shell of electrons. How many spaces do (a) oxygen, (b) nitrogen and (c) carbon atoms have?

② Properties of water

By the end of this topic, you should be able to demonstrate and apply your knowledge and understanding of:

* how hydrogen bonding occurs between water molecules, and relate this, and other properties of water, to the roles of water for living organisms

Water has a number of properties that are essential for life, many of which depend on the polar nature of the water molecule and on hydrogen bonding between water molecules (see previous topic 2.2.1).

Below are listed the properties of water. In each case, the effect of the polarity and hydrogen bonding of water molecules and the reasons why each property is important to living organisms are explained.

Liquid

As in any liquid, water molecules constantly move around. Unlike other liquids, as they move they continually make and break hydrogen bonds (see topic 2.2.1, Figure 4).

The hydrogen bonds between water molecules make it more difficult for them to escape to become a gas. By contrast, other less polar, but similarly-sized molecules (e.g. H_2S) are gases at room temperature. Even with hydrogen bonds, water has quite a low viscosity, which means it can flow easily.

Because it is a liquid at room temperature, water can:

* provide habitats for living things in rivers, lakes and seas
* form a major component of the tissues in living organisms
* provide a reaction medium for chemical reactions
* provide an effective transport medium, e.g. in blood and vascular tissue.

Density

The density of water provides an ideal habitat for living things. If water was less dense, aquatic organisms would find it very difficult to float.

When most liquids get colder, they become more dense. If this were the case with water, then water at the top of a pond would freeze and sink. The water replacing it at the top would do the same, until the pond would be full of ice. But water behaves differently. It becomes more dense as it gets colder until about 4 °C. As it goes from 4 °C to freezing point, because of its polar nature, the water molecules align themselves in a structure which is less dense than liquid water.

Because ice is less dense than water:

* aquatic organisms have a stable environment in which to live through the winter
* ponds and other bodies of water are insulated against extreme cold. The layer of ice reduces the rate of heat loss from the rest of the pond.

Figure 1 The ice on top of the pond is less dense than water and insulates the fish beneath.

Solvent

Water is a good **solvent** for many substances found in living things. This includes ionic **solutes** such as sodium chloride and covalent solutes such as glucose. Because water is polar, the positive and negative parts of the water molecules are attracted to the negative and positive parts of the solute. The water molecules cluster around these charged parts of the solute molecules or ions, and will help to separate them and keep them apart. At this point they dissolve and a solution is formed.

Because water is such a good solvent:

* molecules and ions can move around and react together in water. Many such reactions happen in the cytoplasm of cells, which is over 70% water.
* molecules and ions can be transported around living things whilst dissolved in water.

Cohesion and surface tension

A drop of water on a flat surface does not spread out, but can look almost spherical. This is because hydrogen bonding between the molecules pulls them together. The water molecules demonstrate cohesion.

This happens at the surface of water as well. The water molecules at the surface are all hydrogen-bonded to the molecules beneath them, and hence more attracted to the water molecules beneath than to the air molecules above. This means the surface of the water contracts (because the molecules are being pulled inwards), and it gives the surface of the water an ability to resist force applied to it. This is known as surface tension. Because of cohesion and surface tension:

- columns of water in plant vascular tissue are pulled up the xylem tissue together from the roots.

- insects like pond-skaters can walk on water.

Figure 2 A pond-skater is able to walk on water because of the surface tension resisting the weight of the animal.

High specific heat capacity

Water temperature is a measure of the kinetic energy of the water molecules. Water molecules are held together quite tightly by hydrogen bonds. Therefore, you have to put in a lot of heat energy to increase their kinetic energy and temperature. The amount of heat energy required is known as the specific heat capacity (4.2 kJ of energy to raise the temperature of 1 kg of water by 1 °C). This means that water does not heat up or cool down easily. Because the main component of many living things is water, and many organisms live in water, its high specific heat capacity is important:

- Living things, including prokaryotes and eukaryotes, need a stable temperature for enzyme-controlled reactions to happen properly.

- Aquatic organisms need a stable environment in which to live.

High latent heat of vaporisation

When water evaporates, heat energy, known as the latent heat of vaporisation, helps the molecules to break away from each other to become a gas. Because the molecules are held together by hydrogen bonds, a relatively large amount of energy is needed for water molecules to evaporate. Therefore, water can help to cool living things and keep their temperature stable. For example, mammals are cooled when sweat evaporates, and plants are cooled when water evaporates from mesophyll cells.

> **LEARNING TIP**
> Do not confuse latent heat of vaporisation with specific heat capacity.

Reactant

Water is also a reactant in reactions such as photosynthesis, and in hydrolysis reactions such as digestion of starch, proteins and lipids. Its properties as a reactant do not directly draw on its polarity, but its role as a reactant is extremely important for digestion and synthesis of large biological molecules.

Questions

1. Why is it important for water to be liquid over a broad range of temperatures?

2. Why is water referred to as polar, and how does its polarity affect its solvent properties?

3. Explain how water's thermal properties enable living things to survive.

4. Explain the difference in the molecular structure of liquid water and ice.

5. What is the difference between the latent heat of vaporisation and the specific heat capacity of water?

(3) Carbohydrates 1: Sugar

By the end of this topic, you should be able to demonstrate and apply your knowledge and understanding of:

* the ring structure and properties of glucose as an example of a hexose monosaccharide and the structure of ribose as an example of a pentose monosaccharide

* the chemical elements that make up biological molecules

* the synthesis and breakdown of a disaccharide and polysaccharide by the formation and breakage of glycosidic bonds

> **KEY DEFINITIONS**
>
> **carbohydrates:** a group of molecules containing C, H and O.
> **glycosidic bond:** a bond formed between two monosaccharides by a condensation reaction.

What are carbohydrates for?

Carbohydrates contain carbon, hydrogen and oxygen. Carbohydrates are 'hydrated carbon', which means that for every carbon there are two hydrogen atoms and one oxygen atom. The functions of carbohydrates are three-fold. They act as a source of energy (e.g. glucose), as a store of energy (e.g. starch and glycogen) and as structural units (e.g. cellulose in plants and chitin in insects). Some carbohydrates are also part of other molecules, such as nucleic acids and glycolipids. There are three main groups of carbohydrates: monosaccharides, disaccharides and polysaccharides. The common monosaccharides and disaccharides all have names ending in -ose.

Monosaccharides

Monosaccharides are the simplest carbohydrates. They are particularly important in living things as a source of energy. They are well suited to this role because of the large number of carbon–hydrogen bonds. They are sugars, which taste sweet, are soluble in water and are insoluble in non-polar solvents. Monosaccharides can exist as straight chains or in ring or cyclic forms. They have a backbone of single-bonded carbon atoms, with one double-bonded to an oxygen atom to form a carbonyl group. Different sugars can have different numbers of carbon atoms (see Table 1): hexose sugars have six carbon atoms, pentose sugars have five carbon atoms and triose sugars have three carbon atoms. Monosaccharide hexose sugars, like glucose, are the **monomers** of more complex carbohydrates, and they bond together to form **disaccharides** or **polysaccharides**.

In solution, triose and tetrose sugars exist as straight chains. However, pentoses and hexoses are more likely to be found in a ring or cyclic form (shown in Table 1).

In both forms (straight-chain and cyclic forms), glucose can exist as a number of different isomers (molecules with the same formula, but whose atoms are arranged differently in space). In the straight-chain form, the –H and –OH can be reversed. In a ring shape, isomers can also form. The ring is formed when the oxygen attached to carbon 5 bonds to carbon 1 (see Table 1). Because the –OH and –H on carbon 1 can be above or below the plane of the ring when the ring forms, there are two isomers: α- and β-glucose. This small difference appears insignificant, but it becomes very important when glucose molecules polymerise into starch or cellulose.

> **DID YOU KNOW?**
>
> The word saccharide derives from the Greek word sakkharon, which means sugar.

Disaccharides

Like monosaccharides, disaccharides are sweet and soluble. The most common disaccharides are maltose (malt sugar), sucrose and lactose (milk sugar). Maltose and lactose are **reducing sugars**, whereas sucrose is a **non-reducing sugar**.

Disaccharides are made when two monosaccharides join together. There are lots of different combinations, which determine the disaccharide made:

* α-glucose + α-glucose $\rightarrow$ maltose
* α-glucose + fructose $\rightarrow$ sucrose
* β-galactose + α-glucose $\rightarrow$ lactose
* β-glucose + β-glucose $\rightarrow$ cellobiose

When they join, a condensation reaction occurs to form a **glycosidic bond**. Two hydroxyl groups line up next to each other, from which a water molecule is removed (see topic 2.2.1). This leaves an oxygen atom acting as a link between the two monosaccharide units.

Disaccharides are broken into monosaccharides by a hydrolysis reaction, which requires addition of water. The water provides a hydroxyl group (–OH) and a hydrogen (–H), which help the glycosidic bond to break. For example, cellobiose is obtained by the hydrolysis of the polysaccharide cellulose.

Name of sugar	Displayed formula	Molecular formula	Role in the body	Type of sugar
α-Glucose		$C_6H_{12}O_6$	Energy source. Component of starch and glycogen, which act as energy stores.	Hexose
β-Glucose		$C_6H_{12}O_6$	Energy source. Component of cellulose, which provides structural support in plant cell walls.	Hexose
Ribose		$C_5H_{10}O_5$	Component of ribonucleic acid (RNA), ATP and NAD.	Pentose
Deoxyribose		$C_5H_{10}O_4$	Component of deoxyribonucleic acid (DNA).	Pentose

Table 1 Some monosaccharides.

Like most reactions, when glycosidic bonds are formed and hydrolysed in living things, such reactions are catalysed by enzymes.

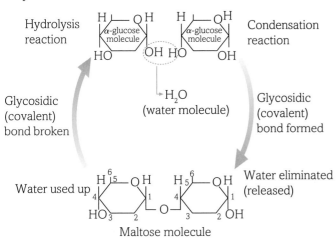

Figure 1 Formation and splitting of maltose from α-glucose. The glycosidic bond is formed between carbon 1 of one α-glucose molecule and carbon 4 of the other α-glucose molecule. It is known as a 1–4 glycosidic bond.

Questions

1. Write down the molecular formula for a triose sugar (write your answer in terms of C, H, O and x).

2. Write down the molecular formula for maltose.

3. To reduce complexity, many writers draw α-glucose like the diagram below. Draw β-glucose in the same way.

4. (a) Draw the displayed formula showing two β-glucose molecules joined together by a condensation reaction.
   (b) Draw the formulae to show what happens when this disaccharide is hydrolysed.

5. Study the formulae of ribose and deoxyribose. State the difference between the two.

6. Draw a trisaccharide of α-glucose.

4 Carbohydrates 2: Polysaccharides as energy stores

By the end of this topic, you should be able to demonstrate and apply your knowledge and understanding of:

* the synthesis and breakdown of a disaccharide and polysaccharide by the formation and breakage of glycosidic bonds

* the structure of starch (amylose and amylopectin), glycogen and cellulose molecules

* how the structures and properties of glucose, starch, glycogen and cellulose molecules relate to their functions in living organisms

Polysaccharides are **polymers** of monosaccharides. They are made of hundreds or thousands of monosaccharide monomers bonded together. Polysaccharides made solely of one kind of monosaccharide are called homopolysaccharides. Those made of more than one monomer are called heteropolysaccharides. Starch is an example of a homopolysaccharide. Hyaluronic acid (in connective tissue) is an example of a heteropolysaccharide.

Energy sources and energy stores

Glucose is a source of energy, as it is a reactant in **respiration**. The energy released is used to make ATP, which is the energy currency of the cell:

$$\text{Glucose} + \text{Oxygen} \xrightarrow{\text{releasing energy which is used to form ATP}} \text{Carbon dioxide} + \text{Water}$$

If you join lots of glucose molecules together into polysaccharides, you can create a *store* of energy, and this is exactly what living things do. Plants store energy as starch in chloroplasts and in membrane-bound starch grains, and humans store energy as glycogen in cells of the muscles and liver.

Why are polysaccharides good energy stores?

The structure of some polysaccharides lends itself to energy storage. **Glycogen** in *animals* and starch in *plants* (comprising **amylose** and **amylopectin**) occur within cells in the form of large granules. They form good stores of monosaccharides for the following reasons:

* Glycogen and starch are compact, which means they do not occupy a large amount of space. They both occur in dense granules within the cell.

* Polysaccharides hold glucose molecules in chains, so they can be easily 'snipped off' from the end of the chain by hydrolysis when required for respiration. Hydrolysis reactions are always catalysed by enzymes.

* Some chains are unbranched (amylose) and some are branched (amylopectin and glycogen). Branched chains tend to be more compact, but also offer the chance for lots of glucose molecules to be snipped off by hydrolysis at the same time, when lots of energy is required quickly. The enzyme amylase is responsible for hydrolysing 1–4 glycosidic linkages, and glucosidase is responsible for hydrolysing 1–6 glycosidic linkages. A 1–4 glycosidic linkage is one between carbon 1 of one glucose and carbon 4 of the other.

Figure 1 Starch grains in potato cells (×80).

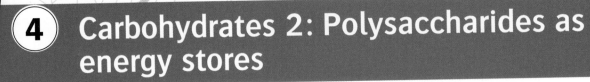

Figure 2 Starch is hydrolysed at 1–4 linkages to produce a mixture of glucose and maltose.

- Polysaccharides are less soluble in water than monosaccharides. If many glucose molecules did dissolve in the cytoplasm, the **water potential** (see topic 2.5.3) would reduce, and excess water would diffuse in, disrupting the normal workings of the cell. Polysaccharides are less soluble because of their size, but also because regions which could hydrogen-bond with water are hidden away inside the molecule. Sometimes the amylose molecule may form a double helix, which presents a hydrophobic external surface in contact with the surrounding solution.

Type of polysaccharide	Detailed structure	
Amylose (in plants): this molecule is a long chain of α-glucose molecules. Like maltose, it has glycosidic bonds between carbons 1 and 4.	Amylose coils into a spiral shape, with hydrogen bonds holding the spiral in place. Hydroxyl groups on carbon 2 are situated on the inside of the coil, making the molecule less soluble and allowing hydrogen bonds to form to maintain the coil's structure.	chains formed by 1,4-glycosidic bond
Amylopectin (in plants): this is like amylose, with glycosidic bonds between carbons 1 and 4 but in addition it has branches formed by glycosidic bonds between carbons 1 and 6.	Amylopectin also coils into a spiral shape, held together with hydrogen bonds, but with branches emerging from the spiral.	branch formed by condensation reaction between carbons 1 and 6; chains formed by 1,4-glycosidic bond
Glycogen (in animals): this molecule is like amylopectin with glycosidic bonds between carbon 1 and 4, and branches formed by glycosidic bonds between carbon 1 and 6.	The 1–4 bonded chains tend to be smaller than in amylopectin, so glycogen has less tendency to coil. However, it does have more branches, which makes it more compact. And it is easier to remove monomer units as there are more ends.	chains formed by 1,4-glycosidic bond; branch formed by condensation reaction beween carbons 1 and 6

Table 1 Energy storage in plants and humans.

DID YOU KNOW?
- The core of a glycogen or amylopectin molecule is resistant to hydrolysis by enzymes. Such a unit is called a limit dextrin, and is one of the major components of mucus.
- Glycogen is stored in the liver and in muscle cells. It may form up to 7% of the mass of the liver.

LEARNING TIP
Remember that glucose is an energy source when used in respiration, and any polymer of glucose therefore provides an energy store.

Questions

1 Copy and complete the following table:

	Amylose	Amylopectin	Glycogen
Do you find it in humans or plants?			
Where in the organism is it stored?			
Is it a form of starch?			
What is the monomer?			
Is it branched?			
Is it soluble?			
Which glycosidic bonds does it have: 1–4 or 1–6?			
Is it spiralled?			
Are hydrogen bonds important in holding the structure in place?			

2 Glycogen is more branched than starch. Explain why this difference is important to animals.

3 Why is it important that hydroxyl groups on carbon 2 of glucose are found on the inside of the amylose spiral?

4 What kind of bonding holds the coiled structure of amylose together?

5 What are the advantages of a polysaccharide having a branched chain? Explain your answer.

5 Carbohydrates 3: Polysaccharides as structural units

By the end of this topic, you should be able to demonstrate and apply your knowledge and understanding of:

* the structure of starch (amylose and amylopectin), glycogen and cellulose molecules

* how the structures and properties of glucose, starch, glycogen and cellulose molecules relate to their functions in living organisms

Cellulose

Cellulose is found in plants, forming the cell walls. It is a tough, insoluble and fibrous substance. Cellulose is a homopolysaccharide made from long chains of up to 15 000 β-glucose molecules, bonded together through **condensation** reactions to form glycosidic bonds.

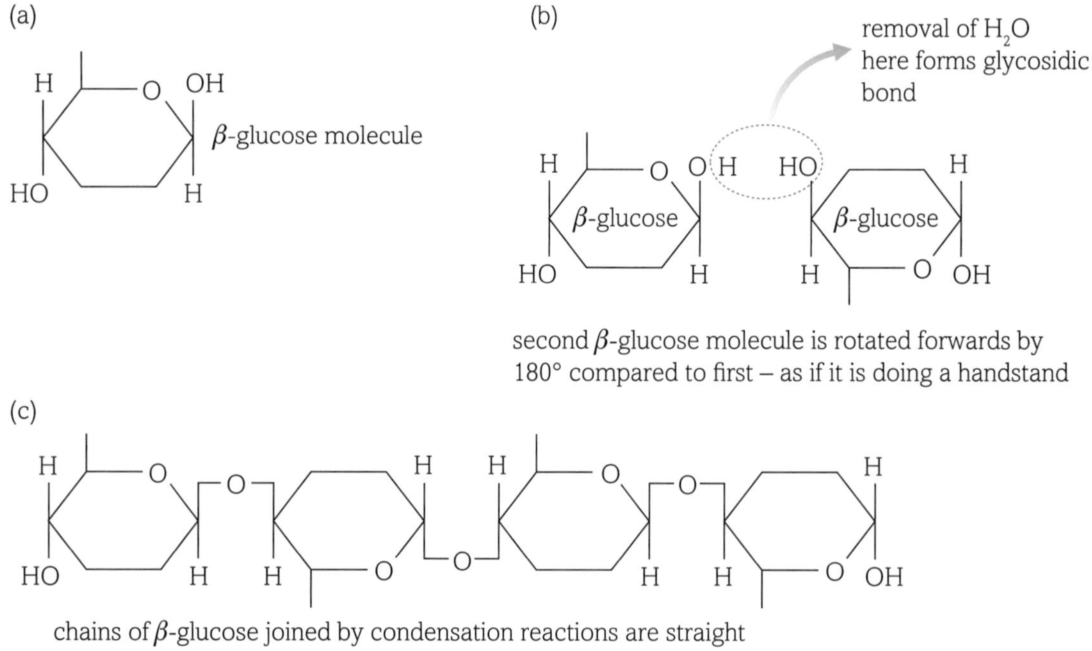

(a)

β-glucose molecule

(b)

removal of H_2O here forms glycosidic bond

β-glucose β-glucose

second β-glucose molecule is rotated forwards by 180° compared to first – as if it is doing a handstand

(c)

chains of β-glucose joined by condensation reactions are straight

Figure 1 Cellulose is made up of a chain of β-glucose molecules.

> **DID YOU KNOW?**
> Cellulose is the most common polysaccharide in the world.

Rather than spiralling like chains of α-glucose, cellulose chains are straight and lie side by side. This difference in structure is a direct result of bonding:

* Hydrogen and hydroxyl groups on carbon 1 are inverted in β-glucose (as compared with α-glucose). This means that every other β-glucose molecule in the chain is rotated by 180 degrees. This and the β-1–4 glycosidic bond help to prevent the chain spiralling.

* Hydrogen bonding between the rotated β-glucose molecules in each chain also gives the chain additional strength, and stops it spiralling.

* Hydrogen bonding between the rotated β-glucose molecules in different chains gives the whole structure additional strength. The hydroxyl group on carbon 2 sticks out, enabling hydrogen bonds to be formed between chains.

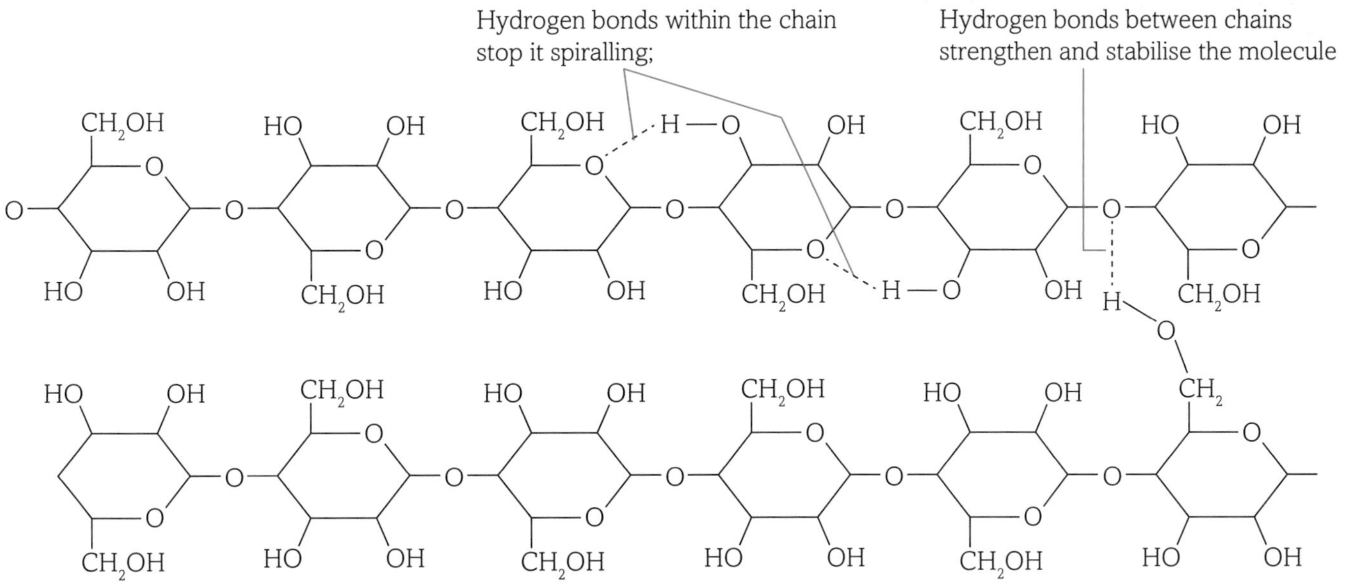

Hydrogen bonds within the chain stop it spiralling;

Hydrogen bonds between chains strengthen and stabilise the molecule

Figure 2 Straight cellulose molecules lie side by side. Hydrogen bonding within and between the chains of β-glucose molecules gives the structure strength.

When 60 to 70 cellulose chains are bound together in this way, they form microfibrils, which are 10–30 nm in diameter. These then bundle together into macrofibrils containing up to 400 microfibrils, which are embedded in pectins (like glue!) to form plant cell walls.

Arrangement of cellulose chains in the formation of cell wall macrofibrils. The macrofibrils are embedded in pectins to form the wall. Macrofibrils run in all directions criss-crossing the wall for extra strength.

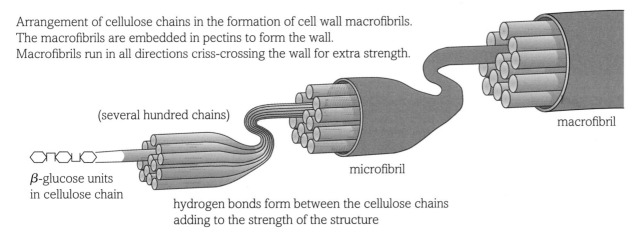

(several hundred chains)

β-glucose units in cellulose chain

macrofibril

microfibril

hydrogen bonds form between the cellulose chains adding to the strength of the structure

Figure 3 Relative structures of cellulose chains, microfibrils and a macrofibril (fibre).

Structure and function of plant cell walls

Cellulose is an excellent material for plant cell walls:

- Microfibrils and macrofibrils have very high tensile strength, both because of the strength of the glycosidic bonds but also because of the hydrogen bonds between chains. Macrofibrils are stronger than steel wire of the same diameter.

- Macrofibrils run in all directions, criss-crossing the wall for extra strength.

- It is difficult to digest cellulose because the glycosidic bonds between the glucose molecules are less easy to break. Indeed, most animals do not even have an enzyme to catalyse the reaction.

These key features help the plant cell wall to do its job:

- Because plants do not have a rigid skeleton, each cell needs to have strength to support the whole plant.

- There is space between macrofibrils for water and mineral ions to pass on their way into and out of the cell. This makes the cell wall fully permeable.

- The wall has high tensile strength, which prevents plant cells from bursting when they are turgid, again helping to support the whole plant. Turgid cells press against each other, supporting the structure of the plant as a whole. The wall also protects the delicate cell membrane.

- The macrofibril structure can be reinforced with other substances for extra support or to make the walls waterproof. For example, cutin and suberin are waxes that block the spaces in the cell wall, and make it waterproof. Lignin (a polymer of phenylpropane units) performs the same function for xylem vessels (see topic 3.3.2). In the woody part of tree trunks, cell walls are extra thick to withstand the weight.

> **DID YOU KNOW?**
> The only vertebrates to use cellulose as food are cattle and other ruminants, like sheep, goats, camels and giraffes. Microorganisms which live in their digestive system produce an enzyme called cellulase, which helps to break down cellulose.

The structural strength of cellulose has been exploited by humans for some time. Cotton is 90% cellulose. Cellophane and celluloid (which used to be used in photographic film) are also derived from cellulose. One of the main components of paper is cellulose. Rayon (viscose) is a semi-synthetic fibre produced from cellulose. It has similar properties to those of silk.

Other structural polysaccharides

Bacterial cell walls

Bacteria also have cell walls, but they are not made of cellulose (see also topic 2.1.7). The whole structure surrounding the cell is called a peptidoglycan, made from long polysaccharide chains that lie in parallel, cross-linked by short peptide chains (made of amino acids).

Exoskeletons

Insect and crustacean exoskeletons are made of chitin. It differs from cellulose because it has an acetylamino group ($NH.OCCH_3$) rather than a hydroxyl group on carbon 2. It forms cross-links between long parallel chains of acetylglucosamine, in a similar way to cellulose.

Questions

1. Explain why plant cell walls need to be strong.

2. Suggest what would happen to plant cell walls if α-glucose was the monomer of cellulose. Explain your answer, highlighting the difference between α- and β-glucose, and discussing the effect of hydrogen bonding.

3. Copy and complete the following table using information from the three topics on carbohydrates. You may include as many examples of mono-, di- and polysaccharides as you like.

Carbohydrate	Example	Displayed formula (to include any glycosidic and hydrogen bonds)	What is its role in living things?	Why is it good at fulfilling this role?
Monosaccharide				
Disaccharide				
Polysaccharide				

4. Explain, using diagrams, why every other β-glucose is rotated in cellulose, and what advantages this gives the molecule.

5. Suggest why a plant cell wall should be (a) fully porous to water and minerals, (b) waterproofed in the xylem vessel by lignin deposition.

6. Plant cells and bacterial cells have walls with long polysaccharide chains that are cross-linked together. Write down one advantage of such cross-links.

(6) Lipids 1: Triglycerides

By the end of this topic, you should be able to demonstrate and apply your knowledge and understanding of:

* the structure of a triglyceride and a phospholipid as examples of macromolecules

* the synthesis and breakdown of triglycerides by the formation (esterification) and breakage of ester bonds between fatty acids and glycerol

* how the properties of triglyceride, phospholipid and cholesterol molecules relate to their function in living organisms

KEY DEFINITIONS

lipids: a group of substances that are soluble in alcohol rather than water. They include triglycerides, phospholipids, glycolipids and cholesterol.
macromolecule: a very large, organic molecule.
phospholipid: molecule consisting of glycerol, two fatty acids and one phosphate group.

What are lipids?

Lipids contain large amounts of carbon and hydrogen, and smaller amounts of oxygen. They are insoluble in water because they are not polar, and so do not attract water molecules, but do dissolve in alcohol. The three most important lipids in living things are triglycerides, **phospholipids** and steroids. These are not **polymers**, but they do have different components bonded together. They are examples of **macromolecules**.

Triglyceride structure

Triglycerides are made up of **glycerol** and **fatty acids**. There are many different types of fatty acid. We can make many of them in our bodies, but some must be ingested 'complete'. These are called essential fatty acids.

Glycerol

Glycerol has three carbon atoms. It is an alcohol, which means it has free –OH groups. There are *three* –OH groups, which are important to the structure of *tri*glycerides.

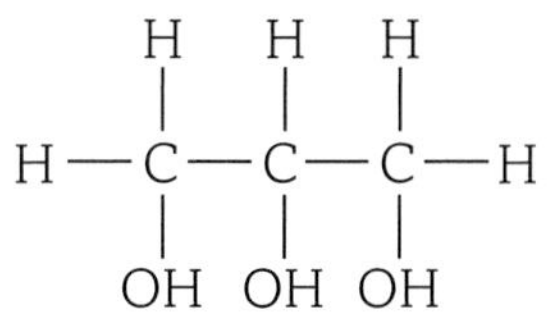

glycerol, a 3-carbon molecule with three OH groups

Figure 1 Molecular structure of glycerol.

Fatty acids

Fatty acids have a carboxyl group (–COOH) on one end, attached to a **hydrocarbon** tail, made of only carbon and hydrogen atoms. This may be anything from 2 to 20 carbons long. The carboxyl group ionises into H^+ and a –COO⁻ group. This structure is therefore an acid because it can produce free H^+ ions.

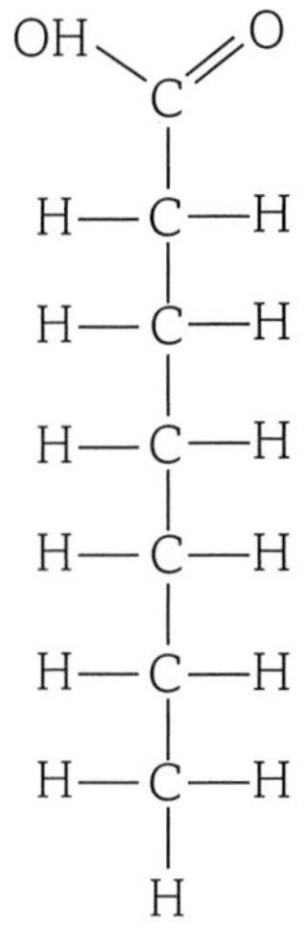

Figure 2 Molecular structure of a saturated fatty acid.

* If a fatty acid is saturated (Figure 2) this means that there are no C=C bonds in the molecule. If a fatty acid is unsaturated, there is a double bond between two of the carbon atoms instead, which means that fewer hydrogen atoms can be bonded to the molecule.

* A single C=C bond makes a fatty acid monounsaturated, e.g. oleic acid. More than one C=C bond makes it polyunsaturated, e.g. linoleic acid.

* Having one or more C=C bonds changes the shape of the hydrocarbon chain, giving it a kink where the double bond is. Because these kinks push the molecules apart slightly, it makes them more fluid. Animal lipids contain lots of saturated fatty acids, which are often solid at 20 °C. If there are more unsaturated fatty acids, the melting point is lower.

Ester bonds

A triglyceride consists of one glycerol molecule bonded to three fatty acids. A **condensation** reaction happens between the –COOH group of the fatty acid and the –OH group of the glycerol. Because there are three –OH groups, three fatty acids will bond, hence the name *tri*glyceride. Because it is a condensation reaction, a water molecule is produced, and the covalent bond formed is known as an **ester bond**. In some cases, the same type of fatty acid may bond to each –OH group, or the fatty acids may be different.

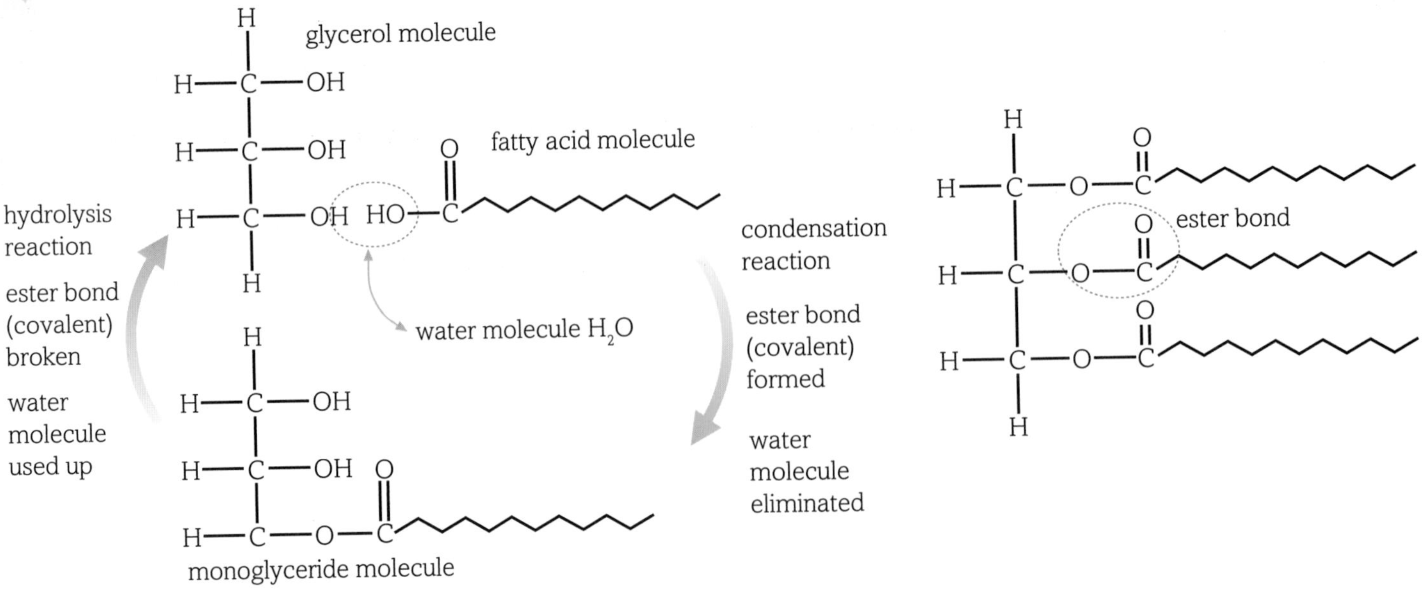

Figure 3 Formation of a triglyceride.

DID YOU KNOW?

Insects and leaves are covered in wax to help make them waterproof. Waxes are a type of lipid formed by combining fatty acids with alcohol molecules much larger than glycerol.

LEARNING TIP

Do not confuse the glycosidic, peptide and ester bonds. They all involve condensation, but each is specific to the particular biomolecules they link together.

Functions of triglycerides

- **Energy source**. Triglycerides can be broken down in **respiration** to release energy and generate ATP. The first step is to hydrolyse the ester bonds, and then both glycerol and the fatty acids can be broken down completely to carbon dioxide and water. Respiration of a lipid produces more water than respiration of a sugar.
- **Energy store**. Because triglycerides are insoluble in water, they can be stored without affecting the water potential of the cell (see topic 2.5.3). Mammals store fat in adipose cells under the skin. One gram of fat releases twice as much energy as 1 g of glucose. This is because lipids have a higher proportion of hydrogen atoms than carbohydrates, and almost no oxygen atoms.
- **Insulation**. Adipose tissue is a storage location for lipid in whales ('blubber'), acting as a heat insulator. Lipid in nerve cells acts as an electrical insulator. Animals preparing for hibernation store extra fat.
- **Buoyancy**. Because fat is less dense than water, it is used by aquatic mammals to help them stay afloat.
- **Protection**. Humans have fat around delicate organs, such as their kidneys, to act as a shock absorber. The peptidoglycan cell wall of some bacteria is covered in a lipid-rich outer coat.

Questions

1. Explain why triglycerides are not polymers.
2. Camels do not store water in their hump, but the contents of the hump are involved in providing water. Suggest how.
3. If you pour oil into water, the two liquids separate into two layers. Explain why.
4. Explain the difference between saturated, monounsaturated and polyunsaturated fatty acids. What effect does the different level of saturation have on their properties?
5. Compare and contrast the structure of triglycerides and carbohydrates.
6. Compare and contrast the glycosidic, peptide and ester bond.

⑦ Lipids 2: Phospholipids and cholesterol

By the end of this topic, you should be able to demonstrate and apply your knowledge and understanding of:

* the structure of a triglyceride and a phospholipid as examples of macromolecules

* how the properties of triglyceride, phospholipid and cholesterol molecules relate to their functions in living organisms

Phospholipids

> **LEARNING TIP**
> Learn about triglycerides in the previous topic, and then learn about phospholipids – it makes it easier to understand their structure.

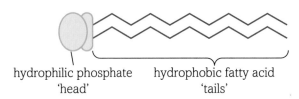

Figure 2 Simplified diagram of a phospholipid molecule.

Structure

Phospholipids have the same structure as triglycerides (see topic 2.2.6), except that one of the fatty acids is replaced by a phosphate group. A condensation reaction between an OH group on a phosphoric acid molecule (H_3PO_4) and one of the three –OH groups on the glycerol forms an ester bond.

Most of the fatty acids found in phospholipids have an even number of carbon atoms (often 16 or 18). Commonly one of these chains is saturated and one of them is unsaturated.

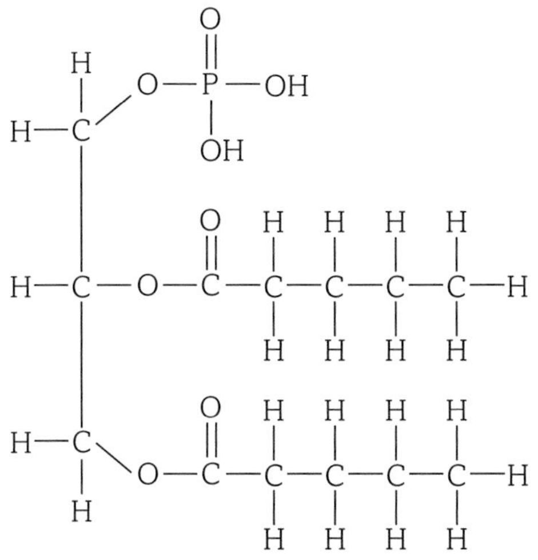

Figure 1 A phospholipid molecule consists of a glycerol molecule, two fatty acids and a phosphate group.

Behaviour in water

When surrounded by water, the phosphate group has a negative charge, making it polar (attracted to water). However, the fatty acid tails are non-polar and so are repelled by water. It is common to refer to the 'head' as **hydrophilic** and the 'tail' as **hydrophobic**, which means that the phospholipid molecule is amphipathic. Membrane lipids tend to be amphipathic, whereas those involved in storage are not.

Phospholipids that are amphipathic have very distinct properties in water. They may form a layer on the surface of the water with heads in the water and tails sticking up out of the water. They may also form micelles – tiny balls with the tails tucked away inside, and the heads pointing outwards into the water.

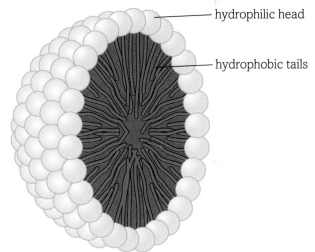

Figure 3 Behaviour of phospholipids in water. This kind of structure is called a micelle.

The phospholipid bilayer

Amphipathic phospholipids are excellent at forming membranes around cells and organelles (see topic 2.5.1). Inside and outside a cell membrane is an aqueous solution. The phospholipids form a bilayer, with two rows of phospholipids, tails pointing inwards and heads pointing outwards into the solution. Between 20 and 80% of membranes in plant and animal cells are made of phospholipids. Bacterial membranes tend to contain a greater proportion of protein.

* The individual phospholipids are free to move around in their layer, but will not move into any position where their hydrophobic tails are exposed to water. This gives the membrane some stability.

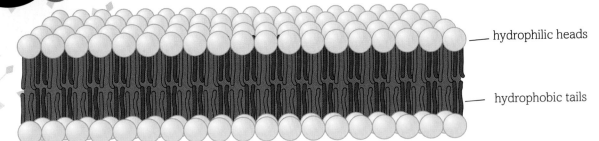

hydrophilic heads

hydrophobic tails

Figure 4 A phospholipid bilayer.

- The membrane is selectively permeable. It is only possible for small and non-polar molecules to move through the tails in the bilayer, such as oxygen and carbon dioxide. This lets the membrane control what goes in and out of the cell, and keeps it functioning properly.

Cholesterol

Cholesterol (chole-sterol) is a steroid alcohol (or sterol) – a type of lipid which is not made from glycerol or fatty acids. It consists of four carbon-based rings or isoprene units (Figure 5). Cholesterol is a small and hydrophobic molecule, which means it can sit in the middle of the hydrophobic part of the bilayer. It regulates the fluidity of the membrane, preventing it from becoming too fluid or stiff.

Cholesterol is mainly made in the liver in animals. Plants also have a cholesterol derivative in their membranes. It is called stigmasterol, and is different from cholesterol in only one respect: it has a double bond between carbon 22 and carbon 23.

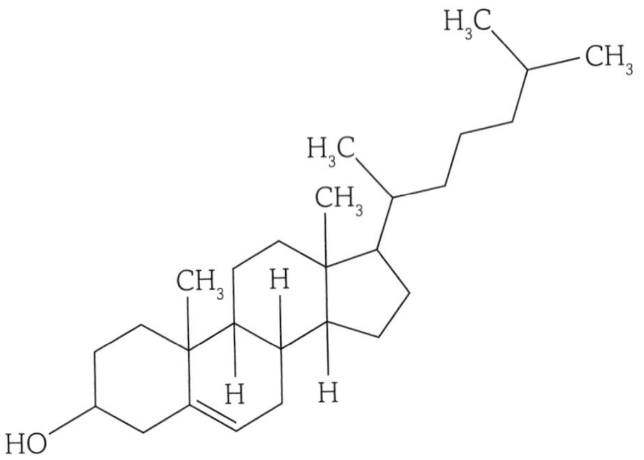

Figure 5 Molecular structure of cholesterol.

The steroid hormones testosterone, oestrogen and vitamin D are all made from cholesterol. Because they are small and hydrophobic, they can pass through the hydrophobic part of the cell membrane and any other membrane inside the cell. Steroids are also abundant in plants, and on ingestion and absorption some can be converted into animal hormones.

> **DID YOU KNOW?**
>
> Digitalis – a cardiac poison – is made from steroids.

> **DID YOU KNOW?**
>
> Plants have another class of lipid called terpenes. These are like steroids, being constructed of isoprene units. Examples include gibberellins (a plant growth substance), carotenoids (a photosynthetic pigment) and phytol (a component of chlorophyll).

Questions

1. Explain the difference between a triglyceride and a phospholipid.

2. How does the presence of a phosphate group affect the properties of a glyceride?

3. What kind of reaction happens when a phosphoric acid molecule reacts with glycerol?

4. Explain why phospholipids will form a bilayer or micelle when exposed to water.

5. Copy and complete the table below, using information from topics 2.2.6 and 2.2.7.

Lipid	Structure	Role(s) in living organisms	How does the structure help it perform its roles in living organisms?
Triglyceride			
Phospholipid			
Cholesterol			

(8) Proteins 1: Amino acids

By the end of this topic, you should be able to demonstrate and apply your knowledge and understanding of:

* the general structure of an amino acid
* the synthesis and breakdown of dipeptides and polypeptides, by the formation and breakage of peptide bonds

KEY DEFINITIONS

amino acids: monomers of all proteins, and all amino acids have the same basic structure.
peptide bond: a bond formed when two amino acids are joined by a condensation reaction.

Proteins

Proteins are large polymers comprised of long chains of **amino acids**. The properties of proteins give them a variety of functions:

*   They form structural components of animals in particular. For example, muscles are made of protein.
*   Their tendency to adopt specific shapes makes proteins important as enzymes (see Chapter 2.4), antibodies (see Chapter 4.1) and some hormones.
*   Membranes have protein constituents that act as carriers and pores for active transport across the membrane and facilitated diffusion (see topics 2.5.2 and 2.5.4).

Both plants and animals need amino acids to make proteins. Animals can make some amino acids, but must ingest the others (called essential amino acids). Plants can make all the amino acids they need, but only if they can access fixed nitrogen (such as nitrate).

Structure of amino acids

Each amino acid contains the elements carbon, hydrogen, oxygen and nitrogen. Some amino acids contain sulfur.

There are over 500 different amino acids, but only 20 of them are proteinogenic, which means that they are found in proteins. Each protein chain of amino acids has an amine group ($-NH_2$) at one end, and a carboxyl ($-COOH$) group at the other end.

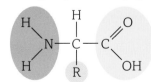

Figure 1 Molecular structure of an amino acid.

The R group in Figure 1 does not stand for a particular element, but is different in each amino acid. In glycine, it is simply an H atom.

However, in other amino acids, the R group can be a lot more complicated. For example, in alanine it is CH_3, whereas in cysteine

it is CH_3S. The name of almost all amino acids ends in –ine. The only exceptions are those which have an acidic R group, such as aspartic acid and glutamic acid. R groups can vary by size, by charge and by polarity, with some being hydrophobic and some being hydrophilic.

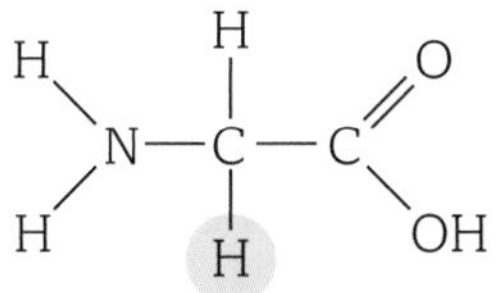

Figure 2 Glycine is the simplest amino acid.

DID YOU KNOW?

Amino acids can act as buffers
When dissolved in water, the amine group and carboxyl group can ionise. This means that the amine group can accept an H^+ ion to change from NH_2 to NH_3^+. The carboxyl group can give up an H^+ ion to change from COOH to COO^-.

$$-COOH \rightleftharpoons -COO^- + H^+$$

The carboxyl group acts as an acid, in producing H+ ions.

$$-NH_2 + H^+ \rightleftharpoons -NH_3^+$$

The amine group acts as a base, in accepting H^+ ions:
*   At low pH (where there are lots of H^+ ions in solution), the amino acid will accept H^+ ions.
*   At high pH (where there are fewer H^+ ions in solution), the amino acid will release H^+ ions.

This means an amino acid has acidic and basic properties, and so is known as amphoteric. In a long chain of amino acids, you will find amine and carboxyl groups on each end, but there are also many on the R-groups of different amino acids.

Protein chains can be affected by this amphoteric nature. By accepting and releasing H^+ ions, amine acids can help to regulate changes in pH. This is known as 'buffering'. A buffer is a substance which helps to resist large changes in pH.

The peptide bond

Amino acids are joined together by covalent bonds called **peptide bonds**. Just like the glycosidic bond and the ester bond, making a peptide bond involves a condensation reaction, and breaking a peptide bond involves hydrolysis. Enzymes catalyse these reactions. Protease enzymes in the intestines break peptide bonds during digestion. They also break down protein hormones so that their effects are not permanent.

All amino acids join together in the same way, whatever R group they may have. Two amino acids joined together are known as a dipeptide. Joining a longer chain of amino acids together forms a **polypeptide**. A protein may consist of a single polypeptide chain, or more than one chain bonded together.

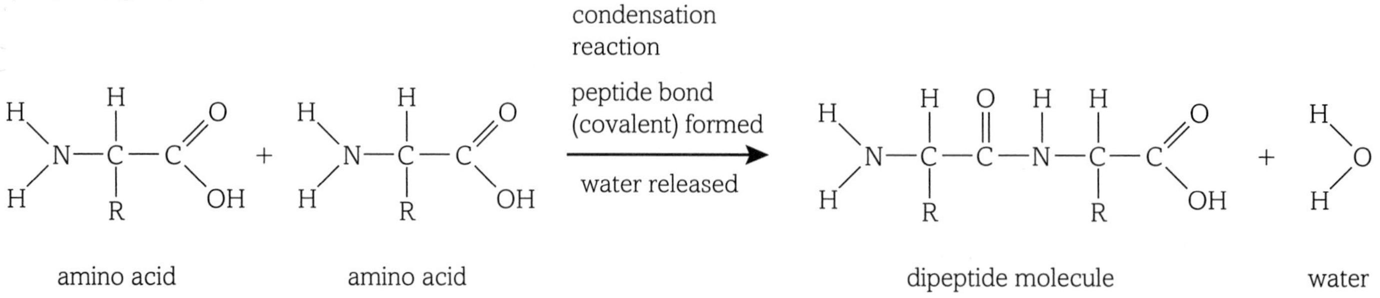

Figure 3a When joining two amino acids together, a condensation reaction is used to form a peptide bond.

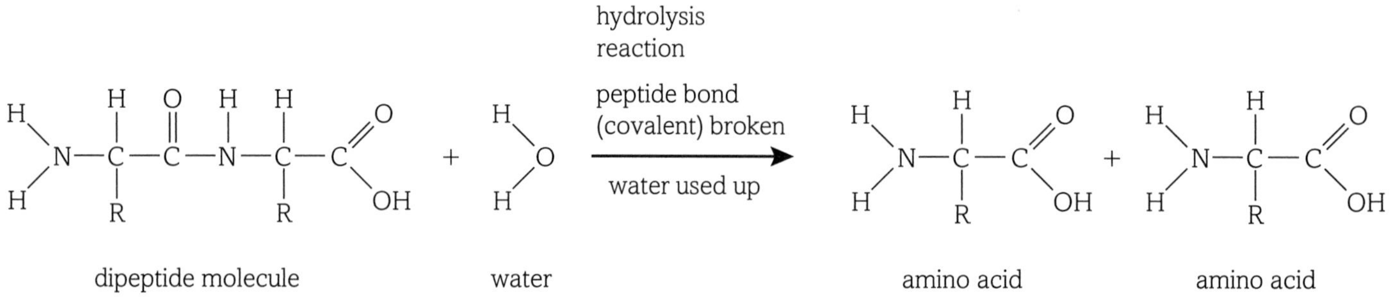

Figure 3b When breaking two amino acids apart, a hydrolysis reaction is used to break the peptide bond.

DID YOU KNOW?

The peptide bond is depicted as a single bond, but due to the electron arrangement around the bond, it has some of the properties of a double bond. This means that the bond is shorter than a conventional C–N bond. It also inhibits rotation around the peptide bond. This makes the polypeptide chain relatively stiff and rigid.

Questions

1. The spine of a molecule formed from three glycine molecules is N–C–C–N–C–C–N–C–C:
   (a) Draw this molecule showing each individual atom, including all carbon, hydrogen and oxygen atoms.
   (b) Label the peptide bonds.
   (c) How many condensation reactions happened to form this chain?
   (d) Explain why we can call this molecule a tripeptide.

2. A buffer solution is a solution which can resist changes in the pH of the solution. Explain why amino acids have some of the properties of a buffer.

3. List the similarities and differences between a peptide bond and a glycosidic bond.

4. Each polypeptide chain has a carboxyl group at one end and an amine group at the other end. Suggest what happens to the carboxyl and amine group in (a) acidic conditions, and (b) alkaline conditions.

⑨ Proteins 2: Protein structure and bonding

By the end of this topic, you should be able to demonstrate and apply your knowledge and understanding of:

* the levels of protein structure

> **KEY DEFINITIONS**
> **primary structure:** the sequence of amino acids found in a molecule.
> **quaternary structure:** protein structure where a protein consists of more than one polypeptide chain. For example, insulin has a quaternary structure.
> **secondary structure:** the coiling or folding of an amino acid chain, which arises often as a result of hydrogen bond formation between different parts of the chain. The main forms of secondary structure are the helix and the pleated sheet.
> **tertiary structure:** the overall three-dimensional shape of a protein molecule. Its shape arises due to interactions including hydrogen bonding, disulfide bridges, ionic bonds and hydrophobic interactions.

Primary structure

The sequence of amino acids in a protein chain is called its **primary structure**. The number and order of amino acids in a protein chain is important, as changing just one amino acid can alter the function of the protein. Because there are 20 amino acids, at every point in the chain there are 20 alternatives. Given that most proteins are at least 100 amino acids long, this gives an enormous number of different proteins that could be formed. There are 20^{100} possible ways of ordering 100 amino acids.

The function of a protein is determined by its structure. The order of amino acids in the primary structure will determine the shape of the protein molecule, through its secondary, tertiary and quaternary structure.

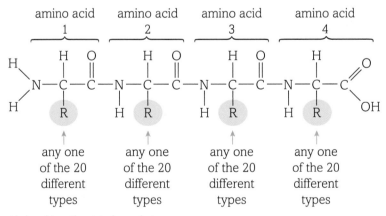

Figure 1 Amino acids bond together into long chains.

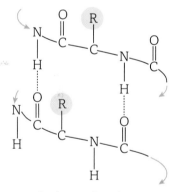

⋯⋯⋯ hydrogen bonds

Secondary structure

The chain of amino acids is not straight, but twists into a shape called the **secondary structure**. Some chains coil into an α-helix, with 36 amino acids per 10 turns of the helix. The helix is held together by hydrogen bonds between the –NH group of one amino acid and the –CO group of another four places ahead of it in the chain.

Other chains fold very slightly in a zig-zag structure. When one such chain folds over on itself, this produces a β-pleated sheet. Hydrogen bonds between the –NH group of one amino acid and the –CO group of another further down the strand hold the sheet together.

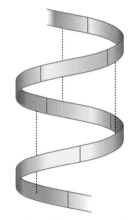

Figure 2 α-helix structure of a polypeptide chain.

Although hydrogen bonds are relatively weak, many are formed, which makes both the α-helix and the β-pleated sheet stable structures at optimal temperature and pH. Some chains do not adopt any regular structure, and some chains may have more than one secondary structure at different ends of the chain.

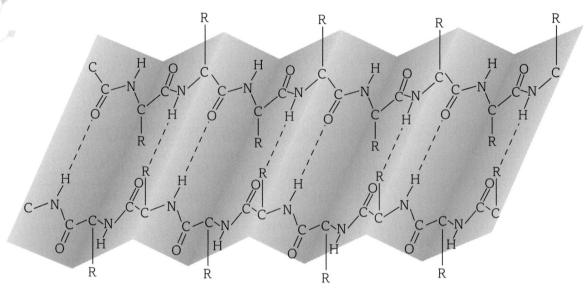

Figure 3 β-Pleated sheet of a polypeptide chain.

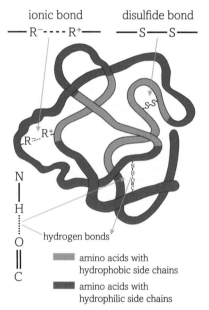

Figure 4 Tertiary structure of a globular protein.

Tertiary structure

When these coils and pleats themselves start to fold, along with areas of straight chains of amino acids, this forms the **tertiary structure**. The tertiary structure is a very precise shape which is held firmly in place by bonds between amino acids which lie close to each other. The tertiary structure may adopt a supercoiled shape (e.g. in fibrous proteins) or a more spherical shape (in globular proteins), as you will learn in topic 2.2.10.

Quaternary structure

Many proteins are made up of more than one polypeptide chain. The **quaternary structure** describes how multiple polypeptide chains are arranged to make the complete protein molecule. This may also be held together with the same types of bond that hold the tertiary structure together. See the quaternary structure of haemoglobin shown in topic 2.2.10.

Protein bonding

The primary structure of proteins, the chain of amino acids, is held together by peptide bonds which are covalent bonds, and hence very strong. Other types of bond form between amino acids in different parts of the polypeptide chain. These hold together the secondary, tertiary and quaternary structures. The secondary structure is primarily held together by hydrogen bonds, but the tertiary structure and quaternary structure are held together by hydrogen bonds and many others.

LEARNING TIP

Remember that the terms primary, secondary and tertiary structure refer to a single polypeptide chain. The quaternary structure describes the association between two or more polypeptide chains.

Hydrogen bonds

Like in water (see topic 2.2.2), **hydrogen bonds** form between hydrogen atoms with a slight positive charge and other atoms with a slight negative charge. In amino acids, these form in hydroxyl, carboxyl and amino groups.

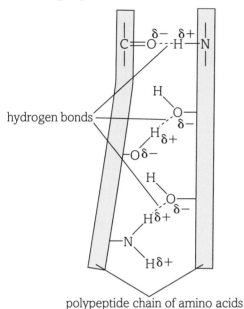

Figure 5 Hydrogen bonds in amino acids.

For example, hydrogen bonds may form between the amino group of one amino acid and the carboxyl group of another. They may also form between polar areas of the R groups on different amino acids. These in particular are involved in keeping the tertiary and quaternary structure of the protein in the correct shape. The presence of multiple hydrogen bonds can give protein molecules a lot of strength.

Ionic bonds

Ionic bonds can form between those carboxyl and amino groups that are part of R groups. These ionise into NH_3^+ and COO^- groups. Positive and negative groups like this are strongly attracted to each other to form an **ionic bond**.

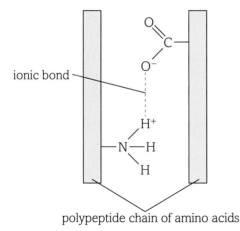

Figure 6 Ionic bonds.

Disulfide links

The R group of the amino acid cysteine contains sulfur. Disulfide bridges are formed between the R groups of two cysteines. These are strong covalent bonds.

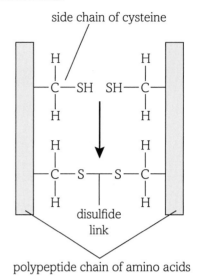

Figure 7 Disulfide link.

Hydrophobic and hydrophilic interactions

Hydrophobic parts of the R groups tend to associate together in the centre of the polypeptide to avoid water. In the same way, hydrophilic parts are found at the edge of the polypeptide to be close to water. Hydrophobic and hydrophilic interactions cause twisting of the amino acid chain, which changes the shape of the protein. These interactions can be a very important influence, given that most proteins are to be found surrounded by water inside a living organism.

Questions

1 Describe the difference between the primary structure and the secondary structure.

2 Describe the difference between the tertiary structure and the quaternary structure.

3 Put the following bonds into order, with strongest first and weakest last: Disulfide. Ionic. Hydrogen.

(10) Proteins 3: Fibrous and globular proteins

By the end of this topic, you should be able to demonstrate and apply your knowledge and understanding of:

* the structure and function of globular proteins including a conjugated protein
* the properties and functions of fibrous proteins

KEY DEFINITIONS

fibrous protein: has a relatively long, thin structure, is insoluble in water and metabolically inactive, often having a structural role within an organism.
globular protein: has molecules of a relatively spherical shape, which are soluble in water, and often have metabolic roles within the organism.
prosthetic group: a non-protein component that forms a permanent part of a functioning protein molecule.

The three-dimensional tertiary and quaternary structure of proteins falls into two main categories. Fibrous and globular proteins do different jobs, and so have different properties:

* **Fibrous proteins** have regular, repetitive sequences of amino acids, and are usually insoluble in water. These features enable them to form fibres, which tend to have a structural function. Examples include collagen and elastin (in connective tissue) and keratin (in hair).

* **Globular proteins** tend to roll up into an almost spherical shape. Any hydrophobic R groups are turned inwards towards the centre of the molecule, while hydrophilic groups are on the outside. This makes the protein water soluble, because water molecules can easily cluster round and bind to them. They often have very specific shapes, which helps them to take up roles as enzymes, hormones (such as insulin) and haemoglobin.

Properties and functions of some fibrous proteins

Collagen

The function of collagen is to provide mechanical strength:

* In artery walls, a layer of collagen prevents the artery bursting when withstanding high pressure from blood being pumped by the heart.
* Tendons are made of collagen and connect muscles to bones, allowing them to pull on bones.

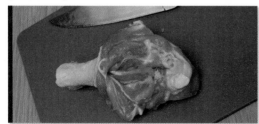

Figure 1 Tendons in a chicken leg.

* Bones are made from collagen, and then reinforced with calcium phosphate, which makes them hard.
* Cartilage and connective tissue are made from collagen.

Keratin

Keratin is rich in cysteine so lots of disulfide bridges form between its polypeptide chains. Alongside hydrogen bonding, this makes the molecule very strong.

Keratin is found wherever a body part needs to be hard and strong. It is found in finger nails, hair, claws, hoofs, horns, scales, fur and feathers. It provides mechanical protection, but also provides an impermeable barrier to infection and, being waterproof, also prevents entry of water-borne pollutants.

Figure 2 This animal horn contains the protein keratin.

Elastin

Cross-linking and coiling make the structure of elastin strong and extensible. It is found in living things where they need to stretch or adapt their shape as part of life processes.

Skin can stretch around our bones and muscles because of elastin. Without elastin, skin would not go back to normal after being pinched.

Elastin in our lungs allows them to inflate and deflate, and in our bladder helps it to expand to hold urine.

Like collagen, elastin helps our blood vessels to stretch and recoil as blood is pumped through them, helping maintain the pressure wave of blood as it passes through.

Structure, properties and function of some globular proteins

Haemoglobin

The **quaternary structure of haemoglobin** is made up of four polypeptides: two α-globin chains and two β-globin chains. Each of these has its own tertiary structure, but when fitted together they form one haemoglobin molecule. The shape of the molecule is held together by the bonds described in the previous topic. The interactions between the polypeptides give the molecule a very specific shape.

At one position on the outside of each chain, there is a space in which a haem group is held. Groups like this are called **prosthetic groups** (see also topic 2.4.2, on enzyme cofactors). They are an essential part of the molecule, without which it could not function, but they are not made of amino acids. The haem group contains an iron ion. A protein associated with this kind of group is called a **conjugated protein**.

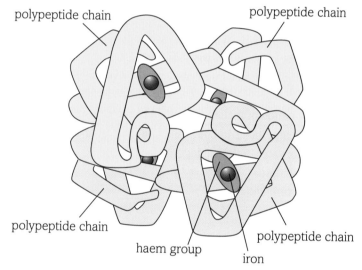

polypeptide chain polypeptide chain polypeptide chain polypeptide chain haem group iron

Figure 3 Structure of haemoglobin.

The function of haemoglobin is to carry oxygen from the lungs to the tissues. In the lungs, an oxygen molecule binds to the iron in each of the four haem groups in the haemoglobin molecule. When it binds, haemoglobin turns from a purple red colour to bright red. The oxygen is released by the haemoglobin when it reaches the tissues. You can find out more about haemoglobin in topic 3.2.7.

Insulin

Insulin is made of two polypeptides chains. The A chain begins with a section of α-helix, and the B chain ends with a section of β-pleat. Both chains fold into a tertiary structure, and are then joined together by disulfide links. Amino acids with hydrophilic R groups are on the outside of the molecule, which makes it soluble in water. Insulin binds to glycoprotein receptors on the outside of muscle and fat cells to increase their uptake of glucose from the blood, and to increase their rate of consumption of glucose.

Pepsin

Pepsin is an enzyme that digests protein in the stomach. The enzyme is made up of a single polypeptide chain of 327 amino acids, but it folds into a symmetrical tertiary structure. Pepsin has very few amino acids with basic R groups (only four), whereas it has 43 amino acids with acidic R groups. This helps to explain why it is so stable in the acidic environment of the stomach, as there are few basic groups to accept H^+ ions, and therefore there can be little effect on the enzyme's structure. The tertiary structure is also held together by hydrogen bonds and two disulfide bridges.

Computer modelling of protein structure

Being able to predict the shape of a protein molecule from its primary structure can be incredibly useful in biochemistry. For example, predicting the occurrence of biologically active binding sites on a protein molecule can help in identifying new medicines.

Scientists can predict protein shapes using computer modelling techniques. As techniques for prediction of secondary structure developed, they were based upon the probability of an amino acid, or a sequence of amino acids, being in a particular secondary structure. Such probabilities were derived from 'already-known' protein molecular structures.

Prediction of tertiary structure is perhaps more interesting, as it is usually the tertiary structure of a protein molecule which contributes directly to its bioactive function. There are two broad approaches:

- *Ab initio* **protein modelling**. In this approach, a model is built based on the physical and electrical properties of the atoms in each amino acid in the sequence. With this technique, there can be multiple solutions to the same amino acid sequence, and other methods sometimes need applying to reduce the number of solutions.

- **Comparative protein modelling**. One approach is protein threading, which scans the amino acid sequence against a database of solved structures and produces a set of possible models which would match that sequence.

Questions

1 Write down the similarities and differences between collagen and (a) cellulose, and (b) haemoglobin.

2 What would be the consequences for keratin if cysteine did not exist?

3 Draw out and complete a table to show the similarities and differences between (a) the fibrous proteins mentioned in this unit, and (b) the globular proteins mentioned in this unit.

4 Describe three ways in which the shape of globular proteins contributes to the maintenance of life.

5 What is the difference between *ab initio* protein modelling and comparative protein modelling?

(11) Inorganic ions

By the end of this topic, you should be able to demonstrate and apply your knowledge and understanding of:

* the key inorganic ions that are involved in biological processes

You have learnt so far in this chapter about the organic molecules that are essential for life. This topic is a fact file on the specific inorganic cations (positive ions) and anions (negative ions) that are involved in key biological processes.

Inorganic ions are essential constituents of skeletal structures, are involved in the maintenance of osmotic pressure and are structural constituents of soft tissue. They are also important for nerve impulse transmission and muscle contraction, and play a vital role in maintaining the pH balance of the body. They serve as essential components and activators of enzymes, vitamins and hormones.

Cations

Calcium	Ca^{2+}	• Increases rigidity of bone, teeth and cartilage and is a component of the exoskeleton of crustaceans • Important in clotting blood and muscle contraction • Activator for several enzymes, such as lipase, ATPase and cholinesterase • Stimulates muscle contraction and regulates transmission of nerve impulses • Regulates permeability of cell membranes • Important for cell wall development in plants, and formation of middle lamella between cell walls
Sodium	Na^+	• Involved in regulation of osmotic pressure, control of water levels in body fluid and maintenance of pH • Affects absorption of carbohydrate in the intestine, and water in the kidney • Contributes to nervous transmission and muscle contraction • Constituent of vacuole in plants which helps maintain turgidity
Potassium	K^+	• Involved in control of water levels in body fluid and maintenance of pH • Assists active transport of materials across the cell membrane • Involved in synthesis of glycogen and protein, and breakdown of glucose • Generates healthy leaves and flowers in flowering plants • Contributes to nervous transmission and muscle contraction • Component of vacuoles in plants, helping to maintain turgidity
Hydrogen	H^+	• Involved in photosynthesis and respiration • Involved in transport of oxygen and carbon dioxide in the blood • Involved in regulation of blood pH
Ammonium	NH_4^+	• A component of amino acids, proteins, vitamins and chlorophyll • Some hormones are made of proteins, e.g. insulin • An essential component of nucleic acids. • Involved in maintenance of pH in the human body • A component of the nitrogen cycle

Anions

Nitrate	NO_3^-	• A component of amino acids, proteins, vitamins and chlorophyll • An essential component of nucleic acids • Some hormones are made of proteins, which contain nitrogen, e.g. insulin • A component of the nitrogen cycle
Hydrogencarbonate	HCO_3^-	• Involved in regulation of blood pH • Involved in transport of carbon dioxide into and out of the blood

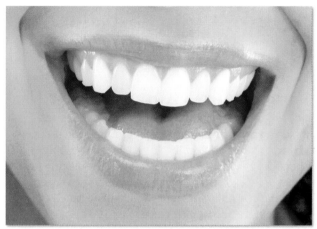

Figure 1 Calcium and phosphate are needed for healthy teeth.

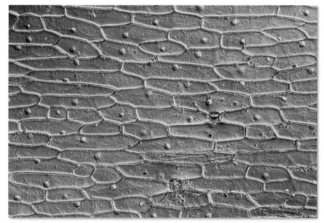

Figure 2 Calcium is needed for development of the plant cell wall.

Figure 3 Metal ions are particularly important for healthy plant growth.

Deficiency

In humans and plants, some ions are required in large amounts (macronutrients or main elements) and some in small amounts (micronutrients or trace elements). Both humans and plants can display deficiency symptoms if they do not consume enough of a particular ion. For example, deficiency of the trace element cobalt causes anaemia, while deficiency of copper in plants causes young shoots to die back.

Questions

1. Which inorganic ions are involved in regulation of blood pH?

2. Which inorganic ions make things hard and rigid?

3. Which inorganic ions are involved as components of protein?

4. Create a table with one row for each ion, and one column for each different function. Place ticks where an ion has a particular function.

5. Decide which two ions have the most overlap in their function. Try to explain why these particular ions undertake the same roles.

(12) Practical biochemistry 1: Qualitative tests for biological molecules

By the end of this topic, you should be able to demonstrate and apply your knowledge and understanding of:

* * how to carry out and interpret the results of the following chemical tests:
  * the biuret test for proteins
  * Benedict's test for reducing and non-reducing sugars
  * reagent test strips for reducing sugars
  * iodine test for starch, and
  * emulsion test for lipids

Each one of the tests in this topic is used simply to determine the presence or absence of the particular biological molecule – hence they are known as qualitative tests. They rely on the biological molecules in the sample passing into solution. Therefore you will need to grind and squash the food samples first, and mix them vigorously with a small volume of water (or in the case of lipids, alcohol – see below). Wear eye protection when carrying out all of these tests.

Testing for carbohydrates

Starch

To test for starch, add iodine solution (in potassium iodide) to a sample. If starch is present, you will see a colour change of yellow-brown to blue-black. When dissolved in potassium iodide, the iodine (I_2) forms a triiodide ion I_3^-, which slips into the middle of the amylose helix. This causes the colour change.

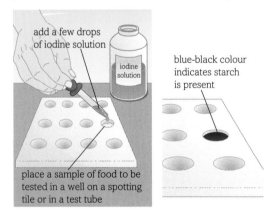

Figure 1 Testing for starch.

DID YOU KNOW?

The iodine test is used in brewing to check that any chains of glucose in the fermentation mix are relatively short (less than 9 molecules long in straight chains, and less than 60 in branched chains). If longer chains go into the fermentation mix, they can give the final beer a haze, which looks unappetising. Longer chains yield a blue-black result with iodine, whereas shorter chains give no colour change, or a much less intense red-purple colour.

Reducing sugars

These include all monosaccharides and some disaccharides. They are known as **reducing sugars** because they can reduce, or give electrons to, other molecules. If you heat a reducing sugar with Benedict's solution (alkaline copper (II) sulfate), there is a colour change from blue to green to yellow to orange-red. Benedict's solution contains Cu^{2+} ions, which are reduced to Cu^+ ions, forming orange-red copper (I) oxide (Cu_2O). This is called a precipitate because it comes out of solution and forms a solid, suspended in the reaction mixture.

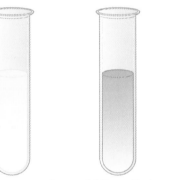

place a sample of food to be tested in a boiling tube

add Benedict's solution then heat in a water bath at 80°C for 3 minutes

orange-red precipitate indicates a reducing sugar is present (when low levels of sugar are present, the contents of the boiling tube may appear yellow or green)

Figure 2 Benedict's test for reducing sugars.

If you use Benedict's solution in excess, the intensity of the red colour is proportional to the concentration of sugar. The reaction mix will appear green if only a little precipitate is formed, and fully orange-red if a lot of precipitate is formed. See the next topic on quantitative testing for a reducing sugar.

It is also possible to use commercially manufactured test strips to test for reducing sugars. Here you simply dip the strip into the test solution, and compare the colour with the calibration card supplied. This tells you whether reducing sugar is present or absent from your solution. These are often used to test for glucose in the urine of diabetic patients.

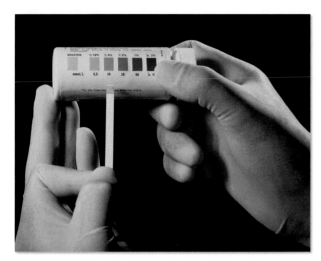

Figure 3 Glucose test strip being held against calibration chart.

Non-reducing sugars

To test for a non-reducing sugar, we have to hydrolyse the bond first, to 'free up' these 'reducing groups', and then test for reducing sugars as normal:

- First, test a sample for reducing sugars to check there are none there in the first place.
- Take a separate sample and boil it with hydrochloric acid to hydrolyse the sucrose into glucose and fructose.
- Cool the solution and use sodium hydrogencarbonate solution to neutralise it.
- Test for reducing sugars again.

A positive result (green-yellow-orange-red) indicates that non-reducing sugar (e.g. sucrose) was present in the original sample.

In some cases, a sample may contain reducing and non-reducing sugars. If you have a positive test for reducing sugars from your first sample, you can go on to test for non-reducing sugars in an equal-sized second sample. If present, the precipitate from this second sample will have more mass than the precipitate from the first sample. You can extract the precipitate from the mixture by filtration.

Testing for lipids

The emulsion test is used to test for the presence of lipids:

- Take a sample and mix it thoroughly with ethanol. Any lipid will go into solution in the ethanol (remember that lipids are not soluble in water).
- Filter.
- Pour the solution into water in a clean test tube.
- A cloudy white emulsion indicates the presence of lipids. This is made up of tiny lipid droplets that come out of solution when mixed with water.

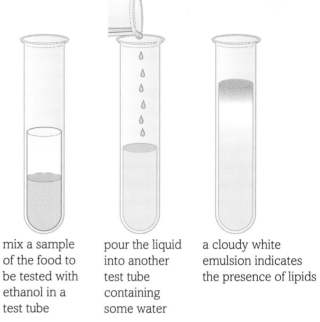

mix a sample of the food to be tested with ethanol in a test tube

pour the liquid into another test tube containing some water

a cloudy white emulsion indicates the presence of lipids

Figure 4 The emulsion test for lipids.

Testing for proteins

For this you use the biuret test. If protein is present, the colour changes from light blue to lilac. You may find the reagents are supplied to you separately as biuret A (sodium hydroxide), which you add first, and biuret B (copper sulfate), which you add next. The colour is formed by a complex between the nitrogen atoms in a peptide chain and Cu^{2+} ions, which is why this test really detects the presence of peptide bonds.

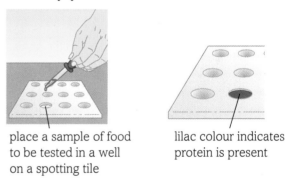

place a sample of food to be tested in a well on a spotting tile

lilac colour indicates protein is present

Figure 5 The biuret test for proteins.

Questions

1. Why is sucrose called a non-reducing sugar?

2. Explain why you have to check there is no reducing sugar in a solution before testing for the presence of non-reducing sugar.

3. What is meant by the terms precipitate and emulsion?

4. A student wants to digest protein into amino acids. Explain how and why the biuret test can help him monitor when that reaction is complete.

5. Why do lipids form an emulsion in water?

(13) Practical biochemistry 2: Quantitative tests for biological molecules

By the end of this spread, you should be able to demonstrate and apply your knowledge and understanding of:

* quantitative methods to determine the concentration of a chemical substance in a solution

Quantitative testing for reducing sugar

Benedict's reagent detects the presence of reducing sugars. If there is more sugar present:

* the amount of precipitate will increase.

* the amount of copper (II) ions remaining in solution will decrease.

We can try to quantify the concentration of sugar in the original sample by assessing how these two variables change, using a technique called **colorimetry**.

Using a colorimeter

A colorimeter works by shining light through a sample. In this case, we would use a centrifuge to separate the precipitate and any excess Benedict's solution (the supernatant).

Using a pipette, we can take the supernatant and place it in a cuvette (a small vial), which is then placed into the colorimeter. The cuvette is commonly made of glass or plastic. Ensure you do not leave a greasy fingerprint on the surface of the cuvette, as it could affect transmission of light.

Colour filters are often used for greater accuracy. By using a red filter in this case, we can shine red light through the solution, and detect how much passes through (percentage transmission). The solution reflects blue light but absorbs red light:

* If there is a lot of unreacted copper sulfate, the supernatant is still quite blue, absorption of red light is high and percentage transmission is low.

* If there is little unreacted copper sulfate, the supernatant is less blue, absorption of red light is low and percentage transmission is high.

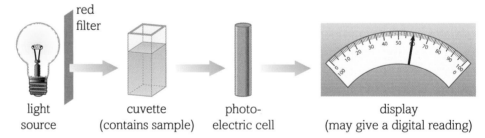

Figure 1 Using a colorimeter.

* When using a colorimeter, the device is usually zeroed between each reading by placing an appropriate 'blank' sample to reset the 100% transmission/absorption. In this case, the blank used would be water.

Creating a calibration curve

Using a colorimeter gives us a semi-quantitative test for sugar, as we can compare how much sugar is contained in different samples. To find the exact amounts, we need to create a calibration curve:

1. First, take a series of known concentrations of reducing sugar.

2. Using a sample of each, carry out Benedict's test.

3. Use a colorimeter to record the percentage transmission of light through each supernatant.

4. Plot a graph to show 'transmission of light' against the concentration of reducing sugar. This provides a calibration curve, which you can use with other 'unknown' samples to determine the concentration of sugar in the original sample. For example, using the calibration curve in Figure 3, if a sample of glucose had a transmission of 92%, we can conclude that it contains $12\,g\,dm^{-3}$ of glucose.

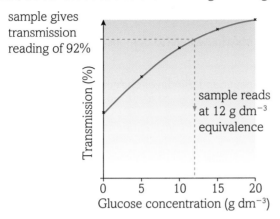

Figure 3 Calibration curve for known concentrations of glucose solution vs transmission of light.

Use of biosensors

Use of colorimetry in this way provides a good introduction to how **biosensors** work. They take a biological or chemical variable which cannot easily be measured, and convert it to an electrical signal.

Figure 4 is a generic diagram of how biosensors work, with the specific parts labelled.

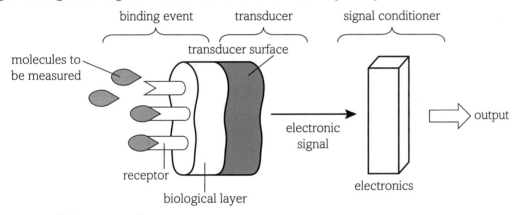

Figure 4 How do biosensors work?

Biosensors have many other applications. For example, they can be used to detect contaminants in water, and pathogens and toxins in food. They can even be used to detect airborne bacteria, for example in counter-bioterrorism programmes.

DID YOU KNOW?

One of the first examples of a biosensor was carrying a canary into a mine. As toxic gases built up, the canary died, providing the miners with an early warning to evacuate the mine.
A new nanocanary has now been produced which aims to assess the toxicity of bioengineered nanomaterials on living cells.

Questions

1. (a) Outline how you would compare the reducing sugar concentration in apple juice and orange juice. (b) Consider the fact that the two liquids have different colours. How would you account for this when using the colorimeter?

2. What are the similarities and differences between a colorimeter and a biosensor?

3. Explain why the biosensor and colorimeter need to be calibrated.

4. Before testing a solution in the colorimeter, we usually use a blank solution, commonly distilled water. Suggest why it is important to use this 'blank' to 'zero' the colorimeter before testing the solution of interest.

14 Practical biochemistry 3: Chromatography

By the end of this topic, you should be able to demonstrate and apply your knowledge and understanding of:

* the principles and uses of paper and thin layer chromatography to separate biological molecules/compounds

* practical investigations to analyse biological solutions using paper and thin layer chromatography

Principles of chromatography

The aim of chromatography is to separate a mixture into its constituents: in this case biological molecules. There are two key components, known as the stationary phase and the mobile phase.

Stationary phase

This is either the chromatography paper or a thin-layer chromatography (TLC) plate. The paper is made of cellulose. The TLC plate is often a sheet of plastic, coated with a thin layer of silica gel or aluminium hydroxide. In each case, there are free –OH groups pointing outwards, in contact with the mobile phase.

Mobile phase

This is the solvent for the biological molecules. At a simple level, we can use water (for polar molecules) or ethanol (for non-polar molecules). The mobile phase flows through and across the stationary phase, carrying the biological molecules with it.

INVESTIGATION

Chromatography practical

Figure 1 shows you how to set up thin layer chromatography, but the protocol would be the same for paper chromatography. As you set it up, please take note of the following points:

Wear eye protection.

Draw the line in pencil and put a tiny dot on the line to show you where to place your solution mixture. If you draw it in ink, the pigments in the ink will also separate.

Spot the solution mixture onto the pencil dot several times by using capillary tubing. Wait for the spot to dry before putting on the next spot, and try to make the spot as thin as possible. When it is completely dry, lower it into the solvent. Ensure the level of the solvent at the start is below the pencil line.

Cover the beaker with a watch glass, or glass plate.

Let the apparatus 'run' until the solvent has reached a point just underneath the top of the paper/TLC plate. Then remove it from the solvent, and lay it on a white tile to dry.

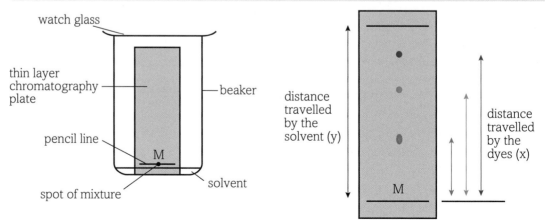

Figure 1 The set-up for thin layer chromatography.

Figure 2 Thin layer chromatography plate with pigments separated out.

What happens?

As the solvent travels up the paper or plate, the components of the solution mixture travel with it. In the example in Figure 2, which involved separating pigments in an ink, you can see that they travel at different speeds. By the time the solvent has reached the top of the plate, some are travelling slowly and some quickly, and so are at different positions on the plate.

You can use the relative distance travelled to help identify the pigments. Calculate the R_f value by measuring the distance from the pencil line to the centre of a spot of pigment (x), and the distance from the pencil line to the solvent front (y) (do not forget to do this before the solvent dries, so you can still see it!).

$$R_f = \frac{x}{y}$$

If you repeat the investigation under the same conditions, each pigment will always have the same R_f value. If you know the R_f values of particular pigments under these conditions, this allows you to identify them. Exactly the same is true of biological molecules.

Sometimes with colourless molecules, you cannot see where they finish. Using thin-layer chromatography, there are solutions:

1. **Ultraviolet light**. Thin layer chromatography plates have a chemical which fluoresces under UV light. If you look at the plate under UV light, most of it will glow, except those places where the spots have travelled to. They mask the plate from the UV light.

2. **Ninhydrin**. To see amino acids, allow the plate to dry, and then spray it with ninhydrin. This binds to the amino acids which are then visible as brown or purple spots.

3. **Iodine**. Allow the plate to dry, and place in an enclosed container with a few iodine crystals. The iodine forms a gas, which then binds to the molecules in each of the spots.

How does it work?

The speed at which molecules move along the paper or TLC plate depends on their solubility in the solvent, and their polarity. In the case of paper chromatography, it may also depend on their size.

Exposed –OH groups make the surface of the paper or plate very polar, and allow it to form hydrogen bonds with the molecules, alongside other dipole interactions. A highly polar solute will tend to stick to the surface (it is adsorbed), and hence move more slowly. A non-polar solute will travel very quickly up the plate.

If two molecules travel at exactly the same speed, it will be difficult to separate them. In this case, you could try using a different solvent, or changing the pH.

How is chromatography used?

Thin layer chromatography is commonly used to monitor the progress of reactions, because it works relatively quickly. It is also used for urine testing of athletes for illegal drugs, analysing drugs for purity of components, and analysis of foods to determine the presence of contaminants.

Questions

1 Explain why different solutes may move at different speeds along a TLC plate.

2 Which of the following will show least adsorption to a TLC plate: saturated fatty acid, glucose, glycine (a simple amino acid)? Explain your answer.

3 You are interested in finding out which amino acids are contained in a particular polypeptide. You are provided with a solution of the polypeptide, a protease enzyme, and pure solutions of all 20 amino acids. Write down an experimental protocol which would allow you to list all the amino acids contained within the polypeptide chain.

BIOLOGICAL MOLECULES

Since the 1950s, dieticians have recommended diets low in fats and particularly low in saturated fats. This has not prevented an epidemic of obesity in Western countries. Experts are now questioning whether it is good advice. The excerpt below comes from a national newspaper.

'WE GOT IT ALL WRONG', CLAIMS EXPERT

Current advice about diet and heart disease is wrong and is putting people's health and lives at risk, claims Professor Joseph Mercer of the Institute of Dietary Research, which is funded by the Dairy Council. 'Current advice is based on research from the 1950s which was flawed' he claims. 'Low-fat diets do not protect against heart disease or make us live longer. The real enemy in our diet is sugar – especially refined sugar' says Professor Mercer, a leading scientist in dietary research.

Writing in the journal *Diet and Cardiovascular Health*, Professor Mercer states: 'There is no conclusive evidence that a diet low in fats has any positive effects on cardiovascular health. The research in the 1950s collected data from 16 countries, but then drew its conclusions from only six of those countries. A review of the literature reveals that there is no real effect of a reduced fat diet. Saturated fats were demonised by public-health campaigns in the 1970s and '80s, we now need one to say that we got it wrong.'

'The previous campaigns were so successful that the public now fear that saturated fat in their diet raises cholesterol, however, this is completely unfounded. Experts also believed that a low-fat diet would lead to less obesity and less diabetes – but the exact opposite is true.' he added.

Last year, experts claimed that the interpretation of scientific studies had been faulty and had perpetuated a myth that a high-fat diet is bad for the heart. For years people have been advised to reduce fat intake and to ensure that only 30 per cent of total energy intake comes from fats. However, modern research fails to show any link between fat intake and risk of cardiovascular disease; in fact, modern research suggests that saturated fat is actually protective. Professor Mercer said: 'From these data it is clear that the recent epidemic of dietary related illnesses including atherosclerosis, heart disease, diabetes and obesity results from a diet high in sugars rather than one high in fats. The best advice we can give now to maintain the health of the heart is to eat more fat and less sugar.'

However, Professor David Leach, head of diabetes research at the University of London, suggests that Professor Mercer has not carried out a full review of the literature available. 'He has misinterpreted the evidence, advising people to eat more fat is not helpful.'

William Wilson, Director of the Department of Nutritional Analysis at the Oxford School of Public Health, warns that the conclusions are seriously misleading. He warns that the analysis contains major errors and omissions. 'This work is bound to cause confusion. The key issue is what replaces saturated fat in the diet if it is reduced. The energy content of the saturated fat is essential. If this is replaced by eating more refined sugars, which are probably now the largest source of energy in many diets, then the risk of heart disease remains the same. This is particularly true if the person eats foods such as biscuits and cakes, which actually contain many saturated fats and trans-fats. However, if saturated fat is replaced with polyunsaturated fat or monounsaturated fat such as olive oil, nuts and other plant oils, there is a lot of evidence to show that the risk of heart disease is reduced.'

Alistair Steadman, Chair of Public Health UK, said: 'It is reasonable to conclude that a reduction in saturated fat intake will lower blood cholesterol, which may reduce the risk of developing heart disease'.

Where else will I encounter these themes?

1.1   2.1   2.2   YOU ARE HERE   2.3   2.4   2.5

Let's start by considering the nature of the writing in the article.

1. Evaluate the article and decide whether it provides a balanced view. Explain your decision. Consider how many people have contributed, who they work for, and what they have said. You may also wish to consider whether these are just statements or whether they are backed up by data.

Now we will look at the biology in, or connected to, this article. Don't worry if you are not ready to give answers to these questions yet. You may like to return to the questions once you have covered other topics later in the book. Use the timeline at the bottom of the page to help you put this work in context with what you have already learned and what is ahead in your course.

2. State three functions of fats in the body.
3. Describe the structure of a fat molecule.
4. State three functions of cholesterol in the body.
5. Professor Mercer talks generally about fats in the diet, whilst William Wilson and Alistair Steadman are more specific. Explain the difference between saturated fats and unsaturated fats.
6. Compare the properties of saturated and unsaturated fats.
7. State three ways that sugars can be used in the body.
8. Using diagrams to show the structural formulae, explain the differences between monosaccharides, disaccharides, oligosaccharides and polysaccharides.

> Remember that a well-annotated diagram is a good way to describe structures.

Activity

Trans-fats are not common in nature, but they can be manufactured easily. They have properties that are considered desirable in certain types of food. Carry out some research to find out more about trans-fats.

Write a leaflet that explains the following points:

- What is meant by the term 'trans-fats'?
- What are the desirable properties of trans-fats?
- What concerns are there over the use of trans-fats in food?
- Should we be using trans-fats in food?

Your leaflet should cover two sides of A4 paper. You should decide whether it is best to avoid trans-fats if possible, or whether you think they are safe to use in food. Create a balanced argument and provide suitable evidence in the form of data from secondary sources to convince the reader.

> This activity could be modified:
> - by ensuring the leaflet has a specific audience: teenage readers concerned about diet and health/parents concerned about their children's health/food mass-manufacturer/cake shop/healthfood shop owner, etc.
> - by considering how it might be written differently depending on the agenda of the author: manufacturers of trans-fats/lobby groups for organic and wholefoods/obesity campaigners, etc.

1. A student carries out a Benedict's test on an unknown solution. The reaction mixture remains blue at the end of the test. Which of the following provides the best interpretation of these results? [1]

   A. There is no sugar in the solution.
   B. There is non-reducing sugar in the solution.
   C. There is reducing and non-reducing sugar in the solution.
   D. There is no reducing sugar in the solution.

2. The following bonds help to hold together protein tertiary structure. Which order below reflects their relative strength (with the weakest on the left and the strongest on the right)? [1]

   A. Hydrogen bond – ionic bond – covalent bond – hydrophobic interactions.
   B. Hydrophobic interactions – hydrogen bond – disulfide bridge – ionic bond.
   C. Ionic bond – disulfide bridge – hydrophobic interactions – hydrogen bond.
   D. Ionic bond – hydrophobic interactions – hydrogen bond – disulfide bridge.

3. Which of the following is a disaccharide? [1]

   A. $C_6H_{12}O_6$
   B. $C_5H_{10}O_5$
   C. $C_{12}H_{22}O_{11}$
   D. $C_3H_6O_3$

4. Which of the following formulae of fatty acids represents a saturated fatty acid? [1]
   (i) Palmitic acid, $C_{15}H_{31}COOH$
   (ii) Oleic acid, $C_{17}H_{33}COOH$
   (iii) Linoleic acid, $C_{17}H_{31}COOH$

   A. (i), (ii) and (iii)
   B. Only (i) and (ii)
   C. Only (ii) and (iii)
   D. Only (i)

   [Total: 4]

5. (a) Lipids form an important part of a balanced diet but if too many lipids are consumed this can result in obesity. What is meant by the term *balanced diet*? [2]

   (b) (i) Lipids are used for energy storage and as a respiratory substrate. List **three** other roles of lipids in the human body. [3]
   (ii) Other than obesity, outline why a diet high in lipids might have a negative effect on the health of an individual. [3]

   (c) Two examples of lipid molecules are triglycerides and phospholipids. Identify **two** differences and **two** similarities in the **structures** of triglycerides and phospholipids. Write your answers in the appropriate boxes in the table below. [4]

	Triglyceride	Phospholipid
Difference		
Difference		
Similarity		
Similarity		

   (d) It is possible to test for the presence of lipids in a food sample.
   (i) Name the test used to identify the presence of lipids. [1]
   (ii) Describe how you would carry out this test on a food sample. [3]
   (iii) State the expected result if lipid is present in the food sample. [1]

   [Q4, F212 January 2013]
   [Total: 17]

6. (a) Amino acids form part of the structure of proteins.
   (i) State the name given to the sequence of amino acids in a protein molecule. [1]
   (ii) Draw the **general structure** of an amino acid molecule. [3]

   (b) Collagen is an important fibrous protein which forms part of the walls of blood vessels.
   (i) State **one** property of collagen that makes it a useful component of blood vessel walls. [1]
   (ii) Describe the **structure** of the collagen molecule. [6]

   [Q1, F212 June 2013]

   (c) Another protein that is important in mammals is haemoglobin.
   (i) State **one** function of haemoglobin. [1]
   (ii) Haemoglobin contains a prosthetic group known as haem. Collagen does not contain a prosthetic group. Describe **three** other ways in which the structure of haemoglobin differs from that of collagen. [3]

   [Total: 15]

7. A number of different biological molecules are represented in Figure 1.

A
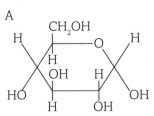

B
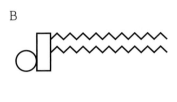

C

D
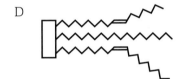

E
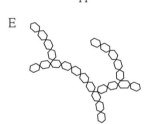

F

Figure 1

(a) (i) State the letter of the molecule shown in Figure 1 which represents:
   a triglyceride
   a monosaccharide
   a protein [3]

   (ii) State the letter of the molecule shown in Figure 1 that contains:
   phosphate
   glycosidic bonds
   peptide bonds
   disulfide bonds [4]

(b) Molecule E shown in Figure 1 is part of the carbohydrate molecule glycogen. Explain why glycogen makes a good storage molecule. [3]

(c) (i) When glycogen is hydrolysed, molecule A, shown in Figure 1, is produced. State the **precise name** of molecule A. [1]

   (ii) State **one** function of molecule A. [1]

   (iii) State the letter of a molecule shown in Figure 1, other than molecule E, that is used as a storage molecule. [1]

(d) Cellulose is a carbohydrate molecule found in plants. Complete the table below to give three **differences** in the structure of glycogen and cellulose. One difference has been done for you. [3]

glycogen	cellulose
no hydrogen bonding	hydrogen bonding

[Total: 16]
[Q3, F212 June 2011]

8. Biological molecules are held together by a variety of bonds.

(a) The diagram in Figure 2 represents an amino acid.

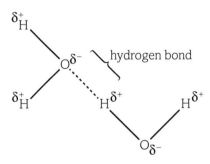

Figure 2

   (i) One of the atoms that makes up the amino acid has been replaced by the letter X. State the chemical symbol of the atom represented by the letter X in Figure 2. [1]

   (ii) Name the polymer formed from a chain of amino acids. [1]

   (iii) Name the bond that is formed when two amino acids are joined together. Describe the formation of this bond. [2]

(b) Figure 3 shows a hydrogen bond between two water molecules.

Figure 3

   (i) Many of the physical properties of water arise as a result of these hydrogen bonds.
   Describe ways in which the physical properties of water allow organisms to survive over a range of temperatures.
   In your answer you should make clear links between the properties of water and the survival of organisms. [9]

   (ii) List **three other** examples of where hydrogen bonds are found in biological molecules. [3]

[Total: 16]
[Q1, F212 January 2012]

Foundations in biology

NUCLEIC ACIDS

Introduction

Recent advances in gene therapy have led to the successful restoration of sight for about 40 people suffering from hereditary blindness. By introducing a gene, which is a length of DNA, the cells in the retina of the eye can now make a protein that previously they could not. Scientists are also developing therapies for some genetic diseases by using pieces of RNA that bind to transcribed mRNA, preventing translation and effectively silencing the gene.

Much research into nucleic acids has been carried out since James Watson and Francis Crick, with valuable information from research by Maurice Wilkins and Rosalind Franklin, elucidated the structure of DNA in 1953.

In this section you will learn about the structure of the large DNA molecule and how it replicates itself before a cell divides so that each new cell contains all the genetic information that was present in the parent cell. You will also learn how in eukaryotic cells DNA molecules are condensed and tightly wound into chromosomes and that each chromosome contains many genes, lengths of DNA, each of which codes for a specific protein. You will discover how the genetic code is made up of just four nucleotide bases read in groups of three. You will see how this code, which never leaves the nucleus, is transcribed or copied into a messenger molecule, RNA, which carries this code to the ribosomes, where amino acids are assembled into proteins, according to the instructions in the mRNA, which are translated. You will compare the structure of DNA and RNA.

All the maths you need

To unlock the puzzles in this section you need the following maths:

- Powers of 3
- Units of measurement
- The relationship between speed, time and rate

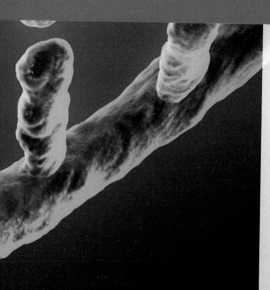

chloroplasts

- The DNA of prokaryotic cells is within the cytoplasm
- The DNA molecule is large and a double helix
- A chromosome is a large molecule of DNA and contains smaller lengths of DNA called genes
- Different versions of genes are called alleles or gene variants
- Genes code for proteins, which are very important molecules; some are structural, some are enzymes, others are receptors on cell surface membranes, some are signalling molecules, some are antibodies and some – carriers and channels in cell surface membranes – allow substances into and out of cells
- Proteins are assembled at ribosomes
- The copying of genetic material at cell division ensures that each new cell receives all the genetic information in the parent cell
- Sexual reproduction brings together different alleles of genes to produce genetic variation in offspring

What will I study later?

- Enzymes – each of which has a specific function that depends on its structure and shape of the active site, which is determined by the coded information in the gene that codes for it (AS)
- How specific genetic disorders may be transmitted in families from parents to offspring (AL)
- How organisms can be genetically modified (AL)
- How gene therapy works (AL)
- How organisms can be cloned to give genetically identical copies of themselves (AL)
- More about how mutations and natural selection contribute to evolution (AS)
- How coordination and control mechanisms in living organisms depend on correctly functioning proteins, whose structure is determined by genes (AL)
- How the functioning of all systems in all living organisms depends on correctly functioning proteins whose structure is determined by genes (AS and AL)

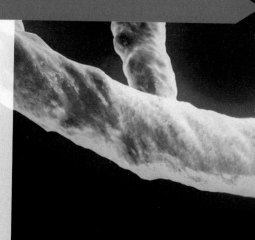

What will I study in this chapter?

- More details about the structure of DNA
- Nucleotide bases and complementary base pairing
- The structure of RNA and how it differs form that of DNA
- The role of DNA in carrying the genetic code to determine the structure of specific proteins
- How DNA replicates (with experimental evidence to support this) before cells divide
- How DNA is transcribed into the messenger RNA which carries the genetic code from the nucleus to the ribosomes
- How the code carried by the mRNA is translated at ribosomes, into proteins

You will also be able to carry out some practical activities involving DNA.

5
6
7
8

(1) DNA – deoxyribonucleic acid

By the end of this topic, you should be able to demonstrate and apply your knowledge and understanding of:

* the structure of a nucleotide as the monomer from which nucleic acids are made
* the structure of ADP and ATP as phosphorylated nucleotides
* the structure of DNA (deoxyribonucleic acid)
* the synthesis and breakdown of polynucleotides by the formation and breakage of phosphodiester bonds
* practical investigations into the purification of DNA by precipitation

KEY DEFINITIONS

double helix: shape of DNA molecule, due to coiling of the two sugar–phosphate backbone strands into a right-handed spiral configuration.
monomer: molecule that when repeated makes up a polymer. Amino acids are the monomers of proteins. Nucleotides are the monomers of nucleic acids.
nucleotide: molecule consisting of a five-carbon sugar, a phosphate group and a nitrogenous base.
polynucleotide: large molecule containing many nucleotides.

Nucleotides

Nucleotides are biological molecules that participate in nearly all biochemical processes (see Figure 1). They are phosphate esters of pentose sugars, where a nitrogenous base is linked to the C_1 (carbon atom 1) of the sugar residue, and a phosphate group is linked to either the C_5 (carbon atom 5) or C_3 (carbon atom 3) of the sugar residue, by covalent bonds formed by condensation reactions. See topic 2.2.1 for more about condensation reactions and covalent bonds.

Nucleotides:

* form the **monomers** of nucleic acids, DNA and RNA. In RNA, the nucleotide pentose sugar is ribose. In DNA, the nucleotide pentose sugar is deoxyribose

* become phosphorylated nucleotides when they contain more than one phosphate group; for example, ADP (adenosine diphosphate) and ATP (adenosine triphosphate). ATP is an energy-rich end-product of most energy-releasing biochemical pathways, and it is used to drive most energy-requiring metabolic processes in cells (see Figure 2)

and coenzyme A (both also involved in respiration). You will learn more about photosynthesis and respiration in the second year of your course.

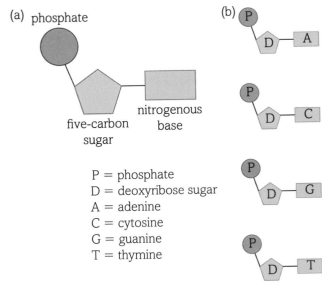

(a) phosphate

five-carbon sugar

nitrogenous base

P = phosphate
D = deoxyribose sugar
A = adenine
C = cytosine
G = guanine
T = thymine

Figure 1 (a) General structure of a single nucleotide. (b) The four DNA nucleotides.

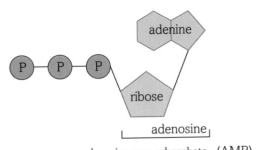

adenine

ribose

adenosine

DNA is a nucleic acid

DNA is found in the nuclei of all eukaryotic cells, within the cytoplasm of prokaryotic cells and is also inside some types of viruses. It is the hereditary material and carries coded instructions used in the development and functioning of all known living organisms. DNA is one of the important **macromolecules** that make up the structure of living organisms, the others being proteins, carbohydrates and lipids.

Structure of DNA

- DNA is a polymer as it is made up of many repeating monomeric units called nucleotides.
- A molecule of DNA consists of two **polynucleotide** strands.
- The two strands run in opposite directions, so they are described as antiparallel.
- Each DNA nucleotide consists of a phosphate group, a five-carbon sugar called deoxyribose, and one of four nitrogenous bases: adenine, guanine, thymine or cytosine (see Figure 3).
- The covalent bond between the sugar residue and the phosphate group in a nucleotide is also called a phosphodiester bond. These bonds are broken when polynucleotides break down and are formed when polynucleotides are synthesised.
- DNA molecules are long and so they can carry a lot of encoded genetic information.

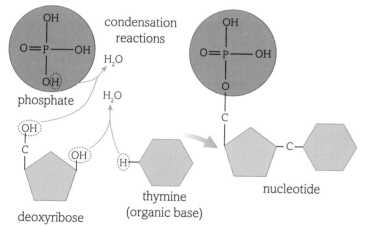

Figure 3 A phosphate group, the five-carbon sugar deoxyribose, and a base, in this case thymine, are joined by condensation reactions to form a DNA nucleotide.

Purines and pyrimidines

DNA consists of just four types of nucleotide. In each nucleotide the phosphate and sugar groups are the same but the organic (nitrogenous) base differs. It may be either a purine – adenine or guanine (two rings) – or a pyrimidine – thymine or cytosine (one ring).

The importance of hydrogen bonds

The two antiparallel DNA strands are joined to each other by hydrogen bonds between the nitrogenous bases. You have already learnt about hydrogen bonds in topic 2.2.1.

- Adenine always pairs with thymine, by means of two hydrogen bonds.
- Guanine always pairs with cytosine, by means of three hydrogen bonds.
- A purine always pairs with a pyrimidine, giving equal-sized 'rungs' on the DNA ladder (see Figure 4). These can then twist, like twisting a rope ladder around an imaginary axis, into the **double helix** (coil). This gives the molecule stability.
- Hydrogen bonds allow the molecule to unzip for transcription and replication.

Nucleotides with adenine as the base can make two hydrogen bonds with nucleotides with thymine as the base.

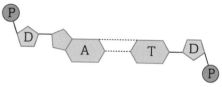

Nucleotides with guanine as the base can make three hydrogen bonds with nucleotides with cytosine as the base.

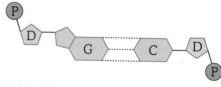

Figure 4 Complementary base pairing. A purine always pairs with a pyrimidine, and the resulting rungs of the DNA ladder are of the same width.

The antiparallel sugar–phosphate backbones

- The upright part of the large DNA molecule that resembles a ladder is formed by the sugar–phosphate backbones of the antiparallel polynucleotide strands (see Figure 5).
- The 'opposite directions' of the two strands refers to the direction that the third and fifth carbon molecules on the five-carbon sugar, deoxyribose, are facing.
- The 5' end of the molecule is where the phosphate group is attached to the fifth carbon atom of the deoxyribose sugar.
- The 3' end is where the phosphate group is attached to the third carbon atom of the deoxyribose sugar.
- The rungs of the ladder consist of the complementary base pairs, joined by hydrogen bonds.
- The molecule is very stable, and the integrity of the coded information within the base sequences is protected.

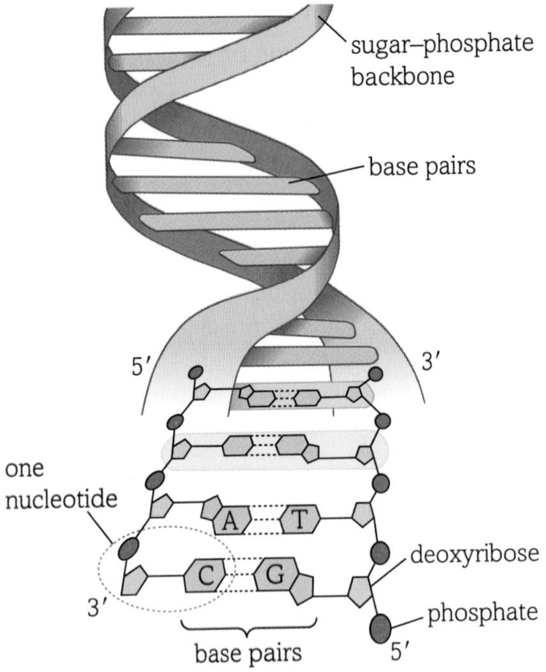

sugar–phosphate backbone

base pairs

5'　3'

one nucleotide

A — T

C - G

3'

deoxyribose

phosphate

base pairs　5'

Figure 5 Part of a DNA molecule, consisting of two antiparallel sugar–phosphate backbones, with complementary pairs of nucleotide bases held together by hydrogen bonds.

How DNA is organised in cells

Eukaryotic cells:

- The majority of the DNA content, or the genome, is in the nucleus.
- Each large molecule of DNA is tightly wound around special histone proteins into chromosomes. Each chromosome is therefore one molecule of DNA (see Figure 6). You will learn more about genes and DNA in the second year of your course. There is more about chromosomes in topic 2.6.1 in the sections about cell division (mitosis and meiosis).
- There is also a loop of DNA, *without* the histone proteins, inside mitochondria and chloroplasts.

Prokaryotic cells:

- DNA is in a loop and is within the cytoplasm, not enclosed in a nucleus.
- It is *not* wound around histone proteins, and is described as naked. See topic 2.1.7 for more information on prokaryotic cells.

Viruses that contain DNA also have it in the form of a loop of naked DNA.

Figure 6 How a molecule of DNA is organised into a chromosome in a eukaryotic cell.

Questions

1　Explain why DNA is described as (a) a macromolecule and (b) a polynucleotide.

2　(a) Describe the general structure of a nucleotide.
(b) Explain how nucleotides are important to cell metabolism, *other than* being part of nucleic acids.

3　Explain how purines and pyrimidines differ from each other.

4　(a) Explain what is meant by 'complementary base pairing'.
(b) A length of DNA was analysed and 23% of the nucleotide bases were adenine. What percentage of the bases were cytosine?

5　Explain why the two strands of each DNA molecule are described as 'antiparallel'.

6　Compare how DNA is organised within eukaryotic cells and prokaryotic cells.

7　What is the relationship between chromosomes and DNA?

8　On one strand of a length of DNA, the nucleotide base sequence is CAGTTCTAGGGTAAT. Write down the sequence of complementary bases on the opposite strand of this portion of the molecule.

9　Write down all the ways in which the structure of the DNA molecule enables it to carry out its function.

How DNA replicates

By the end of this topic, you should be able to demonstrate and apply your knowledge and understanding of:

* semi-conservative DNA replication

KEY DEFINITIONS

DNA polymerase: enzyme that catalyses formation of DNA from activated deoxyribose nucleotides, using single-stranded DNA as a template.
helicase: enzyme that catalyses the breaking of hydrogen bonds between the nitrogenous pairs of bases in a DNA molecule.
semi-conservative replication: how DNA replicates, resulting in two new molecules, each of which contains one old strand and one new strand. One old strand is conserved in each new molecule.

DNA is a self-replicating molecule

All the DNA within a cell (the **genome**), and within every cell of an organism, carries the coded instructions to make and maintain that organism. Every time a cell divides, the DNA has to be copied so that each new daughter cell receives the full set of instructions. Each molecule of DNA replicates. This replication takes place during interphase, before the cell actually divides. In eukaryotes, this results in each chromosome (a chromosome being one molecule of DNA) having an identical copy of itself. At first they are joined together, at the centromere, forming two sister chromatids. See topics 2.6.2 and 2.6.3 for more information on the details of cell division, mitosis and meiosis in eukaryotic organisms.

The DNA within mitochondria and chloroplasts also replicates each time these organelles divide, which is just before the cell divides.

Semi-conservative replication

To make a new copy of itself (see Figure 1), each DNA molecule:

- unwinds – the double helix is untwisted, a bit at a time, catalysed by a gyrase enzyme
- unzips – hydrogen bonds between the nucleotide bases are broken. This is catalysed by DNA **helicase**, and results in two single strands of DNA with exposed nucleotide bases.

Next:

- free phosphorylated nucleotides, present in the nucleoplasm within the nucleus, are bonded to the exposed bases, following complementary base-pairing rules
- the enzyme **DNA polymerase** catalyses the addition of the new nucleotide bases, in the 5' to 3' direction, to the single strands of DNA; it uses each single strand of unzipped DNA as a template
- the leading strand is synthesised continuously, whereas the lagging strand is in fragments (discontinuous) that are later joined, catalysed by ligase enzymes

- hydrolysis of the activated nucleotides, to release the extra phosphate groups, supplies the energy to make phosphodiester bonds between the sugar residue of one nucleotide and the phosphate group of the next nucleotide.

The product of the replication is two DNA molecules, identical to each other and to the parent molecule. Each of these molecules contains one old strand and one new strand, and so this is termed **semi-conservative replication**.

The loops of DNA in prokaryotes, and inside mitochondria and chloroplasts, also replicate semi-conservatively. A bubble sprouts from the loop and this unwinds and unzips, and then complementary nucleotides join to the exposed nucleotides. Eventually the whole loop is copied.

DID YOU KNOW?

Evidence for semi-conservative replication
In the 1950s scientists knew that DNA was a self-replicating molecule but did not know how the molecule made copies of itself. There were three theories:
- conservative – the original molecule acts as a template and a new molecule is made
- dispersive – the original molecule breaks up into nucleotides, each one joins to a complementary nucleotide and new ones join up again
- semi-conservative – the new molecule consists of one original strand and one newly formed strand.

Two scientists, Meselson and Stahl, carried out an experiment in 1958 which showed that DNA replication is semi-conservative.
- They grew bacteria, *E. coli*, for 14 generations in a medium containing the heavy isotope of nitrogen, ^{15}N. This contains an extra neutron in every atomic nucleus. After 14 generations, most of the DNA in the bacteria would be heavy, as it contains ^{15}N.
- They then transferred some of these bacteria into a medium containing the normal ^{14}N isotope of nitrogen, and left them for long enough to undergo one replication.
- The DNA from these bacteria after one division was found to be hybrid DNA. This showed that DNA does not replicate conservatively, as that would have produced two bands of DNA, one heavy and one light.
- The bacteria were allowed to divide once more and their DNA was extracted and centrifuged. This produced two bands of DNA, one hybrid and one light, showing that replication is semi-conservative and not dispersive.

Mutations

During DNA replication, errors may occur and the wrong nucleotide may be inserted. This is estimated to occur in 1 in 10^8 base pairs. This could change the genetic code, and is an example of a point mutation. During the replication process there are enzymes that can proofread and edit out such incorrect nucleotides, reducing the rate that mutations are produced.

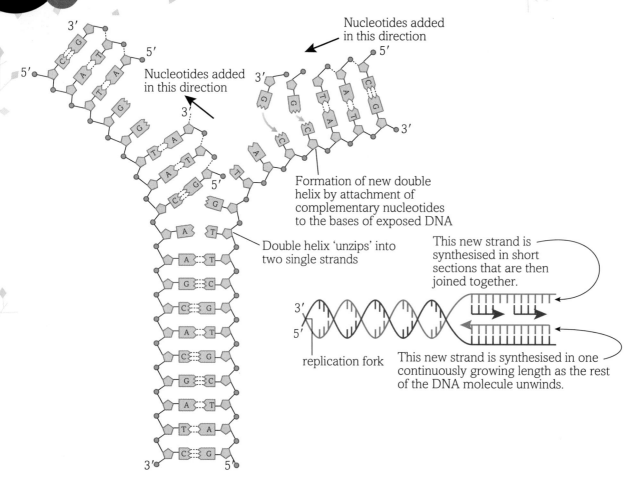

Figure 1 How DNA replicates. The addition of the new DNA nucleotides is catalysed by the enzyme DNA polymerase.

However, many genes have such changes to their nucleotide sequence. Different versions of a particular gene are called *alleles* or *gene variants*.

Not all mutations are harmful. Some appear to give neither advantage nor disadvantage (for example, whether or not you can roll your tongue) and some can be advantageous – for example, a white coat in an animal during winter when snow is on the ground. You will learn more about mutations, alleles and natural selection in Chapter 4.3 and in the second year of your course in Chapter 6.2.

Questions

1. Copy and complete the following table to show the functions of some of the enzymes involved in DNA replication:

Enzyme	Function
gyrase	
helicase	
DNA polymerase	

2. Explain why the replication of DNA is described as 'semi-conservative'.

3. In the experiment by Meselson and Stahl, described above:
   (a) If the bacteria were allowed to divide again, whilst in the same medium, what proportion of the DNA in the fourth generation would be hybrid, and what proportion would be light?
   (b) If the replication of DNA was dispersive, what would have been the results of the DNA composition after a second division of the bacteria?
   (c) In which parts of the DNA molecule is the ^{15}N incorporated?

4. Suggest where the free nucleotides that are in the nucleoplasm of nuclei in your cells have come from.

5. When, during the cell cycle, is mutation – an alteration to the nucleotide base sequence in a molecule of DNA – most likely to happen?

How DNA codes for polypeptides

By the end of this topic, you should be able to demonstrate and apply your knowledge and understanding of:

* the nature of the genetic code

* transcription and translation of genes resulting in the synthesis of polypeptides

---

KEY DEFINITIONS

gene: a length of DNA that codes for a polypeptide or for a length of RNA that is involved in regulating gene expression.

polypeptide: a polymer made of many amino acid units joined together by peptide bonds. Insulin is a polypeptide of 51 amino acids.

protein: a large polypeptide of 100 or more amino acids. However, the terms are often used synonymously, and insulin may be described as a small protein.

transcription: the process of making messenger RNA from a DNA template.

translation: formation of a protein, at ribosomes, by assembling amino acids into a particular sequence according to the coded instructions carried from DNA to the ribosome by mRNA.

RNA

RNA is structurally different from DNA in a number of ways:

* the sugar molecule in each nucleotide is *ribose*

* the nitrogenous base *uracil* (see Figure 1), which is a pyrimidine, replaces the pyrimidine base thymine

* the polynucleotide chain is usually single-stranded

* the polynucleotide chain is shorter

* there are three forms of RNA – messenger RNA (mRNA), transfer RNA (tRNA) and ribosomal RNA (rRNA).

Genes and the genetic code

Genes

You have learnt that each chromosome in a eukaryotic cell nucleus includes a molecule of DNA. On each chromosome, there are specific lengths of the DNA called **genes**. Each gene contains a code that determines the sequence of amino acids in a particular polypeptide or protein.

Protein accounts for 75% of an organism's dry mass. Some proteins are structural, such as the cytoskeleton threads inside cells or the proteins in the cell membrane; others make up the cell's tool-kit, such as enzymes, and these may catalyse the formation of non-protein molecules such as lipids and carbohydrates.

Within each gene there is a sequence of DNA base triplets that determines the amino acid sequence, or primary structure, of a

polypeptide. As long as this primary structure of a **polypeptide** is correct, it will then fold correctly and be held in its tertiary structure or shape, enabling it to carry out its function.

For example:

* the shape of the active site of an enzyme molecule must be complementary to the shape of the substrate molecule

* part of an antibody molecule must have a shape complementary to that of the antigens on the surface of an invading pathogen

* a receptor on a cell membrane must have a shape complementary to the shape of the cell-signalling molecule, such as a hormone or a drug, that it must detect

* an ion-channel protein must have hydrophilic amino acids lining the inside of the channel, and lipophilic amino acids on the outside portion that will be next to the lipid bilayer of the plasma membrane.

Genes are inside the cell nucleus but proteins are made in the cytoplasm, at ribosomes (see topic 2.1.6).

As the instructions inside the genes, on chromosomes, cannot pass out of the nucleus, a copy of each gene has to be **transcribed** (copied) into a length of mRNA. In this form, the sequence of base triplets, now called codons, can pass out of the nucleus to the ribosome, ensuring that the coded instructions are **translated** and the protein is assembled correctly from amino acids.

The nature of the genetic code

* The genetic code is near *universal*, because in almost all living organisms the same triplet of DNA bases codes for the same amino acid.

* The genetic code is described as *degenerate*, because, for all amino acids, except methionine and tryptophan, there is more than one base triplet. This may reduce the effect of point mutations, as a change in one base of the triplet could produce another base triplet that still codes for the same amino acid.

* The genetic code is also *non-overlapping*, and it is read starting from a fixed point in groups of three bases. If a base is added or deleted then it causes a frame shift, as every base triplet after that, and hence every amino acid coded for, is changed.

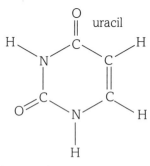

Figure 1 The pyrimidine base uracil.

Transcription and translation

Transcription

Figure 2 summarises the process of **transcription** of a gene into a length of mRNA.

- A gene unwinds and unzips.

- Hydrogen bonds between complementary nucleotide bases break.

- The enzyme RNA polymerase catalyses the formation of temporary hydrogen bonds between RNA nucleotides and their complementary unpaired DNA bases. A bonds with T; C with G; G with C; and U with A, on one strand of the unwound DNA. This DNA strand is called the *template strand*.

- A length of RNA that is complementary to the template strand of the gene is produced. It is therefore a copy of the other DNA strand – the *coding strand*.

- The mRNA now passes out of the nucleus, through the nuclear envelope, and attaches to a ribosome.

Ribosomes are made in the nucleolus, in two smaller subunits. These pass separately out of the nucleus, through pores in the nuclear envelope, and then come together to form the ribosome. Magnesium ions help to bind the two subunits together. Ribosomes are made of ribosomal RNA and protein in roughly equal parts.

Translation

Transfer RNA molecules are made in the nucleolus (see Figure 3) and pass out of the nucleus into the cytoplasm. They are single-stranded polynucleotides, but can twist into a hairpin shape. At one end is a trio of nucleotide bases that recognises and attaches to a specific amino acid. At the loop of the hairpin is another triplet of bases, called an *anticodon*, that is complementary to a specific codon (triplet) of bases on the mRNA.

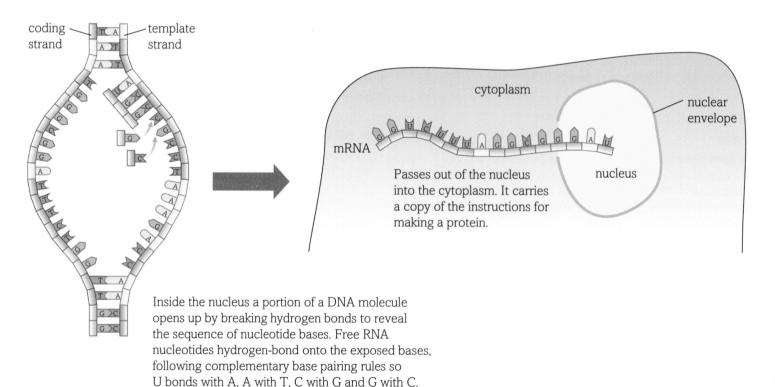

Inside the nucleus a portion of a DNA molecule opens up by breaking hydrogen bonds to reveal the sequence of nucleotide bases. Free RNA nucleotides hydrogen-bond onto the exposed bases, following complementary base pairing rules so U bonds with A, A with T, C with G and G with C.

Passes out of the nucleus into the cytoplasm. It carries a copy of the instructions for making a protein.

Figure 2 How a gene is transcribed into a length of mRNA.

mRNA

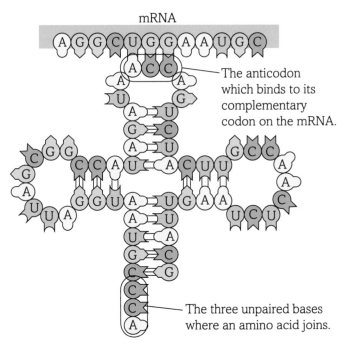

The anticodon which binds to its complementary codon on the mRNA.

The three unpaired bases where an amino acid joins.

Figure 3 A transfer RNA molecule.

Figure 4 Translation at a ribosome.

Ribosomes catalyse the synthesis of polypeptides. Figure 4 shows translation occurring at a ribosome.

- Transfer RNA molecules bring the amino acids and find their place when the anticodon binds by temporary hydrogen bonds to the complementary codon on the mRNA molecule.

- As the ribosome moves along the length of mRNA, it reads the code, and when two amino acids are adjacent to each other a peptide bond forms between them.

- Energy, in the form of ATP, is needed for polypeptide synthesis.

- The amino acid sequence for the polypeptide is therefore ultimately determined by the sequence of triplets of nucleotide bases on the length of DNA – the gene.

- After the polypeptide has been assembled, the mRNA breaks down. Its component molecules can be recycled into new lengths of mRNA, with different codon sequences.

- The newly synthesised polypeptide is helped, by chaperone proteins in the cell, to fold correctly into its 3D shape or tertiary structure, in order to carry out its function.

LEARNING TIPS

- A triplet of bases on a DNA molecule is called a base triplet.

- A triplet of bases on a length of mRNA is called a codon.

- A triplet of bases on a tRNA molecule, complementary to the mRNA codon, is called an anticodon.

INVESTIGATION

Using methyl green–pyronin stain to show the distribution of DNA and RNA within cells

DNA takes up the methyl green and RNA takes up the pyronin.
You can make slides of root tips of bean or onion roots and stain them to show that DNA is in the nuclei and RNA is in the cytoplasm (and nucleolus). *The cut tips of bean or onion roots have been fixed in acetic ethanol or absolute ethanol for at least 30 minutes.*
Wear eye protection.

1. Use a clean sharp scalpel or razor blade and cut thin longitudinal sections of the tip of the fixed root. Each section should be about 3 mm long.

2. Place the sections on a microscope slide and cover with methyl green–pyronin stain for 30 minutes.

3. Use a pipette to remove the stain, and then add distilled water to the root tips.

4. Change the water several times to wash the root-tip sections.

5. Add a drop of distilled water and a coverslip.

6. Observe cells in the sections under low power and then under high power.

7. You should see DNA stained blue-green in the cells' nuclei and RNA stained red in the cytoplasm.

Questions

1 During which physiological process is the ATP, which is needed for protein synthesis, made in cells?

2 List all the organelles used within a cell that is carrying out protein synthesis, and for each one briefly describe its role in the process.

3 Copy and complete the table below to compare the structure of DNA with that of RNA.

Feature	DNA	RNA
Type of sugar		
Nucleotide bases		
Number of polynucleotide strands		
Number of different types		
Location in eukaryotic cells		
Shape of molecule	double helix	not a double helix

4 Name the pyrimidine nucleotide bases found in RNA.

5 Where in a eukaryotic cell is mRNA made?

6 Suggest why the two subunits of ribosomes pass out of the nucleus separately before joining together in the cytoplasm.

7 Suggest why ribosomes may be described as catalysts.

8 Make a table comparing the processes of DNA replication and transcription.

THE RNA REVOLUTION

Long-perceived as a mere cellular housekeeper, RNA is now seen to have great potential for medical treatment. By manipulating new forms of RNA, scientists have the potential to develop new treatments for cancer, infectious disease and some chronic illnesses. Some types of RNA can direct specialised proteins to block or silence some cellular processes and these pathways may be adapted to develop more precise medical treatments. Recently, a new form of gene editing tool – called CRISPR – has been attracting attention; this has the potential to make precise changes to target genes.

GENE EDITING COMES OF HEALTHCARE AGE

The view of RNA, developed by biologists in the 1950s and 1960s, as just a messenger molecule acting as go-between from nucleus to cytoplasm, held for several decades.

Three types of RNA – messenger, transfer and ribosomal – were recognised. In the 1990s two other types, micro RNAs and small interference RNAs, were discovered and in 2003 long non-coding RNAs were found.

Genetic engineering first became possible following the development, during the 1970s, of recombinant DNA technology. However, this has been a fairly hit and miss process, as researchers have had little control over where the new gene was inserted into the genome.

Recently, gene editing technology has made it possible to make precise changes by adding or subtracting DNA at specific regions in the genome, turning target genes on or off.

In 2012 the CRISPR (clustered regularly interspaced short palindromic repeats) system gained prominence as one such precision gene editing tool. CRISPR (pronounced 'crisper') has two main components, both derived from naturally-occurring DNA cutting and repair systems used by bacteria to deal with invading viruses. One is a homing device (guide RNA)

that targets a specific length of DNA. The other is enzymatic 'scissors' (Cas9 nuclease) that cut the DNA there.

In a separate process, small pieces of corrective DNA can be inserted at the same location.

Several biotech companies are already exploiting CRISPR for drug discovery and improved clinical outcomes, with several competing patent claims pending. Big pharma has also picked up the technology for in-house research labs and, in January 2015, Astra Zeneca announced what the company says is the industry's most far-reaching CRISPR programme so far.

Collaboration with the Wellcome Trust Sanger Institute in Cambridge, UK, will delete specific genes relevant to cancer, as well as to cardiovascular, metabolic, respiratory, autoimmune and inflammatory diseases and to regenerative medicine, so as to understand the precise roles of genes in these conditions. Sanger researcher Kosuke Yusa says "CRISPR has transformed the way we study the behaviour of cells."

Source
- Cookson, C. Gene editing comes of healthcare age. *Financial Times*, 16 February 2015.

Let's start by considering the nature of the writing in the article.

1. This extract is adapted from the article 'Gene editing comes of healthcare age' by Clive Cookson, in the newspaper Financial Times. What features of the writing indicate that this article is aimed at a general audience, who may have some fairly basic scientific knowledge?

2. Is it clear from this article that CRISPR RNA and the Cas proteins occur naturally in bacterial cells?

3. Why do you think this article devotes several lines of text to the biotech and pharmaceutical companies involved?

Now let's look at the biological concepts underlying the information in this article. Remember to use knowledge acquired at GCSE level, as well as what you have learnt so far in this course, about cells, nucleic acids, proteins and viruses.

4. Many medicines work by blocking proteins' functions by association with their active sites. However, some protein targets involved in diseases have no suitable drugs either because they have no active site (e.g. cytoskeleton threads) or because their active sites are hidden (e.g. in ion transport channels within cell membranes).
   a. Explain how blocking a protein's active site may affect the functioning of that protein.
   b. Explain why proteins such as cytoskeleton threads do not have an active site.

5. Suggest how the guide RNA, used in CRISPR technology, finds its specific sequence of target DNA.

6. The source refers to enzymatic scissors called Cas (CRISPR-associated) proteins. Which types of bonds would be broken to cut a length of DNA in two?

7. RNA molecules used as medical treatments need to be introduced into cells. They may be wrapped in fat molecules, making liposomes which can pass through the plasma membrane and through the nuclear envelope. Suggest why these relatively large liposomes are able to pass through cell membranes.

8. Gene therapy, developed in the 1990s, introduced a copy of a functioning gene (also wrapped in fat molecules as liposomes) into a patient's cells and it was hit or miss as to whether the functioning gene inserted itself into a patient's genome. Explain how CRISPR is potentially a more precise form of gene therapy.

9. Twenty years ago scientists thought that much of our DNA was non-coding as it was not involved in making proteins. Some of it was described as 'junk DNA'. Scientists now know that most of our genes are transcribed and, although many are not involved in making proteins, their products are RNA of some sort. Since many of our genes are transcribed but the resulting RNA does not get translated into proteins, should the definition of a gene as 'a length of DNA that codes for a protein' be revised? Suggest a possible alternative definition of a gene.

> Although you usually associate 'active site' with enzymes, the specifically-shaped regions of other protein molecules such as antibodies and receptors are also active sites.

Activity 1

Micro RNAs (miRNAs), discovered in 1993, attach to mRNA (messenger RNA) at ribosomes and halt translation. They have the potential to prevent the hepatitis C virus from replicating in infected human cells. More people are infected by hepatitis C than are infected by the human immunodeficiency virus (HIV) and hepatitis C is the leading cause of liver cancer in humans. Develop an annotated diagram that could be used in an A level Biology textbook to explain the role of miRNAs.

Activity 2

Small interfering RNAs (siRNAs) were discovered in 1998. Each type is complementary to a specific length of mRNA. The siRNA can be taken up by specific proteins that can then cut a length of mRNA. The siRNA that has been synthesised to block the production, in infected cells, of a protein needed for the Ebola virus to replicate has been tested on monkeys and given to some healthy humans, producing no side effects. However it has not been fully trialled. The Ebola virus kills up to 90% of those infected. What are the ethical issues for giving this treatment to Ebola patients during an outbreak, such as that of 2014–2015?

DID YOU KNOW?
The Human Genome Project has shown us that almost all DNA in the genome is transcribed, not just genes that code for proteins. Much DNA is transcribed only to RNA of some sort which acts to regulate expression of other genes.

3.1   3.2   3.3   4.1   4.2   4.3

Practice questions

1. What are the monomers of nucleic acids? [1]
   A. pyrimidines
   B. purines
   C. nucleotides
   D. base triplets

2. Which of the following are not derived from nucleotides? [1]
   A. DNA and RNA
   B. ATP and ADP
   C. NAD and FAD
   D. ADH and FSH

3. DNA is a self-replicating molecule. What term is used to describe the way in which it replicates itself? [1]
   A. conservative
   B. dispersive
   C. semi-conservative
   D. discontinuous

4. The sequence of base triplets on the coding strand of part of a DNA molecule is TGACCGTTAGCG.

   Which of the following options shows the correct corresponding tRNA molecules? [1]

   A. ACT   GGC   AAT   CGC
   B. UGA   CCG   UUA   GCG
   C. UGA   GGC   AAU   CGC
   D. ACU   GGC   AAU   CGC

5. Read the following statements:
   (i) Each chromosome includes one molecule of DNA.
   (ii) The genetic code is overlapping and near-universal.
   (iii) The two ribosomal sub-units are joined in the nucleus and then pass into the cytoplasm.

   Which statement(s) is/are true? [1]
   A. (i), (ii) and (ii)
   B. Only (i) and (ii)
   C. Only (ii) and (iii)
   D. Only (i)

   [Total: 5]

In questions 6–9, you may need to use your knowledge of other topics, such as biological molecules, cell structure, enzymes, membranes, cell diversity and classification.

Bear in mind that when you sit your exam at the end of your course, you will have covered and learned about all the topics in the specification and should be able to make links between them.

6. Figure 1 is a diagram of part of a DNA molecule.

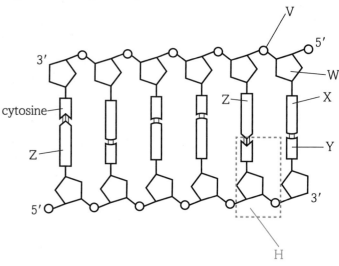

Figure 1

(a) Look at the labels on the diagram.
   (i) Name the parts labelled V–Z. [5]
   (ii) What do the lines between X and Y represent? [1]
   (iii) Name an element found in the part labelled X that is not in V or W. [1]
   (iv) Name the structure H. [1]

(b) A section of DNA was analysed and 24% of its bases were found to be adenine. Calculate the percentage of bases that were cytosine, in this piece of DNA. Show your working. [2]

(c) State three structural differences between DNA and RNA. [3]

[Total: 13]

7. The following questions relate to DNA replication.
   (a) Name the stage in the cell cycle in which DNA replication occurs. [1]
   (b) Explain why, during DNA replication, each molecule should be replicated exactly. [1]

   Figure 2 represents part of a DNA molecule.

Figure 2

(c) Show, by means of annotated diagrams, how this piece of DNA is replicated. Distinguish clearly between the original and new strands. [4]

(d) Outline the functions of the following enzymes during DNA replication.
   (i) helicase [1]
   (ii) DNA polymerase [1]
   (iii) ligase. [1]

(e) Errors may occur during DNA replication, resulting in a change to the sequence of base triplets. Explain how a change in the base sequence of a length of DNA may affect the structure and function of the protein encoded by this gene. [5]

[Total: 14]

8. The housefly is an insect. Its eggs hatch into larvae (maggots) that each grow before changing into a pupa and then into an adult fly. The larva and adult fly feed on similar types of food.

Scientists extracted amylase and lysozyme enzymes from homogenised tissues from the guts of larval and adult flies. Amylase digests starch to maltose, and lysozyme breaks bonds between residues in the peptidoglycan of bacterial cell walls.

- Both larval and adult amylase digested starch but the enzymes had a different protein structure.
- Both adult and larval lysozyme had the same protein structure, but the amounts produced in larva and adult differed.

(a) Explain, using your knowledge of genes (DNA), how the amylase enzymes in the fly larva and the adult fly can have a different protein structure. [4]

(b) Suggest how different amounts of lysozyme are produced by the cells of adult and larval flies. [1]

(c) Suggest the advantage to adult and larval flies of producing lysozyme enzymes in their guts. [2]

(d) Some new-born human babies have a deficiency of lysozyme. If they are fed on formula feed, then they may suffer from diarrhoea. However, if they are breast fed, then they are far less likely to suffer from diarrhoea. Suggest why breast milk helps to protect these babies from diarrhoea. [2]

[Total: 9]

9. The table opposite shows the standard (coding strand) DNA triplet codes for the 20 amino acids involved in protein synthesis. A section of DNA template strand is shown below.

5'-CATCCAAATTGTTGCCCG–3'

(a) Write down the sequence of amino acids formed when this section of DNA is transcribed and translated. [1]

(b) The standard (coding strand) base triplets TAA, TAG and TCA do not correspond with an amino acid.
   (i) Write down the corresponding mRNA codons for these base triplets. [3]

(ii) Suggest what happens when a ribosome reaches one of these codons during translation. [1]

(c) With reference to the base triplets shown in the table below:
   (i) Explain what is meant by 'The genetic code is degenerate.' [2]
   (ii) Explain why the genetic code is degenerate. [2]
   (iii) Explain how the code being degenerate is advantageous to living organisms. [1]

(d) Why is the genetic code described as universal (or near-universal)? [1]

[Total: 11]

DNA triplet codes						amino acid
TTT	TTC					phenylalanine
TTA	TTG	CTT	CTC	CTA	CTG	leucine
ATT	ATC	ATA				isoleucine
GTT	GTC	GTA	GTG			valine
TCT	TCC	TCA	TCG	AGT	AGC	serine
CCT	CCC	CCA	CCG			proline
ACT	ACC	ACA	ACG			threonine
GCT	GCC	GCA	GCG			alanine
TAT	TAC					tyrosine
CAT	CAC					histidine
CAA	CAG					glutamine
AAT	AAC					asparagine
AAA	AAG					lysine
GAT	GAC					aspartic acid
GAA	GAG					glutamic acid
TGT	TGC					cysteine
TGG						tryptophan
CGT	CGC	CGA	CGG	AGA	AGG	arginine
GGT	GGC	GGA	GGG			glycine
ATG						methionine
TAA	TAG	TCA				

10. The following questions relate to transcription and translation.

(a) Describe the process of transcription. [5]

(b) Describe the process of translation. [5]

(c) In what ways is transcription:
   (i) similar to DNA replication?
   (ii) different from DNA replication? [9]

(d) Briefly outline how you could demonstrate the distribution of RNA and DNA within cells. [2]

[Total: 21]

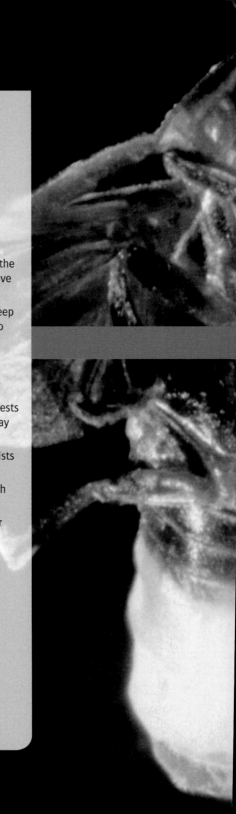

Foundations in biology

ENZYMES

Introduction

Much of the food we eat, such as cheese and biscuits, and drinks such as fruit juice, owe their production to the use of enzymes. If you have contact lenses, the fluid used to clean them contains protease enzymes to remove any proteins on their surface and prevent eye infections.

Without enzymes the metabolic reactions that take place in our bodies would not progress fast enough to keep us alive. Some enzymes break down large molecules, such as the proteins, carbohydrates and fat we eat, into smaller molecules; others, such as DNA polymerase, synthesise large molecules from smaller ones.

Enzymes also help our cells communicate with each other as they are crucial for cell signalling and for regulating the cell cycle.

In 1946 three scientists were awarded the Nobel Prize in Chemistry for work they had carried out in 1930, showing that enzymes were proteins. In 1965 two other scientists crystallised lysozyme, an enzyme that digests the walls of some bacteria and is found in tears, saliva and egg white, and worked out its structure using X-ray crystallography.

Most enzyme molecules are much larger than their substrate molecules but the active site of enzymes consists of only a few amino acids.

All living organisms and viruses have enzymes. Enzymes that do not function properly, because the gene with instructions for their synthesis is mutated, may cause severe diseases.

Enzymes are used in many manufacturing processes, including the food and drinks industries, brewing, paper making, biofuels, detergents and in medicine and biotechnology.

All the maths you need

To unlock the puzzles in this section you need the following maths:

- Units of measurement
- Powers of 10
- Division and multiplication, for example when making serial dilutions
- Ratios, percentages and fractions
- Use an appropriate number of significant figures when handling data from investigations
- Calculate percentage error
- Calculate the rate of change from a graph

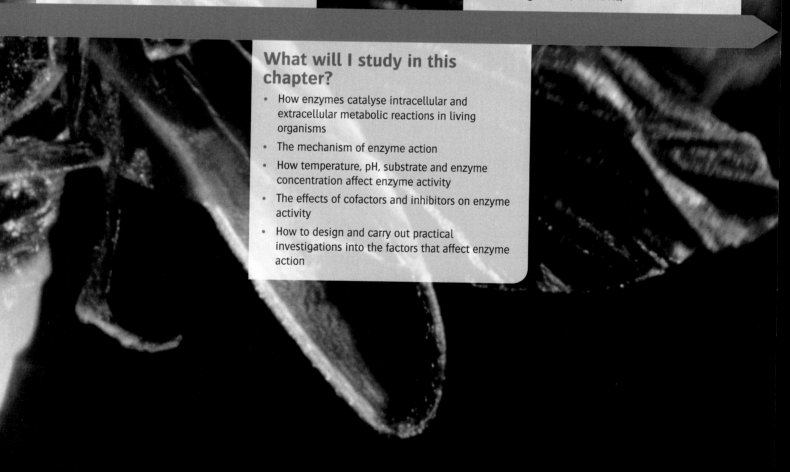

What have I studied before?

- Enzymes are catalysts, made of protein
- Enzymes are crucial for speeding up chemical reactions involved in metabolic activities such as photosynthesis, digestion and protein synthesis
- Substrate molecules fit into an enzyme's active site
- Enzymes are specific
- Enzyme action is affected by temperature and pH

What will I study later?

- How enzymes, such as ATPase, help to actively transport substances across cell membranes (AS and AL)
- The role of enzymes at checkpoints in regulating the cell cycle (AS)
- The importance of enzymes in catalysing protein synthesis for cell division (AS)
- The role of enzymes in all metabolic processes that sustain life (AS and AL)
- The roles of specific enzymes in biotechnology and genetic manipulation techniques (AL)
- How comparing the presence and structures of enzymes in different living organism can help elucidate the evolutionary relationships between those organisms (AS and AL)

What will I study in this chapter?

- How enzymes catalyse intracellular and extracellular metabolic reactions in living organisms
- The mechanism of enzyme action
- How temperature, pH, substrate and enzyme concentration affect enzyme activity
- The effects of cofactors and inhibitors on enzyme activity
- How to design and carry out practical investigations into the factors that affect enzyme action

(1) Enzymes – biological catalysts

By the end of this topic, you should be able to demonstrate and apply your knowledge and understanding of:

* the role of enzymes in catalysing reactions that affect metabolism at a cellular and whole organism level

* the role of enzymes in catalysing both intracellular and extracellular reactions

---

KEY DEFINITIONS

active site: indented area on the surface of an enzyme molecule, with a shape that is complementary to the shape of the substrate molecule.
catalyst: chemical that speeds up the rate of a reaction and remains unchanged and reusable at the end of the reaction.
extracellular: outside the cell.
intracellular: inside the cell.
metabolic/metabolism: the chemical reactions that take place inside living cells or organisms.
product: molecule produced from substrate molecules, by an enzyme-catalysed reaction.
substrate: molecule that is altered by an enzyme-catalysed reaction.

Enzymes are called biological **catalysts** because they speed up **metabolic** reactions in living organisms. Their actions affect both structure and function within cells, tissues and organs.

* Catalysts speed up chemical reactions and remain unchanged at the end of the reaction, able to be used again.

* A small amount of catalyst can catalyse the conversion of a large number of **substrate** molecules into **product** molecules.

* The number of reactions that an enzyme molecule can catalyse per second is known as its *turnover number*.

Why are enzymes so remarkable?

Whereas chemical catalysts usually need very high temperatures, increased pressures and extremes of pH, enzymes speed up metabolic reactions by up to 10^{12} times at lower temperatures, often at neutral pH and at normal pressures. Hence, as biological catalysts, they are able to function in conditions that sustain life.

Enzymes are also *more specific* than chemical catalysts. They do not produce unwanted by-products and rarely make mistakes. The cells in which they are made and/or act can also regulate their production and activity to fit the needs of the cell or organism at the time.

Enzymes are therefore remarkable molecules, and in this chapter we shall see how they work.

Enzyme structure determines function

As with all biological molecules, the structure of enzymes enables them to carry out their functions.

* For enzymes to catalyse some reactions, they may need help from **cofactors** (see topic 2.4.2).

* The instructions for making enzymes are encoded in genes. If the gene has a mutation that alters the sequence of amino acids in the protein, then this may alter the enzyme's tertiary structure and prevent it from functioning.

* If an enzyme that catalyses a metabolic reaction is deficient, then a metabolic disorder results.

* Enzymes also catalyse the formation of the organism's structural components, such as collagen in bone, cartilage, blood-vessel walls, joints and connective tissue. Some genetic disorders cause malformations of connective tissue and can be very harmful, such as 'stone man syndrome' (see the Thinking Bigger spread in Chapter 2.6).

DID YOU KNOW?

Many metabolic disorders are caused by deficient or non-functioning enzymes, for example if the active site is misshapen. The genetic disorder phenylketonuria, PKU, results when the enzyme phenylalanine hydroxylase does not function and cannot convert the essential amino acid, phenylalanine, to another amino acid, tyrosine. As a result, sufferers cannot make melanin (which is made from tyrosine), and the accumulation of phenylalanine in their blood impairs brain development, leading to severe mental impairment.
Because this enzyme deficiency is so severe, all new-born babies are screened for PKU, so that if the result is positive their diet can be adjusted to include only very small amounts of phenylalanine, to prevent the irreversible brain damage.

The active site

Enzymes are large molecules with a specific area, an indentation or cleft on the surface of the molecule, called the **active site**. This consists of just a few – often about 6 to 10 – amino acids.

* The tertiary structure of the active site is crucial, as its shape is complementary to the shape of the substrate molecule.

* So, each type of enzyme is highly specific in its function, as it can only catalyse a reaction involving the particular type of substrate molecule that fits into its active site.

- The shape of the enzyme's active site, and hence its ability to catalyse a reaction, can be altered by changes in temperature and pH, as these affect the bonds that hold proteins in their tertiary structure.

substrate molecules attached to the enzyme's active site

> **LEARNING TIP**
>
> The active site is part of the enzyme molecule, not part of the substrate molecule.

> **LEARNING TIP**
>
> Most enzymes catalyse a reaction in either direction depending on the cell's needs. Hence ATPase can catalyse the formation of ATP or the hydrolysis of ATP. Sometimes an enzyme catalyses two reactions, because it is really a large enzyme complex and has more than one active site. So, the same enzyme complex may be known by different names.

Figure 1 Structure of an enzyme molecule. The protein, made from a specific sequence of amino acids (the primary structure) has some pleats and coils (secondary structure) and is folded into its tertiary structure. The substrate molecule is shown attached to the cleft that forms the active site. This region, consisting of just a few amino acids, has a specific shape that is complementary to that of the substrate molecule.

Where enzymes work

Enzymes catalyse a wide range of **intracellular** as well as **extracellular** reactions.

Intracellular enzymes

In any cell, and within its organelles, there may be up to 1000 metabolic reactions going on at the same time, each being catalysed by a different enzyme. Some of these reactions are part of a metabolic pathway (see Figure 2).

- Each metabolic pathway in a living cell is one of a series of consecutive reactions, every step catalysed by a specific enzyme that produces a specific product.

- The various reactants and intermediates act as substrates for specific enzymes.

- The reactants, intermediates and products are known as metabolites.

- In some metabolic pathways, described as *catabolic*, metabolites are broken down to smaller molecules and release energy.

- In other metabolic pathways, described as *anabolic*, energy is used to synthesise larger molecules from smaller ones.

- Respiration and photosynthesis are examples of complex metabolic pathways, with many enzymes involved.

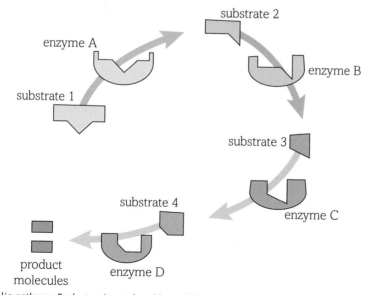

Figure 2 A metabolic pathway. Each step is catalysed by a different enzyme. If one of the enzymes cannot function, then the metabolic pathway cannot run.

Catalase is found in nearly all living organisms that are exposed to oxygen. It is a very important enzyme, as it protects cells from damage by reactive oxygen by quickly breaking down hydrogen peroxide, a potentially harmful by-product of many metabolic reactions, to water and oxygen.

- Catalase consists of four polypeptide chains and contains a haem group with iron (see topic 2.4.2).
- It is the fastest-acting enzyme, having the highest turnover number known, of about 6 million per second.
- In eukaryotic cells, catalase is found inside small vesicles called peroxisomes.
- When white blood cells ingest pathogens they use catalase to help kill the invading microbe.

The optimum pH for human catalase is around pH 7, but for other species it varies between pH 4 and 11. The optimum temperature also varies with species. For humans this is 45 °C and for some thermophilic archaea it is 90 °C.

> **DID YOU KNOW?**
>
> Some people lose hair pigment and 'go grey' earlier in life than others. One reason may be that these people have lowered levels of catalase, and so more hydrogen peroxide bleaches their hair shafts from the inside.

Extracellular enzymes

Some enzymes are secreted from the cells where they are made and act on their substrates, extracellularly. Fungi, such as the bread mould *Mucor*, release hydrolytic enzymes from their thread-like hyphae. The enzymes digest carbohydrates, proteins and lipids in the bread, and the products of digestion – glucose, amino acids, glycerol and fatty acids, are absorbed into the fungal hyphae for use in respiration and growth.

In our digestive system many enzymes are secreted, from cells lining the alimentary canal, into the gut lumen. There they extracellularly digest the large molecules, such as proteins, lipids, carbohydrates and nucleic acids, found in food. The products of digestion are then absorbed, via epithelial cells of the gut wall, into the bloodstream in order to be used for respiration, growth and tissue repair.

- Amylase is produced in the salivary glands, and acts in the mouth to digest the polysaccharide starch to the disaccharide maltose. It is also made in the pancreas, and acts to catalyse the same reaction in the lumen of the small intestine.

- Trypsin is made in the pancreas, and acts in the lumen of the small intestine to digest proteins into smaller peptides by hydrolysing peptide bonds. Its optimum pH is between 7.5 and 8.5.

> **DID YOU KNOW?**
>
> If you had to wait for a meal to be digested in your alimentary canal, without any enzymes, it would take several years. Hence, without enzymes, the speed of chemical reactions inside living organisms could not sustain life. Humans produce about six times as much amylase as chimpanzees do. Chimpanzees eat some meat and a lot of fruit, but very few starchy vegetables.

Questions

1. Explain why enzymes (a) are specific in their action, and (b) are described as biological catalysts.

2. Peptidases are enzymes that catalyse the breaking of peptide bonds within small peptide molecules. Suggest names for the following enzymes: (a) those that catalyse the breaking of glycosidic bonds (between glucose residues in polysaccharides) and (b) those that catalyse the breaking of ester bonds (in triglycerides).

3. Explain what is meant by the 'turnover number' of an enzyme.

4. Explain why a change to the sequence of DNA base triplets in a gene can result in an enzyme that cannot function properly.

5. Suggest why a non-functioning enzyme may cause a serious disease.

② Cofactors

By the end of this topic, you should be able to demonstrate and apply your knowledge and understanding of:

* the need for coenzymes, cofactors and prosthetic groups in some enzyme-controlled reactions

Some enzymes need help

Some enzymes, particularly those involved in catalysing oxidation–reduction reactions, can only work if another small non-protein molecule is attached to them. These small molecules are called **cofactors**.

Prosthetic groups

A cofactor that is permanently bound, by covalent bonds, to an enzyme molecule, is called a **prosthetic group**.

The enzyme carbonic anhydrase contains a zinc ion permanently bound, as a prosthetic group (see Figure 1), to its active site. This enzyme is found in erythrocytes (red blood cells) and catalyses the interconversion of carbon dioxide and water to carbonic acid, which then breaks down to protons and hydrogencarbonate ions.

As with most enzyme-catalysed reactions, this reaction may proceed in either direction, depending on the concentration of substrate or product molecules.

$$CO_2 + H_2O \longleftrightarrow H_2CO_3 \leftrightarrow H^+ + HCO_3^-$$
$$\text{Carbonic anhydrase}$$

The reaction is vitally important, as it enables carbon dioxide to be carried in the blood from respiring tissues to the lungs.

Other cofactors

Whereas the zinc associated with carbonic anhydrase is present in a compound that is permanently bound to the enzyme's active site, some enzymes work better in the presence of ions that are *not* permanently bound to them. These ions are also cofactors.

During an enzyme-catalysed reaction, the enzyme and substrate molecules temporarily bind together to form an **enzyme-substrate complex**. The presence of certain ions that may temporarily bind to either the substrate or the enzyme molecule may ease the formation of such enzyme-substrate complexes and therefore increase the rate of the enzyme-catalysed reaction.

* Some cofactors act as co-substrates – they and the substrate together form the correct shape to bind to the active site of the enzyme.

* Some cofactors change the charge distribution on the surface of the substrate molecule or on the surface of the enzyme's active site, and make the temporary bonds in the enzyme-substrate complex easier to form.

The enzyme amylase digests starch to maltose, and will only function if chloride ions are present.

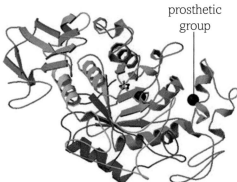

Figure 1 Zinc-based prosthetic group attached to the active site of the enzyme carbonic anhydrase.

Coenzymes

Coenzymes are small organic non-protein molecules that bind *temporarily* to the *active site* of enzyme molecules, either just before or at the same time that the substrate binds. The coenzymes are chemically changed during the reaction, and they need to be recycled to their original state, sometimes by a different enzyme.

Some vitamins are sources of coenzymes

Many coenzymes are derived from water-soluble vitamins. If these vitamins are deficient in the diet of humans, then certain diseases may result (see Table 1).

Vitamin	Coenzyme derived from it	Human deficiency disease
B_{12}	cobalamin coenzymes	pernicious anaemia (progressive and fatal anaemia)
Folic acid	tetrahydrofolate	megablastic anaemia (large, irregularly shaped erythrocytes)
Nicotinamide, B_3	NAD, NADP	pellagra (diarrhoea, dermatitis and dementia)
Pantothenate, B_6	coenzyme A	elevated blood-plasma triglyceride levels
Thiamine, B_1	thiamine pyrophosphate	beriberi (mental confusion, irregular heartbeat, muscular weakness, paralysis and heart failure)

Table 1 Some coenzymes and the vitamins that they are derived from, and the deficiency diseases that result when such vitamins are lacking in the diet.

NAD and NADP (see Table 1) are hydrogen acceptors that you will learn more about when you study respiration and photosynthesis in the second year of your course. They are both derivatives of nucleotides – see topic 2.3.1.

Questions

1. The recommended daily dietary allowance for nicotinamide for adult humans is 18 mg. Suggest why this amount is so low, although a great deal of NAD is used by humans for their metabolism, throughout each day.

2. Explain how metallic ion cofactors enable enzymes to work efficiently.

3. In what ways are metallic ion cofactors similar to and different from organic coenzymes?

4. In what ways are metallic ion cofactors different from zinc-based prosthetic groups?

5. Which type of bonds (a) are present between a prosthetic group and its enzyme? (b) form between metallic ion cofactors and an enzyme's active site? (c) form between coenzymes and the enzyme or substrate molecules?

3 The mechanism of enzyme action

By the end of this topic, you should be able to demonstrate and apply your knowledge and understanding of:

* the mechanism of enzyme action

KEY DEFINITIONS

enzyme-product complex: enzyme molecule with product molecule(s) in its active site. The two are joined temporarily by non-covalent forces.

enzyme-substrate complex: enzyme molecule with substrate molecule(s) in its active site. The two are joined temporarily by non-covalent forces.

The lock-and-key hypothesis

You have already seen (in topic 2.4.1) that a specific indented area on the surface of the enzyme molecule, called the active site, is where the substrate molecules fit. They fit because the tertiary structure of the enzyme's active site gives it a shape that is *complementary* to that of the substrate molecule – rather like the way in which only a specific key will fit into a particular lock. This idea about how enzymes work is described as the lock-and-key hypothesis (see Figure 1). The lock is the enzyme's active site, and the key is the substrate molecule.

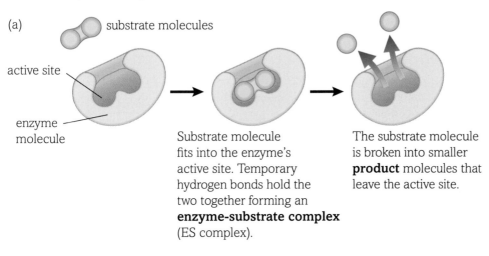

(a) substrate molecules

active site

enzyme molecule

Substrate molecule fits into the enzyme's active site. Temporary hydrogen bonds hold the two together forming an **enzyme-substrate complex** (ES complex).

The substrate molecule is broken into smaller **product** molecules that leave the active site.

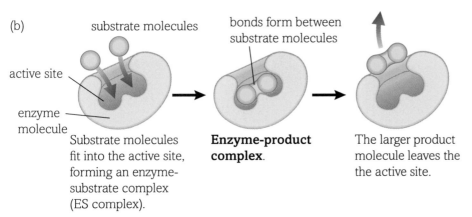

(b) substrate molecules

bonds form between substrate molecules

active site

enzyme molecule

Substrate molecules fit into the active site, forming an enzyme-substrate complex (ES complex).

Enzyme-product complex.

The larger product molecule leaves the the active site.

Figure 1 Lock-and-key mechanism of enzyme action. (a) A large substrate molecule is split into smaller product molecules that are then released from the active site. (b) Two small substrate molecules attach to the active site and bonds form between them, making a larger product molecule, which then leaves the active site. In each case, at the end of the reaction, the enzyme is able to form an ES complex with another substrate molecule and catalyse another reaction.

- The substrate molecules and enzyme molecules each have kinetic energy and are constantly moving randomly.
- If a substrate molecule successfully collides with an enzyme molecule, then an **enzyme-substrate complex** (ES complex) forms as the substrate molecule fits into the complementary-shaped active site on the enzyme molecule.
- The substrate molecules are either broken down or built up into the product molecule(s), and these form an **enzyme-product complex** whilst still in the active site.
- The product molecules leave the active site.
- The enzyme molecule is now able to form another enzyme-substrate complex.
- A small number of enzyme molecules can therefore convert a large number of substrate molecules into product molecules.

DID YOU KNOW?

A scientist, Emil Fischer, discovered in 1894 that glycolytic (sugar-splitting) enzymes could distinguish between sugar molecules that have the same molecular formula but slight differences in the geometry of how the atoms are arranged within the molecules (think of a left-hand glove and a right-hand glove). This led him to form his lock-and-key hypothesis, which explains enzyme specificity.

DID YOU KNOW?

Although scientists have known about enzymes since the late 19th century, it was only in the 1930s that it was established that enzymes are proteins. The first enzyme to have its amino acid sequence worked out was bovine pancreatic ribonuclease, in 1963. In 1965, the first 3D structure of an enzyme, hen egg-white lysozyme, was worked out using X-ray diffraction. Since then, the structures of many enzymes have been elucidated. In the 1990s, scientists discovered that some types of RNA have catalytic properties in cells.

The induced-fit hypothesis

Although the lock-and-key hypothesis explains enzyme specificity, it does not explain how the transition state – namely the ES complex – is stabilised.

In 1958, Daniel Koshland modified the lock-and-key hypothesis by suggesting that the active site of the enzyme is not a rigid fixed structure, but that the presence of the substrate molecule in it *induces* a shape change, giving a good fit (see Figure 2).

He suggested that:

- When the substrate molecules fit into the enzyme's active site, the active site changes shape slightly to mould itself around the substrate molecule. Think of putting on a glove – it will only accept a hand-shaped object, but when you insert your hand the glove moulds around and fits your hand perfectly.

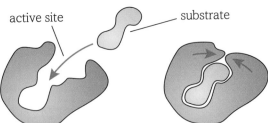

Figure 2 The induced fit hypothesis. Binding of the substrate molecule (yellow) to the active site of the enzyme (blue) results in a change of shape in the enzyme's active site so that it fits around the substrate molecule more closely to give the enzyme-substrate (ES) complex. When the product molecules are still attached to the active site this gives an enzyme-product complex.

- The active site still has a shape complementary to the shape of the substrate molecule. But, on binding, the subtle changes of shape of the side chains (R-groups) of the amino acids that make up the active site give a more precise conformation that exactly fits the substrate molecule.
- This moulding enables the substrate to bind more effectively to the active site.

- An enzyme-substrate complex is formed, and *non-covalent* forces such as hydrogen bonds, ionic attractions, van der Waals forces and hydrophobic interactions, bind the substrate molecule to the enzyme's active site.
- When the substrate molecules have been converted to the product molecules and these are still in the active site, they form an enzyme-product complex.
- As the product molecules have a slightly different shape from the substrate molecule, they detach from the active site.
- The enzyme molecule is now free to catalyse another reaction with another substrate molecule of the same type.
- The equation below outlines how an enzyme catalyses a reaction:

Enzyme + → Enzyme-substrate → Enzyme-product → Enzyme +
Substrate   complex   complex   Product

E + S   →   ESC   →   EPC   →   E + P

Enzymes lower the activation energy of a reaction

Chemical reactions need energy to activate or begin them. Many chemicals can be heated to provide this activation energy and make them react together. This increases the kinetic energy of the molecules so that they move about more, in a random fashion, and are more likely to successfully collide and then react together.

In a living cell, the temperature cannot be raised too much or the proteins within it would denature and lipids would melt. Because enzymes have an active site specific to *only* the substrate molecules, they bring the substrate molecules close enough together to react, without the need for excessive heat. Therefore they lower the activation energy and hence speed up metabolic reactions (see Figure 3).

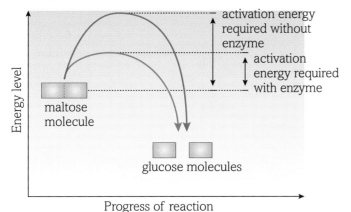

Adding the enzyme maltase reduces the amount of activation energy required for the reaction to take place

Figure 3 How the enzyme maltase lowers the activation energy needed to hydrolyse maltose to glucose.

INVESTIGATION

Enzyme specificity

Wear eye protection.

Yeast, *Saccharomyces cerevisiae*, is a single-celled fungus. It is a facultative anaerobe, which means it can respire with or without oxygen. If oxygen is present, then it will respire aerobically, but if oxygen is absent then it can respire anaerobically.

When yeast respires anaerobically it produces ethanol and carbon dioxide. The carbon dioxide produced during a unit time, such as 10 minutes, can be collected and its volume measured.

Yeast may respire different sugars. Enzymes are specific and each type will catalyse a specific sugar. You can investigate whether yeast can metabolise the monosaccharide sugars glucose, fructose and galactose.

Questions

Use information gained from previous chapters, as well as the information in this chapter, to help you answer these questions.

1. Explain how enzymes lower the activation energy and therefore enable metabolic reactions to proceed quickly at the relatively low body temperatures of living organisms.

2. Explain how a cell can carry out many different types of metabolic reaction at the same time.

3. Describe how the lock-and-key hypothesis of enzyme action differs from the induced-fit hypothesis.

4. Some organisms grow in extreme environments. For example, some archaea and bacteria grow in very hot springs. Suggest how the structure of their enzymes, and other proteins, enables them to withstand such high temperatures.

5. Explain how an enzyme-substrate complex differs from an enzyme-product complex.

6. Are the enzymes involved in respiration intracellular or extracellular? Explain your answer.

7. Explain why a large amount of substrate can be converted by a small amount of enzyme.

The effect of temperature on enzyme activity

By the end of this topic, you should be able to demonstrate and apply your knowledge and understanding of:

* the effect of temperature on enzyme activity
* practical investigations into the effect of temperature on enzyme activity

KEY DEFINITION

Q_{10}: temperature coefficient, calculated by dividing the rate of reaction at $(T + 10)\,°C$ by the rate of the reaction at $T\,°C$.

Heat and kinetic energy

All molecules have kinetic energy and can continuously move around randomly. In so doing, molecules will collide with one another.

If a substance is heated, then:

* The extra energy, in the form of heat, causes the molecules to move faster.
* This increases the rate of collisions between molecules.
* It also increases the force with which they collide, as they are moving faster.

Collisions between enzyme and substrate molecules

You have already seen in topic 2.4.3 that an enzyme-catalysed reaction only takes place if a substrate molecule successfully collides with the active site of an enzyme molecule, forming an enzyme-substrate complex.

If the reactant mixture containing enzyme and substrate molecules is heated, then:

* both types of molecule will gain kinetic energy and move faster
* this will increase the rate (number per second) of successful collisions
* therefore the rate of formation of ES complexes increases, and the rate of reaction increases, increasing the number of enzyme-product complexes per second, up to a point
* at a particular temperature, called the enzyme's **optimum temperature**, the rate of reaction is at its *maximum*.

Heat makes molecules vibrate

As well as making molecules move faster, increasing the temperature also makes them vibrate.

* This may break some of the weak bonds, such as hydrogen and ionic bonds, that hold the tertiary structure of the enzyme's active site.
* As the active site shape begins to change, the substrate molecules will not fit into it so well and the rate of reaction begins to decrease.
* As more heat is applied, the shape of the enzyme's active site completely and *irreversibly* changes so that it is no longer complementary to the shape of the substrate molecule.
* The reaction cannot proceed at all.
* The enzyme is **denatured** (see Figure 1).

Heat does not break the peptide bonds between amino acids, so the enzyme's *primary structure* is not altered.

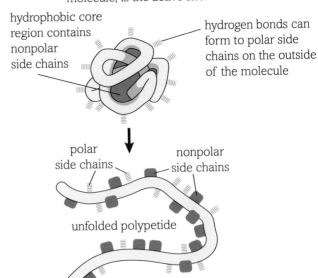

This protein is folded into its correct 3D shape, part of which, in an enzyme molecule, is the active site.

hydrophobic core region contains nonpolar side chains

hydrogen bonds can form to polar side chains on the outside of the molecule

polar side chains

nonpolar side chains

unfolded polypetide

Heat has broken hydrogen bonds so that the protein has unfolded. It no longer has its 3D shape and the active site is not complementary to the shape of the substrate molecule.

Figure 1 Denaturation.

Optimum temperature

This is the temperature at which the enzyme works *best*. It is the temperature at which the enzyme has its *maximum* rate of reaction.

Some enzymes work best at cool temperatures. Bear in mind that some organisms are adapted to living in cold places. For example, there are *psychrophilic* bacteria (see Figure 2(a)), which live in very cold conditions. Their enzymes can work at really low temperatures.

Some organisms, such as *thermophilic* bacteria in hot springs, live at very high temperatures (see Figure 2(b)). Their enzymes (as well as their other proteins) are heat stable. They have more disulfide bonds that do not break with heat and keep the shape of the protein molecules stable. Their enzymes will have high optimum temperatures. You will learn in the second year of your course about an enzyme obtained from a thermophilic bacterium, *Thermophilus aquaticus*, called Taq polymerase, that is used at high temperatures in the polymerase chain reaction (PCR) to amplify fragments of DNA for forensic gene-screening analysis or cloning.

LEARNING TIP
Enzymes are biological molecules but they are not organisms. They are *not* killed; they are denatured.

(a)

(b)

Figure 2 (a) Red algae growing in ice. Their enzymes function at cold temperatures. (b) Thermophilic bacteria in a hot pool. The enzymes of these bacteria function well at high temperatures.

The graph in Figure 3 shows the effect of temperature on the rate of an enzyme-catalysed reaction. In this case, the reaction is the digestion (hydrolysis) of cooked egg-white (the protein albumen) by the protease enzyme pepsin, at pH 2. The rate of reaction was measured by the time taken for the end point (a clear solution) to be reached.

$$\text{Rate of reaction} = \frac{1}{\text{time taken to reach end point}}$$

The units for rate of reaction are s^{-1}.

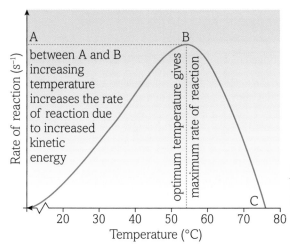

increasing temperature beyond the optimum (B) temperature reduces the rate of reaction due to the breaking of bonds holding the enzyme's tertiary structure in place

at C there is no reaction because the enzyme is denatured

Figure 3 The results of an investigation into the effect of temperature on the action of the protease enzyme pepsin. This enzyme has an optimum temperature of around 50–60 °C, which is higher than the mammalian body temperature.

Temperature coefficient, Q_{10}

The temperature coefficient here refers to the increase in the rate of a process when the temperature is increased by 10 °C.

It is given by the equation:

$$Q_{10} = \frac{\text{rate of reaction at } (T + 10)\,°C}{\text{rate of reaction at } T\,°C}$$

For chemical reactions in a test tube, Q_{10} is approximately 2, which means that for every 10 °C rise in temperature the rate of reaction is doubled.

For metabolic reactions catalysed by enzymes, between temperatures of 0 °C and approximately 40 °C, the rate of these reactions is also roughly doubled for every 10 °C rise in temperature. This is because the increase in temperature provides more kinetic energy, so enzyme and substrate molecules move faster and collide more often.

For enzymes at temperatures above their optimum temperature, the value of the temperature coefficient, Q_{10}, drops. This is because the higher temperatures alter the structure of the active sites of the enzyme molecules so that they are no longer complementary to the shape of the substrate molecules.

WORKED EXAMPLE

The table below shows the results of starch hydrolysis in a solution of Lintner's (soluble) starch and amylase enzyme.

Temperature (°C)	Mass of starch hydrolysed in 10 minutes (mg)
25	50
35	108
45	100

Q_{10} (between 25 and 35 °C) is $\frac{108}{50} = 2.16$

Q_{10} (between 35 and 45 °C) is $\frac{100}{108} = 0.93$

INVESTIGATION

There are many enzymes that you can use to investigate the effect of temperature change. One is phosphatase enzyme. This enzyme breaks down organic phosphates in cells in order to maintain the pool of phosphate ions for use by cells to make chemicals such as ATP, ribulose bisphosphate and NADP.

Effect of temperature on phosphatase enzyme

Wear eye protection.

In this investigation, a chemical substrate phenolphthalein phosphate, (PPP), is used. Phosphatase enzyme breaks this down, liberating free phenolphthalein. When sodium carbonate, which is alkaline, is added in excess, the free phenolphthalein produces a deep-pink colour. The intensity of the colour is proportional to the concentration of free phenolphthalein. This intensity can be measured using a colorimeter with a green filter. If each reaction tube has been given the same length of time, the intensity of the colour gives an indication of the rate of reaction – the darker the pink, the more molecules of PPP were hydrolysed by the enzyme in the set period of time.

Questions

1 Explain why the rate of an enzyme-controlled reaction is lower at 10 °C than it is at 30 °C.

2 Explain why an enzyme solution kept in the fridge for a week will catalyse a reaction at 40 °C, whereas an enzyme solution that has been boiled does not catalyse a reaction at 40 °C.

3 Explain why the rate of an enzyme-catalysed reaction does not keep on increasing as the temperature is increased above 40 or 50 °C.

4 Explain why using living tissue, such as bean sprouts, as a source of enzyme may lead to difficulties in controlling the enzyme concentration.

5 (a) Explain the difference between the independent variable and the dependent variable.

(b) Explain why tubes of reactants and separate tubes of enzyme solutions are placed in the water-bath for 5–10 minutes before they are mixed and the reaction allowed to proceed.

(c) Explain why, in an investigation about enzyme activity, the enzyme solution is always added last.

(5) The effect of pH on enzyme activity

By the end of this topic, you should be able to demonstrate and apply your knowledge and understanding of:

* the effect of pH on enzyme activity
* practical investigations into the effect of pH on enzyme activity

What is pH?

The pH indicates whether a substance is acidic, alkaline or neutral.

pH values of 0–6 are acidic, pH 7 indicates neutral, and pH 8–14 indicates that the solution is alkaline.

Acids such as hydrochloric acid and sulfuric acid dissociate into protons and a negatively charged ion.

$$HCl \rightarrow H^+ + Cl^-$$
$$H_2SO_4 \rightarrow H^+ + HSO_4^-$$

Organic acids are also proton donors:

* lactic acid dissociates into H^+ and lactate
* pyruvic acid dissociates into H^+ and pyruvate

Buffers

In biology, a **buffer** is something that resists changes in pH. There are certain chemicals in your blood that help resist changes in pH so that the blood pH remains within fairly narrow limits close to pH 7.4. These chemicals can donate or accept hydrogen ions. Some proteins, such as haemoglobin, can also donate or accept protons and so act as buffers.

In laboratory investigations, you will use buffer solutions to maintain the desired pH for investigating enzyme action at different pH values or to keep the pH constant whilst you investigate another factor.

LEARNING TIP

* Remember that pH is worked out by the formula $\log \frac{1}{[H^+]}$, so a large concentration of hydrogen ions gives a small value of pH. If the acidity increases, then the pH decreases.
* Remember also that hydrogen ions are protons.

How changes in pH affect bonds within molecules

* A hydrogen ion, which is a proton, has a positive charge, so it is attracted towards negatively charged ions, molecules or parts of molecules. Figure 1 shows how hydrogen bonds and ionic forces between amino acids hold the tertiary structure of an enzyme molecule, particularly the active site, in the correct shape so that the substrate molecule will fit into it.

* Excess hydrogen ions will interfere with these hydrogen bonds and ionic forces, and so the active site of the enzyme molecule will change shape. If the substrate molecule does not fit well into the active site, then the rate of the reaction that the enzyme catalyses will be lowered.

* Increasing the concentration of hydrogen ions will also alter the charges on the active site of enzyme molecules, as more protons will cluster around negatively charged groups (for example amino acid R-groups – see topic 2.2.8) in the active site. This interferes with the binding of the substrate molecule to the active site.

Hydrogen bonds hold structures like an α helix in place in protein molecules.

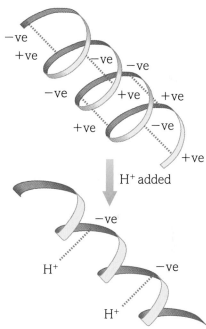

As H^+ is increased in concentration, the positive charges are attracted to the negative charges on the α helix and so 'replace' the hydrogen bonds.

Figure 1 How protons interfere with the hydrogen bonds holding the tertiary structure of an enzyme molecule.

Enzymes work within a narrow range of pH.

* *Small* changes of pH, either side of the **optimum pH**, slow the rate of reaction, because the shape of the active site is disrupted (see Figure 2).

* However, if the normal optimum pH is restored, the hydrogen bonds can re-form and the active site's shape is restored.

* At extremes of pH, the enzyme's active site may be permanently changed. When the enzyme is thus denatured, it cannot catalyse the reaction.

Reducing or increasing the pH away from the optimum pH reduces the rate of reaction because the concentration of H⁺ in solution affects the tertiary structure of the enzyme molecule.

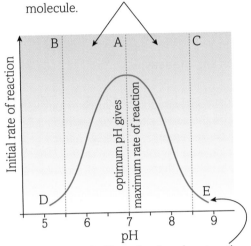

Extremes of pH can lead to denaturation of the enzyme (at points D + E).

Figure 2 The effect of pH on an enzyme-catalysed reaction.

Not all enzymes have the same optimum pH

Enzymes that work intracellularly have an optimum pH that is close to pH 7.

- Enzymes that work extracellularly may have optimum pH values different from pH 7. During digestion, for example, food is first taken into your mouth, and the amylase enzymes that digest starch to maltose there work best at a pH of 6.8.

- As food passes into your stomach, hydrochloric acid is secreted, giving a very low pH. One of the acid's functions is to kill bacteria and other pathogens in the food. The protease enzyme, pepsin, in the stomach works best at very low pH values, of between 1 and 2. It digests large protein molecules into smaller peptide molecules.

- As the partly digested food moves into the small intestine, salts in bile made in the liver neutralise it and raise the pH to around 7.8. This is optimal for the protein-digesting enzymes, trypsin and enterokinase, that catalyse further digestion of peptides to amino acids in the small intestine (Figure 3).

DID YOU KNOW?

It is important that your blood pH is maintained within quite narrow limits, between 7.35 and 7.45. There are many plasma proteins in blood. Changes in blood pH can also cause vasodilation or vasoconstriction. In the blood plasma there are buffers that resist changes in pH. For example, potassium dihydrogenphosphate can donate hydrogen ions. Some proteins act as buffers, as they can donate or accept protons. Hydrogencarbonate ions also act as buffers.

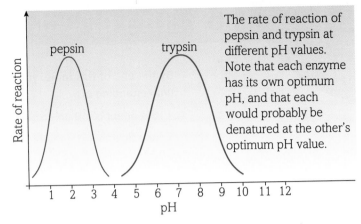

The rate of reaction of pepsin and trypsin at different pH values. Note that each enzyme has its own optimum pH, and that each would probably be denatured at the other's optimum pH value.

Figure 3 Pepsin and trypsin – rates of reaction at different pH values.

INVESTIGATION

The effect of pH changes on the rate of an enzyme-catalysed reaction

Design an investigation to find the effect of changing pH on the activity of the enzyme phosphatase. You can then carry out similar experiments with the enzymes pepsin and trypsin.

In each case, write a plan for your investigation. Include a clearly stated testable hypothesis; a list of apparatus needed; the independent and dependent variables; control variables with details of how and why you need to control them; and how you will deal with your data.

Questions

1. Explain why enzymes only work within narrow ranges of pH.

2. Suggest why blood pH is one of the factors monitored closely in some hospital patients.

3. Explain what is meant by 'a buffer solution'.

4. Explain why both increasing and decreasing the pH away from the optimum pH for a specific enzyme reduces its rate of reaction.

5. Dogs produce a much greater concentration of stomach acid than do humans. Hence they are able to eat food that would make us ill. How would you expect the optimum pH of canine pepsin to differ from that of human pepsin? How could you investigate this experimentally?

6. The proteases pepsin and trypsin are both produced in an inactive form – pepsinogen and trypsinogen, respectively. The stomach acid changes pepsinogen to the active pepsin, and an enzyme enterokinase changes trypsinogen to trypsin. Suggest why these enzymes are first secreted in an inactive form.

(6) The effect of substrate concentration on the rate of enzyme-catalysed reactions

By the end of this topic, you should be able to demonstrate and apply your knowledge and understanding of:

* the effects of substrate concentration on enzyme activity
* practical investigations into the effects of substrate concentration on enzyme activity

KEY DEFINITION

concentration: number of molecules per unit volume.

The effect of changing substrate concentration

If there is no substrate present, then an enzyme-catalysed reaction (see topic 2.4.3) cannot proceed. This is because there are no substrate molecules to fit into the enzyme molecules' active sites, and so no enzyme-substrate complexes can be formed.

As substrate is added and its **concentration** increases, the rate of reaction increases (see Figure 1).

* This is because more enzyme-substrate (ES) complexes can form.
* As a result, more product molecules are formed.
* Substrate concentration is limiting the reaction, because, as it increases, the rate of reaction increases.
* Substrate concentration is the **limiting factor**.

As the concentration of substrate is increased even further, the reaction will reach its *maximum* rate.

* Adding more substrate molecules to increase the substrate concentration will *not* increase the rate of reaction.
* This is because all the enzymes' active sites are occupied with substrate molecules.
* If more substrate molecules are added, then they cannot successfully collide with and fit into an enzyme's active site.

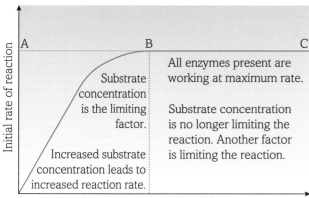

Figure 1 The effect of increasing substrate concentration on the rate of an enzyme-controlled reaction.

INVESTIGATION

The effect of increasing substrate concentration on the rate of an enzyme-catalysed reaction

Urease is a hydrolytic enzyme found in some bacteria, fungi and plants. It breaks the C–N bonds in amides, such as urea.

$$(NH_2)_2CO + H_2O \rightarrow 2NH_3 + CO_2$$

This releases ammonia, which can be used by the organisms as a source of nitrogen.

As ammonia is released, the pH of the solution is increased. The indicator phenol red changes from yellow to pink when alkaline (see Figure 2). Universal Indicator solution may also be used to detect the change of pH due to the evolution of ammonia. The solution will change from green to blue.

Colorimetry can be used to measure absorbance and hence depth of colour, and therefore the amount of ammonia produced in a set time. This gives an indication of the rate of reaction.

The independent variable is concentration of urea. The dependent variable is rate of reaction as measured by absorbance.

Control variables:

* concentration of enzyme solution
* volume of enzyme and substrate solutions
* temperature (use a thermostatically controlled water-bath at a specified temperature within the range 30–40 °C)
* time for reaction, e.g. 20 minutes
* stirring/shaking of reactants
* wavelength of light/filter in the colorimeter.

Wear eye protection.

You will have a stock solution of urea of 0.1 M concentration. Make up a range of urea concentrations as follows:

Volume of 0.1 M urea solution (cm³)	Volume of distilled water (cm³)	Concentration of urea (M)
10	0	0.10
9	1	0.09
8	2	0.08
7	3	0.07
6	4	0.06
5	5	0.05
4	6	0.04
3	7	0.03
2	8	0.02
1	9	0.01
0	10	0.00

Use a 0.05% solution of urease, which has been kept in the water-bath at a specified temperature within the range 30–40 °C.

1 Label 11 test tubes with your initials and the numbers 0–10.
2 Place 5 cm³ of urea solution of differing concentrations, as shown in the table above, into the tubes.
3 Add 1 cm³ of phenol red to each tube.
4 Place these tubes in the water-bath for 10 minutes.
5 Make a table for your results, and set up the colorimeter. Get your cuvettes ready.
6 Add 1 cm³ of urease solution to each of your 11 tubes, in the water-bath. Shake each tube.
7 Leave the tubes for 20 minutes. Shake each one once every 5 minutes.
8 Just before the 20 minutes is up, fill a cuvette with solution from tube 0. Use this for calibration and set the absorbance at 0, using a blue or green filter in the colorimeter.
9 After 20 minutes, read the absorbance of the contents of each tube, 1–10. Make sure that you give each one a final shake before filling the cuvette.
10 Present your data as a graph.
11 Between which concentrations is the concentration of urea the limiting factor?
12 Compare your data with those of others in your class. Is there concordance?

Figure 2 Phenol red indicator changes colour when urease hydrolyses urea to ammonia. Yellow indicates no reaction. Pink develops when urease breaks down urea to ammonia, which reacts with the phenol red indicator. Therefore the depth of pink colour indicates the degree of enzyme activity.

Questions

1 Explain the shape of the graph shown in Figure 1, between points A and B, and between points B and C.

2 Suggest why fungi, plants and some bacteria have the enzyme urease.

3 Suggest limitations in the above-mentioned investigation on the effect of urea concentration on the rate of activity of urease.

4 Suggest sources of errors in the above-mentioned investigation on the effect of urea concentration on the rate of activity of urease.

5 Suggest improvements to the above-mentioned investigation on the effect of concentration of urea.

6 Design an investigation into the effect of substrate concentration on the activity of phosphatase enzyme.

7 When the rate of a reaction, as shown in Figure 1, has reached its maximum, increasing the substrate concentration does not increase the rate of reaction. Suggest why increasing the concentration of enzyme will increase the rate of reaction.

8 Explain why each of the following variables should be controlled in the investigation of urease activity above: (a) temperature; (b) time of reaction; (c) mixing of reactants; and (d) wavelength of light in a colorimeter.

DID YOU KNOW?

Tanning animal skins to make leather is an ancient craft. After the skins were soaked in urine and water, scraped to remove fat and treated with lime to remove the hairs, they were placed in a vat of water and dung to make them softer. Children were often employed to gather dog and pigeon dung from streets. We now know that there are protease enzymes in the dung that partly digest the collagen protein in hides and make them soft and supple. This process is called bating. The hides were de-limed before the dung was added, to make the pH less alkaline.

LEARNING TIP

We refer to concentration of substances in solution, not just 'the amount'. Concentration indicates how many molecules are present in a unit volume. Hence it will always affect the rate of a reaction.

7 The effect of enzyme concentration on the rate of reaction

By the end of this topic, you should be able to demonstrate and apply your knowledge and understanding of:

* the effects of enzyme concentration on enzyme activity

* practical investigations into the effects of enzyme concentration on enzyme activity

Enzyme availability in cells

In a laboratory practical, you may be able to change the concentration of a solution of a particular enzyme. In living cells, the enzyme concentration or availability depends on the rate of synthesis of the enzyme and its rate of degradation.

Each of these rates is directly controlled by the cell.

Enzyme synthesis

Depending on the cell's needs, genes for synthesising particular enzymes can be switched on or off.

Enzyme degradation

The protein component of living cells is constantly being turned over. Cells are continuously degrading old enzyme molecules to their component amino acids and synthesising new enzyme molecules from amino acids.

This might appear to be rather wasteful, but there are advantages:

* the elimination of abnormally shaped proteins that might otherwise accumulate and harm the cell

* the regulation of metabolism in the cell by eliminating any superfluous (surplus to requirements) enzymes.

Therefore, for a cell to regulate its metabolism properly, the control of enzyme degradation is equally as important as the control of enzyme synthesis.

DID YOU KNOW?

One of the longest-lived proteins in humans is haemoglobin, found inside erythrocytes. Each haemoglobin molecule lasts for about 120 days.

Increasing the enzyme concentration in an enzyme-controlled reaction

As enzyme concentration increases (see Figure 1):
* more active sites on the enzyme become available
* more successful collisions between the enzyme and substrate occur

* more enzyme-substrate (ES) complexes can form per unit time, so the rate of reaction increases
* enzyme concentration is the limiting factor – as it increases, so does the rate of reaction.

If substrate concentration is fixed or limited, all the substrate molecules will be occupying an active site, or will have occupied an active site and been released as product molecules. The reaction is at its maximum rate for the fixed substrate concentration, so:

* if the enzyme concentration is increased further, then there will be no increase in the rate of reaction because the active sites of the extra enzyme molecules will not be occupied by substrate molecules
* the enzyme concentration is no longer the limiting factor; as enzyme concentration increases, the rate of reaction does *not* increase
* substrate concentration is now the limiting factor. Lack of substrate molecules is preventing the rate of reaction from increasing.

LEARNING TIP

When you study a graph, describe what is happening in each region and explain why these changes are occurring. Refer to both axes and quote data figures if you can.

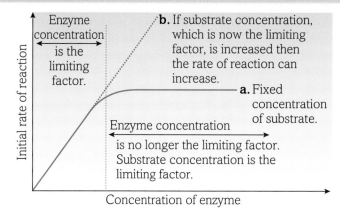

Figure 1 Effect of enzyme concentration on the rate of an enzyme-controlled reaction.

Initial rate of reaction

In any reaction, the initial rate of reaction between the reactants is fastest. The same is true of an enzyme-catalysed reaction.

- At the beginning of the reaction, when enzyme and substrate molecules are first mixed and are then moving randomly, there is a great chance of a substrate molecule successfully colliding with an enzyme's active site.

- As the reaction proceeds, substrate molecules are used up as they are converted to product molecules, so the concentration of substrate drops.

- As a result, the frequency of successful collisions between enzyme and substrate molecules decreases because some enzymes may collide with product molecules, and so the rate of the reaction slows down.

- Thus, the initial reaction rate gives the maximum reaction rate for an enzyme under a particular experimental situation.

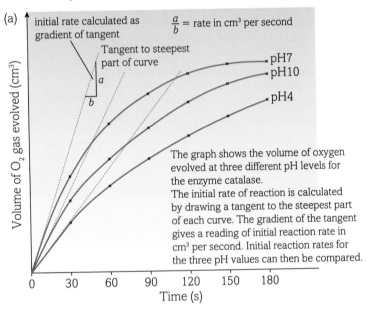

The graph shows the volume of oxygen evolved at three different pH levels for the enzyme catalase.
The initial rate of reaction is calculated by drawing a tangent to the steepest part of each curve. The gradient of the tangent gives a reading of initial reaction rate in cm³ per second. Initial reaction rates for the three pH values can then be compared.

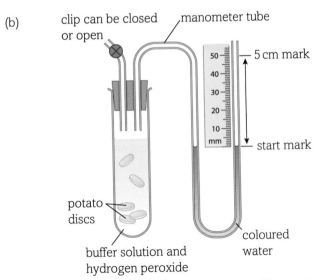

Figure 2 (a) Comparing the initial rates of reaction, at different pH values, for the breakdown of hydrogen peroxide catalysed by the enzyme catalase, using the apparatus shown in (b).

INVESTIGATION

The effect of enzyme concentration on an enzyme-catalysed reaction

Wear eye protection.
You have carried out an investigation into the effect of changing the substrate (urea) concentration on the activity (rate of reaction) of the enzyme urease.
Now think about how changing the enzyme (urease) concentration will affect the rate of the same reaction – the breakdown of urea catalysed by urease.
Write a plan for the investigation to include the following:

1 State your hypothesis and explain it.

2 State your independent and dependent variables.

3 Decide which concentration of substrate (urea) you will use. Do you want the substrate to be in excess?

4 Decide on a suitable range of enzyme concentrations and how you will make up these solutions. Which strength of stock solution of urease will you need?

5 Which variables will you need to control? How will you control them? Why do they need to be controlled?

6 Draw up a table showing the apparatus that you will need and explain why you need each piece.

7 Carry out a risk assessment.

8 Carry out your investigation, repeating each reading so you have three sets of data at each enzyme concentration and can find the mean value. This will also help you identify anomalies and repeat those anomalous readings if necessary.

9 Graph your data.

10 Write a conclusion.

11 Discuss sources of errors and limitations, and suggest improvements to your method.

Questions

1 Suggest why cells regulate the amount of enzymes in them.

2 When cells make new enzymes they may need more of specific types of amino acids than have been made available by enzyme degradation. What is the source/origin of the amino acids that have not come from the degradation of enzymes?

3 Suggest why the number of enzyme molecules within a living cell is kept fairly low.

4 Explain why, if you want to demonstrate the effect of any independent variable on the rate of an enzyme-catalysed reaction, it is best to compare the initial reaction rates at different values for the independent variable.

⑧ Enzyme inhibitors

By the end of this topic, you should be able to demonstrate and apply your knowledge and understanding of:

* the effects of inhibitors on the rate of enzyme-controlled reactions

---

> **KEY DEFINITIONS**
>
> **competitive inhibition:** inhibition of an enzyme, where the inhibitor molecule has a similar shape to that of the substrate molecule and competes with the substrate for the enzyme's active site. It blocks the active site and prevents formation of enzyme-substrate (ES) complexes.
>
> **inhibitor:** a substance that reduces or stops a reaction.
>
> **non-competitive inhibition:** inhibition of an enzyme, where the competitor molecule attaches to a part of the enzyme molecule but not the active site. This changes the shape of the active site, which prevents ES complexes forming, as the enzyme active site is no longer complementary in shape to the substrate molecule.

Inhibitors are substances that reduce the activity of an enzyme. They do this by combining with the enzyme molecule in a way that influences how the substrate binds to the enzyme or affects the enzyme's turnover number. Some may *block* the active site and some *change the shape of* the active site. Both of these actions will inhibit the formation of ES complexes and therefore of product formation.

Competitive inhibitors

Competitive inhibitors are substances whose molecules have a similar shape to an enzyme's substrate molecules. As you can see in Figure 1:

- The competitive inhibitor fits into the active site and so a substrate molecule cannot enter.
- The amount of inhibition depends on the relative concentration of substrate and inhibitor molecules. More inhibitor molecules means more inhibitors collide with active sites and so the effect of inhibition is greater.
- Increasing substrate concentration effectively 'dilutes' the effect of the inhibitor. If enough substrate is added, the inhibitor is unlikely to collide with the enzyme.

(a)

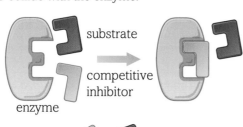

(b)

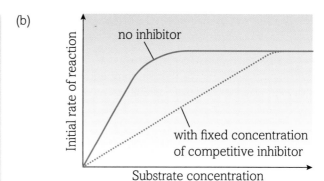

Figure 1 (a) How a competitive inhibitor works. (b) The effect of concentrations of inhibitor and substrate on the rate of an enzyme-catalysed reaction.

- They compete directly with substrate molecules for a position on the enzyme's active site, forming an *enzyme-inhibitor complex* that is catalytically inactive.
- Once on the active site, the inhibitor is not changed by the enzyme, as the normal substrate molecule would be.
- The presence of the inhibitor prevents the substrate molecule from joining to the active site. This reduces the rate of formation of ES complexes and of product molecule formation.
- A competitive inhibitor reduces the number of free enzyme active sites available for the substrate molecules to bind to and form ES complexes.
- Most enzyme inhibition by competitive inhibitors is *reversible*. As collisions between enzyme and substrate or inhibitor molecules are random, increasing the concentration of substrate would reduce the effect of reversible competitive inhibition, as there would be more chance of an enzyme molecule colliding with a substrate molecule than with an inhibitor molecule.
- If the competitive inhibitor binds *irreversibly* to the enzyme's active site it is called an *inactivator*.

Non-competitive inhibition

If the inhibitor molecule binds to the enzyme somewhere *other than* at the active site, it is called **non-competitive inhibition** (see Figure 2).

- Non-competitive inhibitors do not compete with substrate molecules for a place on an enzyme's active site. They attach to the enzyme molecule in a region (known as an allosteric site) away from the active site and, in so doing, they disrupt the enzyme's tertiary structure and change its shape.

- This distortion changes the shape of the active site so that it is no longer complementary to the shape of the substrate molecule, and the substrate molecule *can no longer bind* to the enzyme's active site. ES complexes cannot form.
- The maximum rate of reaction is reduced by the presence of non-competitive inhibitors. Adding more substrate might allow the reaction to attain this new, lower rate, but even very high concentrations of substrate will not allow the rate of reaction to return to its uninhibited maximum. This is shown in Figure 2(b).

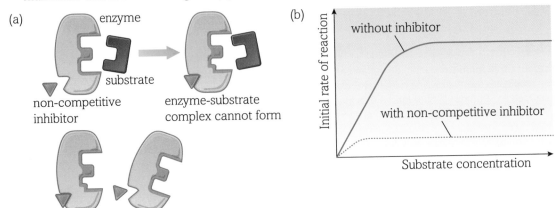

Figure 2 (a) How non-competitive inhibitors work. (b) The rate of an enzyme-catalysed reaction, with and without a non-competitive inhibitor.

- The more inhibitor molecules are present, the greater the degree of inhibition, because more enzyme molecules are distorted and either cannot form ES complexes or cannot complete the catalytic reaction involving ES complexes.
- Some non-competitive inhibitors bind reversibly to the allosteric site. Other non-competitive inhibitors bind irreversibly to the allosteric site.

End-product inhibition

One way in which enzyme-catalysed reactions may be regulated is by end-product inhibition. After the catalysed reaction has reached completion, product molecules may stay tightly bound to the enzyme. In this way, the enzyme cannot form more of the product than the cell needs. Such regulation is an example of *negative feedback*.

> **A LEVEL STUDENTS NEED TO KNOW**
>
> Some enzymes are synthesised and produced in an inactive precursor form. Before they can carry out their function some of their amino acids have to be removed so that their active sites assume the correct shape or are exposed. Many digestive enzymes are produced in this way so that while in cells they do not digest any of the cell's molecules. The proteolytic enzyme, trypsin, is produced in the small intestine in the inactive form trypsinogen and, after a portion of its molecule is removed by another enzyme, it becomes active as trypsin. The proteolytic enzyme pepsin is secreted as inactive pepsinogen and this is converted to active pepsin, by the action of hydrochloric acid, in the stomach.

Control of metabolic sequences

Many metabolic processes, such as photosynthesis and respiration, involve a series of enzyme-catalysed reactions (see Figure 3).

- The product of one enzyme-catalysed reaction becomes the substrate for the next enzyme-catalysed reaction in the metabolic pathway.
- Cells do not need to accumulate too much of the end product, so the product of the last enzyme-catalysed reaction in the metabolic pathway may attach to a part of the first enzyme in the pathway, but not at its active site.
- This binding changes the shape of enzyme 1's active site, preventing the pathway from running. This is *non-competitive inhibition*, but it is *reversible*.
- When the concentration of this product within the cell falls, those molecules will detach from enzyme 1 and allow its active site to resume its normal shape: the metabolic pathway can run again.

An example of a metabolic sequence is shown in Figure 3. The substrate A is converted into product B by the action of the first enzyme (enzyme 1). Product B is the substrate for enzyme 2, converted to product C, and so on.

The end product E can bind to enzyme 1, and acts as a non-competitive inhibitor. This means that the end product E will not build up in the cell, because as E is made, it slows the formation of itself by inhibiting the first enzyme in the sequence.

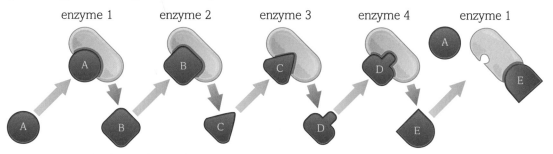

Figure 3 An example of a metabolic pathway. The enzymes in this pathway may operate as one large enzyme complex.

Multi-enzyme complexes increase the efficiency of metabolic reactions without increasing substrate concentration, as they keep the enzyme and substrate molecules in the same vicinity and reduce diffusion time.

Many metabolic reactions are carried out in particular regions or organelles in cells, and this also increases the efficiency of metabolism. Some of the enzymes within organelles are bound into the organelle membranes.

DID YOU KNOW?

Many enzyme-catalysed metabolic pathways are driven in one direction by being coupled to hydrolysis of ATP. DNA helicase is one such enzyme, and will only move in one direction along a molecule of DNA, causing the DNA to unwind.

LEARNING TIP

Remember the shape of the graphs of competitive and non-competitive inhibition.
- If the rate of the enzyme-controlled reaction stays low, even if the substrate concentration is increased, then it is non-competitive inhibition.
- If the rate of the enzyme-controlled reaction increases when the substrate concentration is increased, then it is competitive inhibition.

Questions

1. Explain how competitive inhibitors reduce an enzyme's activity.

2. Explain how non-competitive inhibitors reduce an enzyme's activity.

3. The rate of a reaction catalysed by enzyme Y is reduced when an inhibitor is added. However, if more substrate is added, the rate of the enzyme-catalysed reaction increases. Is this competitive or non-competitive inhibition? Explain your answer.

4. Explain how end-product inhibition may regulate metabolic pathways.

5. How does the presence of organelles within a cell increase the efficiency of metabolic reactions?

(9) Enzyme inhibition: poisons and medicinal drugs

By the end of this topic, you should be able to demonstrate and apply your knowledge and understanding of:

* the effects of inhibitors on the rate of enzyme-controlled reactions

Metabolic poisons that act as enzyme inhibitors

Many **toxins** (poisons) exert their effect because they inhibit or inactivate enzymes. Some examples are described here.

Cyanide

Potassium cyanide (KCN) is highly toxic because it inhibits aerobic respiration. It also inhibits catalase.

- When ingested, KCN is hydrolysed to produce hydrogen cyanide, a very toxic gas that can readily dissociate into H^+ and CN^- ions.
- The CN^- ions bind irreversibly to an enzyme found in mitochondria and inhibit the final stage of aerobic respiration. Because the final stage is inhibited, earlier stages cannot run and aerobic respiration stops.

DID YOU KNOW?

The monk Grigori Rasputin, who seemed to exert quite an influence over the Tsarina Alexandra, grand-daughter of Queen Victoria and wife of Tsar Nicholas II of Russia, was given cyanide but survived. This may be because his stomach acid was unusually low, reducing the formation of hydrogen cyanide from potassium cyanide. To many people potassium cyanide tastes of bitter almonds, but some people, due to a genetic mutation, cannot taste it.

During World War II, British secret agents were issued with cyanide pills so they could commit suicide and avoid torture if captured. In 1954, the mathematician, computer scientist and Enigma code-breaker Alan Turing died after eating an apple poisoned with cyanide.

Snake venom

The **venom** of the green mamba snake (see Figure 1) contains a chemical that inhibits the enzyme acetylcholinesterase (AChE). This enzyme is important at neuromuscular synapses (gaps between neurones and muscles) to break down the neurotransmitter, acetylcholine (ACh). If this enzyme is inhibited, the ACh stays attached to receptors on the muscle membrane and keeps the muscle contracted.

- This causes paralysis, as movement depends on muscles being able to contract and relax alternately. If the muscles involved in breathing are paralysed, then victims die from suffocation.

Figure 1 Green mamba snake, *Dendroapsis angusticeps*, a solitary, tree-dwelling snake from East Africa. Its venom is highly toxic and can be fatal to humans. Snake venom may also contain hydrolytic enzymes, such as hyaluronidase, to break down the connective tissues of the prey and allow the toxin to spread more rapidly through the body.

Medicinal drugs acting by enzyme inhibition

Aspirin

This drug has been used for over 3000 years, and marketed as aspirin since the late 19th century. In 1971, Professor John Vane and his team discovered that salicylic acid binds to enzymes that catalyse the formation of prostaglandins. Thus it prevents the formation of prostaglandins that are cell-signalling molecules produced by cells when tissues are infected or damaged. Prostaglandins make nerve cells more sensitive to pain and increase swelling during inflammation. Aspirin can also reduce the risk of blood clots forming in blood vessels, and many people take a low dose to reduce the risk of strokes. However, children under 12 years old should not take it, as aspirin can damage their stomach lining.

ATPase inhibitors

Extracts from purple foxglove leaves have been used for centuries to treat heart failure and atrial arrhythmia (abnormal beat rate of the atria).

The chemicals are now identified as cardiac glycosides, also known as digitalis, digitoxin, digitalin or digoxin.

They inhibit the sodium potassium pump in the cell membranes of heart-muscle cells, and allow more calcium ions to enter the cells. Calcium ions increase muscle contraction, and this strengthens the heartbeat.

ACE inhibitors

These are medical drugs that inhibit the angiotensin converting enzyme (ACE), which normally operates in a metabolic pathway that ultimately increases your blood pressure. They are used:

- to lower blood pressure in patients with **hypertension** who cannot take beta-blockers

- to treat heart failure – a low dose is given at first, and the patient's blood pressure is checked in case it falls too low

- to minimise risk of a second heart attack or a stroke in patients who have suffered a **myocardial infarction**.

Protease inhibitors

Protease inhibitors, such as amprenavir and ritonavir, are used to treat some viral infections. They prevent the replication of the virus particles within the host cells, by inhibiting protease enzymes so that the viral coats cannot be made. These inhibitors often inhibit viral protease enzymes by *competitive inhibition*.

Nucleoside reverse transcriptase inhibitors

Many of the antiviral drugs, such as zidovudine and abacavir, used to treat patients who are HIV-positive are **nucleoside** reverse transcriptase inhibitors. They inhibit enzymes involved in making DNA using the viral RNA as a template.

LEARNING TIP

There is a difference between venom and poison.

- Venom is introduced into the victim's body by injection – e.g. by a snake's fangs during a bite. An adder is another venomous snake.

- Poisons are toxins that are ingested. The fungus *Amanita phalloides*, the death cap, is poisonous or toxic. The toxin in a puffer fish that has not been properly prepared (to make the Japanese dish *fugu*) is a poison.

Questions

1. During the early stages of cyanide poisoning, blood flow to tissues increases by about seven-fold, causing the patient's skin to appear red. Suggest why blood flow increases at this time.

2. Explain why a substance that inhibits the enzyme acetylcholinesterase is toxic.

3. Explain how digitalin strengthens the heartbeat of patients with heart failure or atrial arrhythmia.

4. Suggest how the hyaluronidase enzymes in snake venom increase the effectiveness of its toxicity.

5. Suggest what type of enzyme inhibition is caused by aspirin.

THINKING BIGGER

THE BITE THAT HEALS

Some venom toxins can cause paralysis of the muscles used for breathing and lead to death in humans in a matter of hours. However, scientists are unlocking the potential of venom to cure some diseases. This article gives some information on the potential medicinal uses of venom toxins.

HOW DEADLY POISONS MAY SAVE LIVES

Whilst on holiday in Mexico, a young man was bitten by a bark scorpion, *Centruroides sculpturatus*, causing fierce pain followed by what felt like electric shocks. He managed to catch the scorpion and his family rushed him to a Red Cross facility, where the scorpion was identified and an antivenom injected. In about 30 hours the pain was gone.

What followed was a complete surprise. Days afterwards, the pain of an autoimmune disease, ankylosing spondylitis (a sort of spinal arthritis) that he had suffered from for eight years, disappeared and he remains pain free and off his medication. The trigger for this disease is unknown, but scorpion toxins appear to selectively block the T cells implicated in many autoimmune diseases.

Venom is a complex mixture of toxic proteins and peptides – short strings of amino acids similar to proteins. Whereas poisons are ingested, venom is injected or sprayed onto the victim's skin. Some of the molecules in venom cause paralysis by blocking messages between nerves and muscles. Others are enzymes that digest connective tissue so that cells and tissues collapse. Some make blood clot and others prevent blood from clotting. Some interfere with enzyme action and some work as enzymes.

More than 100 000 known animals of unrelated species, in groups such as fish, scorpions, spiders, bees, lizards, octopuses, anemone, jellyfish and cone snails, have independently evolved the biochemistry to produce venom, glands to house it and apparatus to expel it. The male duck-billed platypus is one of the few venomous mammals. The World Health Organisation (WHO) estimates that five million bites and stings each year kill 100 000 people, but the real figure is probably higher.

There is evidence, in Sanskrit texts and in ancient Roman literature, of venom-based cures being used about 2000 years ago, and they have long been used in Chinese and Indian traditional medicine. The science of changing venoms into cures took off in the 1960s when a British clinician developed the use of a clot-busting drug derived from pit-viper venom, to treat deep-vein thrombosis. The ACE inhibitor drugs used to treat hypertension are derived from the Brazilian pit viper's venom, after it was observed that Brazilian banana-plantation workers bitten by these snakes collapsed with crashing blood pressure.

Gila monster lizards eat about three big meals a year and store fat in their tails, but their blood-sugar level remains constant. Exenatide, a drug derived from the venom in their saliva, is used to help diabetics produce insulin. A drug for ischaemic stroke victims is based on an anticoagulant protein obtained from the saliva of vampire bats.

National Geographic explorer, herpetologist and toxinologist, Zoltan Takacs, says 'From fewer than one thousand toxins screened, about twelve major drugs have made it to the market, but there could be 20 million venom toxins out there waiting to be screened. The study of venom has opened up a new avenue of pharmacology.' Toxins from venom are also helping scientists find out how proteins control many crucial cellular functions. Studies of the toxin from puffer fish, which is 1000 times more deadly than cyanide, have revealed intricate details of how nerve cells communicate. However, snakes are in decline, and the changing chemistry of oceans due to increased atmospheric carbon dioxide could lead to loss of some potential sources of venom, such as octopuses and cone snails. 'In conserving biodiversity worldwide' says Takacs, 'we should better appreciate molecular biodiversity as these deadly poisons may save lives.'

Source
Holland, Jennifer S. The bite that heals. *National Geographic,* Feb 2013, pp. 64–83.

Where else will I encounter these themes?

1.1   2.1   2.2   2.3   2.4   YOU ARE HERE   2.5

This extract is from the magazine *National Geographic*, which is aimed at the general public.

1. Discuss the general style of writing – are there too many technical terms that some people may not understand?
2. Is the difference between venom and poison made clear in this article?

Now let's look at the biological concepts underlying the information in this article. You have studied enzymes, biological molecules, cells and tissues at this stage of your A level course. At GCSE you have studied heart disease, diabetes, biodiversity, classification, evolution and immunity. There is more information on these topics in Chapters 4.1, 4.2 and 4.3 in this book.

3. Outline the possible mechanisms involved when toxins in venom cause paralysis by 'blocking messages' between nerves and muscles.
4. Explain why toxins in venom that act as hydrolytic enzymes will digest only particular types of molecule (e.g. collagen) in body tissues. *(Think about how enzymes work and the features of their structure that makes them specific in their action.)*
5. 'Most medicines work as molecular keys fitting into specific locks'. Explain what this phrase means. *(Think about complementary shapes, but not just in the context of enzymes.)*
6. Suggest why the saliva of vampire bats contains an anticoagulant.
7. What do you think is meant by 'autoimmune disease'? *(Think about what 'auto' means as in 'autobiography', and you could read the topic on immunity in topic 4.1.6)*
8. Suggest why production of venom has independently evolved in many unrelated species. *(Think about natural selection. See topic 4.3.9)*
9. Why do you think the real number of deaths from venomous bites and stings is probably higher than the estimate by the World Health Organisation?

DID YOU KNOW?

The original anti-wrinkle treatment, Botox™, is derived from a potent bacterial toxin derived from the *Botulinum* bacterium. There is also an anti-wrinkle treatment made from snake venom.

For Question 3, you could start by remembering what you learnt at GCSE about synapses and neurotransmitters as well as revisiting topics 2.4.8 and 2.4.9.

What do you think vampire bats feed on? – the clue is in the name.

Activity

Zoltan Takacs has left his university post and is setting up a World Toxin Bank. He aims to collect blueprints for 'toxin libraries', so that there is information on all venoms.

Research to find out more about the World Toxin Bank. Write a short article (about 500 words) that could appear in a local newspaper in order to inform people about the importance of venom and of having information on many types of venom. Think about your style of writing for your target audience, many of whom may have limited knowledge of biology.

Figure 1 The venomous Geography Cone snail, *Conus geographus*. Inside its proboscis is a harpoon-like structure loaded with venom that it can eject to paralyse its prey, e.g. small fish.

3.1   3.2   3.3   4.1   4.2   4.3

Practice questions

1. Which graph shows the effect of changing the pH on the rate of an enzyme-controlled reaction? [1]

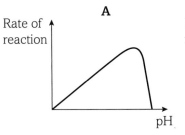

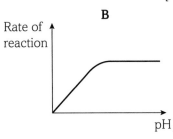

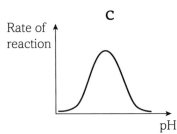

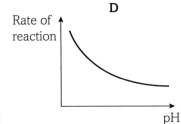

2. Which row correctly shows the substances that would give positive reactions for each reagent? [1]

	Benedict's reagent	Alcohol emulsion test	Biuret reagent
A	maltase	lipid	amylase
B	maltose	lipase	amylase
C	maltose	lipid	amylose
D	maltose	lipid	amylase

3. Read the following statements.
   (i) Some metallic metal ions are enzyme cofactors.
   (ii) Organic coenzymes are cofactors that bind permanently to the enzyme.
   (iii) Enzyme cofactors are made of protein.

   Which statement(s) is/are true? [1]
   A. (i), (ii) and (iii)
   B. Only (i) and (ii)
   C. Only (ii) and (iii)
   D. Only (i)

4. The enzyme catalase contains iron. The enzyme carbonic anhydrase contains zinc. What are these metallic groups called? [1]
   A. coenzymes
   B. EP complexes
   C. ES complexes
   D. prosthetic groups

5. Read the following statements.
   (i) Enzymes only work inside living cells.
   (ii) Enzymes are biological catalysts with greater specificity than chemical catalysts.
   (iii) Enzymes lower the activation energy needed to start a reaction.

   Which statement(s) is/are true? [1]
   A. (i), (ii) and (iii)
   B. Only (i) and (ii)
   C. Only (ii) and (iii)
   D. Only (i)

   [Total: 5]

In questions 6–8 you may need to use your knowledge of other topics such as cell structure, biological molecules, stem cells, cell differentiation and scientific methodology, as well as your knowledge of enzymes.

6. In Central and South America, chica, a type of beer, is made from cassava roots. In some cultures the cassava roots are crushed and boiled and then, after cooling, are chewed and then expectorated (spat out) into a container which is left in the open air for several days. Yeasts from the air infect the brew and ferment the sugar in it to ethanol.

   (a) Cassava roots contain a lot of starch. Explain how the following processes help to produce sugar for the fermentation process outlined above.
      (i) Crushing and boiling the roots [2]
      (ii) Chewing the boiled roots [2]
   (b) Name the type of bonds that are broken when starch is converted to sugar. [1]
   (c) Name the sugar produced when starch is digested by amylase. [1]
   (d) Amylase will only function if chloride ions are present. What term describes the function of these chloride ions? [1]
   (e) Suggest why chewing the roots and then spitting it out digests starch to sugar more effectively than adding saliva to crushed roots in cold water. [2]

   [Total: 9]

7. Aspirin has been used to reduce fever and inflammation for over 100 years. Scientists have recently discovered that aspirin prevents cells from producing prostaglandins by inhibiting cyclooxygenase-1 (Cox-1), one of the enzymes in the metabolic pathway involved in prostaglandin synthesis. (Prostaglandins are 'eicosanoids' (their molecules contain 20 carbon atoms) and act as local hormones. They were at first thought to originate in the prostate gland and, although scientists now know this is not true, the name has been kept.)

Aspirin adds a chemical group (acetyl) permanently to one of the monomers in the enzyme's active site.

Thromboxanes are also eicosanoids. The enzyme thromboxane synthase catalyses their synthesis, within platelets. Thromboxanes promote blood clotting.

Aspirin also inhibits the production of thromboxanes. Platelets have no DNA and cannot synthesise more enzymes.

Some people take a low daily dose of aspirin to reduce their risk of heart attack or stroke.

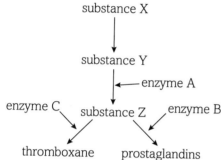

Figure 1 Some stages of the metabolic pathway for prostaglandin and thromboxane synthesis.

(a) What are the monomers that make up the active site of cyclooxygenase? [1]

(b) Suggest which enzyme in the metabolic pathway shown above is inhibited by aspirin. Explain the reasons for your choice. [3]

(c) Explain how aspirin inhibits the formation of product molecules from substrate molecules. [4]

(d) Explain why platelets cannot synthesise more enzymes. [2]

(e) Suggest why people continue to take a low daily dose of aspirin to reduce the risk of blood clotting even though their platelets cannot make enzymes to replace those permanently inhibited by aspirin. [2]

[Total: 12]

8. In an investigation, the enzyme activity of saliva from smokers was compared to that of non-smokers. The saliva samples were mixed with starch solutions in test tubes (*in vitro*) and the production of maltose from starch was monitored over a 10-minute period. The results are shown in Figure 2.

(a) Describe the reaction that is catalysed by salivary amylase. [2]

(b) Briefly outline how the concentration of maltose could be measured. [2]

(c) Calculate the initial rate of reaction of this reaction with non-smokers' salivary amylase. [2]

(d) What effect does smoking appear to have on salivary amylase? [2]

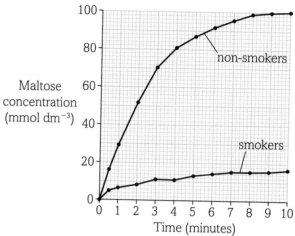

Figure 2

(e) Other studies show that cigarette smoke contains chemicals called aldehydes, which react with a chemical group in protein molecules. These aldehydes inhibit salivary amylase.

In one investigation, amylase activity in 20 smokers and 20 non-smokers was investigated *in vivo* and the results are shown in Figure 3. Standard error bars are shown.

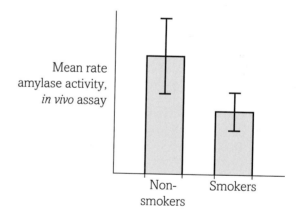

Figure 3

These data were published in a scientific journal and a journalist used them to write an article on smoking. In that article, he said, 'there is no doubt, smoking permanently affects our ability to digest food'.

Do you agree or disagree with the journalist's comment? Explain your answer. [4]

[Total: 12]

Foundations in biology

BIOLOGICAL MEMBRANES

Introduction

Many people in the world still do not have access to clean drinking water. Recently a cheap and easy to use method has been introduced. The dirty water is put into a 25 litre container and by using a hand pump the water is forced through a membrane that contains very small pores of 0.1 nm. This is the same as 0.0001 micron (micrometre) and is 500 000 times less than the diameter of a human hair. Only water molecules can pass through these pores and all impurities including organic particles, dissolved salts, chemicals, bacteria and viruses cannot. After this type of filtering, clean drinking water can be poured from the container.

Our knowledge of such membranes is based upon what we know about the membrane surrounding all cells.

In this section you will learn about the structure of cell surface membranes and how their structure enables them to carry out their functions of regulating what can go into and out of a cell. You will also learn that many of the organelles within eukaryotic cells are also surrounded by a membrane. You will learn in more detail about the processes by which substances enter and leave cells. You will also learn how cells communicate with each other by making signalling molecules or having receptors on their membranes for signalling molecules.

All the maths you need

To unlock the puzzles of this chapter you need the following maths:

- Units of measurement
- Ratios
- How to calculate surface area
- How to calculate volume
- The relationship between area, thickness and volume

What have I studied before?

- All living things are made of cells
- Cells have a membrane that separates the cells' exterior from its interior and controls what passes into and out of the cell
- The structure and functions of proteins
- The structure and functions of lipids and phospholipids
- The structure of plant, animal and prokaryotic cells
- That cells become differentiated and specialised for different functions
- Water passes across cell membrane by osmosis
- Dissolved substances pass across cell membranes by diffusion, down their concentration gradient
- Some substances pass across cell membranes, against their concentration gradient, by active transport
- Glands secrete hormones that help coordinate body functions, growth and maturation. These hormones travel to target tissues in the blood
- Plants respond to light and gravity by tropic growth responses, involving plant hormones
- Drugs alter body chemistry

What will I study later?

- How RNA passes out of the nucleus through the nuclear envelope (AS)
- How gaseous exchange occurs through the membranes of cells in special structures for both plant and animals (AS)
- How the blood system transports substances to cells and how these substances enter cells across their surface membranes (AS)
- How cells excrete their metabolic waste through cell membranes (AL)
- How the membranes of neurones are involved in transmission of electrical impulses (AL)
- The action and effects of hormones (AL)
- The actions and effects of plant growth regulatory substances (AS)
- More about the immune response and the role of cell surface antigens, receptors and signalling molecules (AS)

What will I study in this chapter?

- The detailed structure of cell membranes
- Cell membranes protect the cell from adverse environmental effects
- Cell membranes have receptors that recognise specific chemicals, allowing cells to respond to hormones or to other signals
- Cell membranes have glycoproteins that allow cells to group together to form tissues
- Cell membranes maintain a potential difference across themselves and are crucial for impulse conduction along neurones
- How drugs affect cell signals
- Cell membranes allow attachment of structures such as the cell wall, cytoskeleton and glycocalyx
- Cell membranes contain antigens
- Some cells have extensions of the cell surface membrane, called microvilli, which increase surface area for absorption
- The membranes around organelles are similar but not identical to the cell surface membrane

The structure of cell membranes

By the end of this topic, you should be able to demonstrate and apply your knowledge and understanding of:

* the roles of membranes within cells and at the surface of cells
* the fluid mosaic model of membrane structure and the roles of its components

KEY DEFINITIONS

fluid mosaic model: theory of cell membrane structure with proteins embedded in a sea of phospholipids.
glycolipid: lipid/phospholipid with a chain of carbohydrate molecules attached.
glycoprotein: protein with a chain of carbohydrate molecules attached.
plasma membrane: cell surface membrane.

Cell membranes are partially permeable barriers

Because cell membranes form a barrier and separate the cell contents from the cell's exterior environment, or separate organelles from cytoplasm, they need to allow some molecules through, into or out of the cell. Some organelles also have membranes within them, and these form barriers too.

Permeability refers to the ability to let substances pass through.

* Some very small molecules simply diffuse through the cell membrane, in between its structural molecules.
* Some substances dissolve in the lipid layer and pass through.
* Other substances pass through special protein channels or are carried by carrier proteins.

Because these membranes do not let all types of molecule pass through them, they are described as **partially permeable** barriers. The properties of the component molecules of the cell membrane determine its permeability – i.e. which molecules it allows through.

The roles of membranes
At the surface of cells

The **plasma membrane** (sometimes called the cell surface membrane):

* separates the cell's components from its external environment (in single-celled organisms the environment is its external surroundings; in multicellular organisms, such as humans, each cell's environment is the tissue fluid or cells surrounding it)
* regulates transport of materials into and out of the cell
* may contain enzymes involved in specific metabolic pathways
* has antigens, so that the organism's immune system can recognise the cell as being 'self' and not attack it
* may release chemicals that signal to other cells

* contains receptors for such chemical signals, and so is a site for cell communication or signalling; hormones and drugs may bind to membrane-bound receptors
* may be the site of chemical reactions.

Within cells

The membranes around many organelles present in eukaryotic cells separate the organelle contents from the cell cytoplasm, so that each organelle is a discrete entity and able to perform its function. In some organelles, metabolic processes occur on membranes.

* You have seen in topic 2.1.4 that mitochondria have folded inner membranes, called *cristae*. These give a large surface area for some of the reactions of aerobic respiration and localise some of the enzymes needed for respiration to occur.
* The inner membranes of chloroplasts, called thylakoid membranes, house chlorophyll. On these membranes some of the reactions of photosynthesis occur.
* There are some digestive enzymes on the plasma membranes of epithelial cells that line the small intestine, and these enzymes catalyse some of the final stages in the breakdown of certain types of sugars.

The fluid mosaic model of cell membrane structure

In 1972, Singer and Nicolson proposed a model that allowed the passage of molecules through the membrane. Their structure explained how cell membranes could be more dynamic and interact more with the cells' environment. It was called the **fluid mosaic model**, and proposed that the fabric of the membrane consisted of a phospholipid bilayer (double layer) with proteins floating in it, making up a mosaic pattern. The lipid molecules can change places with each other, and some of the proteins may move, giving fluidity.

Figure 1 explains the fluid mosaic model of cell membrane structure.

The fabric of the membrane is the lipid bilayer made up of two layers of phospholipid molecules. Their hydrophilic heads are in contact with the watery exterior or watery interior (cytoplasm). The hydrophobic tail regions are in the centre of the membrane, away from water.

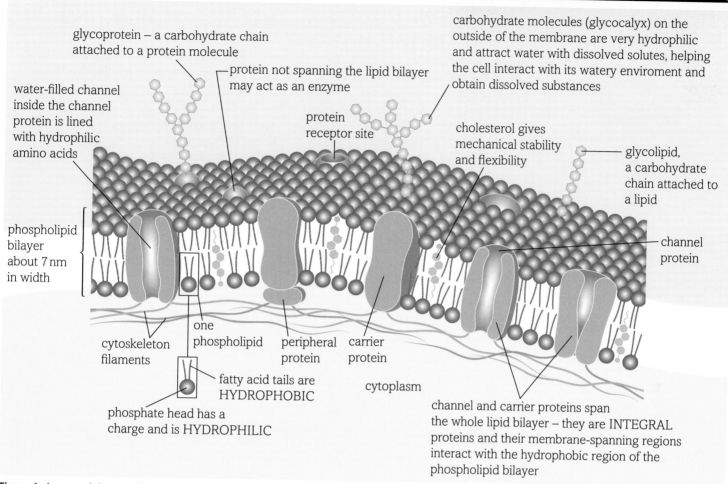

Figure 1 Annotated diagram illustrating the structure of a plasma membrane according to the fluid mosaic model.

Spanning the membrane are various proteins.

- Some of them have pores and act as channels to allow ions, which have an electrical charge and are surrounded by water molecules, to pass through.
- Some proteins are carriers and, by changing their shape, carry specific molecules across the membrane.
- Other proteins may be attached to the carrier proteins and function as enzymes, antigens or receptor sites for complementary-shaped signalling chemicals such as hormones.

Eukaryotic cell membranes contain cholesterol, which is important for helping to regulate the fluidity of the membrane, maintain mechanical stability and resist the effects of temperature changes on the structure of the membrane (see topic 2.5.5).

The total thickness of a cell membrane is between 5 and 10 nm. Outside the membrane is the **glycocalyx** – formed from the carbohydrate chains attached to either lipids (**glycolipids**) or proteins (**glycoproteins**) in the membrane.

You will have read more about phospholipids, proteins and carbohydrates in Chapter 2.2.

Not all cell membranes have the same composition

You will see in topics 2.6.4 and 2.6.5 how cells become differentiated and specialised. Their membranes may have particular distributions of proteins in order to enable them to carry out their specific functions.

- In neurones (nerve cells), the protein channels and carriers in the plasma membrane covering the long axon allow entry and exit of ions to bring about the conduction of electrical impulses along their length.

- Neurones have a myelin sheath formed by flattened cells wrapped around them several times, giving several layers of cell membrane. The membrane forming the myelin sheath is about 20% protein and 76% lipid (see Figure 2).
- The plasma membranes of white blood cells contain special protein receptors that enable them to recognise the antigens on foreign cells, usually from invading pathogens but also from tissue or organ transplants.

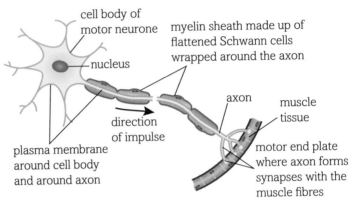

Figure 2 Myelinated motor neurone synapsing with muscle tissue.

- Root hair cells in plants have many carrier proteins to actively transport nitrate ions from the soil into the cells.
- The inner membranes of mitochondria are 76% protein and 24% lipid. This is because their inner membranes contain many electron carriers that are made of protein, and hydrogen ion channels associated with ATP synthase enzymes.

DID YOU KNOW?

As well as cells in your pancreas having receptors to detect sugars, cells lining your intestines also have taste receptors for sugar. For 50 years, scientists were mystified as to why eating glucose produces a quicker release of insulin than if glucose was injected straight into the bloodstream. In 2007, they discovered that cells lining the intestine contain taste receptors for sugars, and when stimulated these cause a cascade of hormones that ultimately ends in release of extra insulin from the pancreas. In fact, these receptors also respond to artificial sweeteners, just as the receptors on your tongue do, so these sweeteners also cause an insulin surge.

LEARNING TIP

Cell membranes are described as *partially* permeable. This means that they allow some, but not all, substances to pass through. Substances may pass from inside the cell to the outside, or from outside the cell into its interior.

Questions

1. Explain why the lipids that are the main fabric of the cell membrane form a bilayer, with heads towards the cytoplasm and external environment and tails to the inside of the bilayer.

2. Explain why the current model to describe cell membrane structure is called the fluid mosaic model.

3. Explain clearly the meanings of the following terms: hydrophilic, hydrophobic, glycoprotein, glycolipids and glycocalyx.

4. Explain why proteins embedded in cell membranes to form channels have hydrophilic amino acids lining the part of the channel that forms the pore.

5. Describe and explain how and why the protein component of axon cell surface membranes will differ from that of the myelin sheath surrounding the axon.

6. Explain why the protein, phospholipid, cholesterol and carbohydrate composition of different cell types varies.

② Diffusion across membranes

By the end of this topic, you should be able to demonstrate and apply your knowledge and understanding of:

* the movement of molecules across membranes
* practical investigations into the factors affecting diffusion rates in model cells

KEY DEFINITIONS

diffusion: movement of molecules from an area of high concentration of that molecule to an area of low concentration; it may or may not be across a membrane; it does not involve metabolic energy (ATP).
facilitated diffusion: movement of molecules from an area of high concentration of that molecule to an area of low concentration, across a partially permeable membrane via protein channels or carriers; it does not involve metabolic energy (ATP).

You have seen from Chapter 2.1 that the biochemical processes that sustain life go on in cells. Cells therefore need to receive raw material or reactants for these reactions. They respire to make ATP, which provides cellular energy, to drive these biochemical reactions. Therefore they need oxygen and glucose. They also need to remove the toxic metabolic waste products, such as carbon dioxide, and they need to export some of the molecules that they make, such as enzymes, hormones or other signalling molecules. Some substances can pass across cell membranes *without* using any of the cell's metabolic energy. These are described as *passive* processes because they use only the **kinetic energy** of the molecules and do *not* use ATP.

DID YOU KNOW?

When you have been active for a long time, a lot of ATP molecules are broken down to adenosine. The build-up in concentration of adenosine eventually acts as a signal to part of your brain and triggers you to feel sleepy.

Simple diffusion

All molecules have kinetic energy and can move freely and randomly within gas or liquid media. This will happen even if the medium is not mixed by stirring or shaking. If there is a high concentration of a certain type of molecule in an area, then the molecules will bump into each other as they randomly move, and eventually they will spread further from each other. More will move to an area where they are in lower concentration, until eventually they will become evenly dispersed (see Figure 1).

When the molecules have moved *down* their concentration gradient, they are still moving randomly but remain evenly dispersed so there is no *net* **diffusion**. They have reached equilibrium.

LEARNING TIP

When they are diffusing, molecules move *down* their concentration gradient.

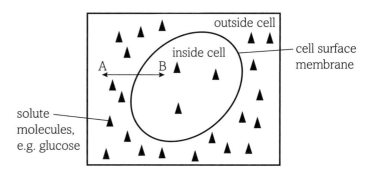

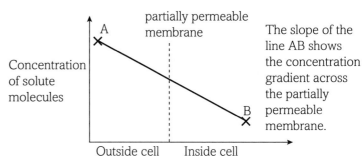

Figure 1 There is a high concentration of the molecules on one side of the membrane and a low concentration on the other side. The difference between them forms a gradient (slope). The steeper the slope, the faster the molecules will diffuse down that gradient.

Some molecules such as oxygen and carbon dioxide, which are small, can pass through cell membranes by simple diffusion. Fat-soluble molecules such as steroid hormones, even if they are larger, can diffuse through cell membranes as they dissolve in the lipid bilayer. They still move down their concentration gradient (see Figure 2).

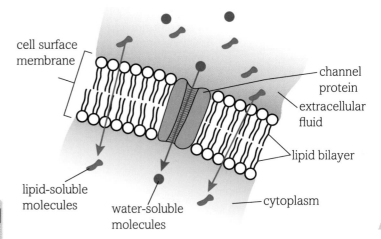

Figure 2 Diffusion of lipid-soluble molecules and water-soluble (lipid-insoluble) molecules across a cell surface membrane (plasma membrane).

Water molecules are a special case. Since they are polar and insoluble in lipid the phospholipid bilayer would seem to be an impenetrable barrier. However, water is present in such great concentrations that significant direct diffusion does happen. In membranes where a very high rate of water movement is required there may indeed be specific water channel proteins known as aquaporins to allow water molecules to cross the membrane without the challenge of moving through a lipid environment.

How the concentration gradient is maintained

Many molecules entering cells then pass into organelles and are used for metabolic reactions; this maintains the concentration gradient and keeps more of the molecules entering the cell.

- Oxygen diffusing into the cytoplasm of respiring cells then diffuses into mitochondria and is used for aerobic respiration.
- Carbon dioxide diffusing into the palisade mesophyll cells of a plant leaf will then diffuse into chloroplasts and be used for photosynthesis.

These are both examples of how the concentration gradient across the cell membrane is maintained.

Factors that affect the rate of simple diffusion

Simple diffusion relies only on the molecules' own kinetic energy, and so factors that alter this kinetic energy will affect the rate of diffusion.

- **Temperature**: as temperature increases, molecules have more kinetic energy, so their rate of diffusion will increase. Conversely, as they lose heat their rate of diffusion will slow down.
- **Diffusion distance**: the thicker the membrane across which molecules have to diffuse, the slower the rate of diffusion.
- **Surface area**: more diffusion can take place across a larger surface area. Cells specialised for absorption have extensions to their cell surface membranes, called microvilli. These increase the surface area.
- **Size of diffusing molecule**: smaller ions or molecules diffuse more rapidly than larger molecules.
- **Concentration gradient**: the steeper this gradient (the more molecules there are on one side of the membrane compared with the other side), the faster the diffusion to the side where there are fewer molecules, down the gradient.

Facilitated diffusion

Small molecules that have polarity (opposite charges at either end of the molecule), such as ions that have an electrical charge, are insoluble in lipid because they cannot interact with the hydrophobic tails of the lipid bilayer. This means that they diffuse through water-filled protein channels (pores) embedded in the membrane. These channels are around 0.8 nm in diameter.

Cholesterol molecules within membranes reduce the permeability of the membranes to small water-soluble molecules. Glucose molecules are too large to diffuse through the water-filled protein channel in a membrane, but they can bind to a transmembrane carrier protein, which then opens to allow the glucose to pass out on the other side of the membrane. There are specific carrier proteins for different types of molecules (see Figure 3).

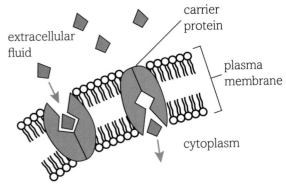

Figure 3 Facilitated diffusion.

Different cell types have membranes with differing proportions of transmembrane protein channels and transmembrane protein carriers. This allows cells to control the types of molecule that pass in or out.

- Neurone plasma membranes have many channels specific to either sodium ions or potassium ions. The diffusion of these ions into and out of the neurone axon is crucial for the conduction of nerve impulses. At synapses (gaps between neurones), there are also calcium ion channels and there may be chloride ion channels.
- The plasma membranes of epithelial cells that line your airways have chloride ion channels, and these play a crucial role in regulating the composition of mucus to trap particles and pathogens.

Questions

1. Explain why single-celled organisms such as *Amoeba* and *Paramecium* can gain enough oxygen for their aerobic respiration through simple diffusion across their cell surface membranes.

2. How would you expect the direction of net diffusion of carbon dioxide into/out of a palisade mesophyll leaf cell to differ (a) at noon on a sunny day compared with (b) at midnight? Explain your answer.

3. How would you expect the direction of net diffusion of oxygen into/out of a palisade mesophyll leaf cell to differ at (a) noon on a sunny day compared with (b) at midnight? Explain your answer.

4. Explain why glucose and amino acids pass into or out of cells by facilitated diffusion.

5. When molecules diffuse they use their kinetic energy. Explain why such diffusion is referred to as passive.

6. List five factors that affect the rate of diffusion, and explain how each alters the rate of diffusion.

(3) Osmosis

By the end of this topic, you should be able to demonstrate and apply your knowledge and understanding of:

* the movement of water across membranes by osmosis and the effects that solutions of different water potential can have on plant and animal cells

* practical investigations into the effects of solutions of different water potential on plant and animal cells

---

KEY DEFINITIONS

osmosis: passage of water molecules down their water potential gradient, across a partially permeable membrane.
water potential: measure of the tendency of water molecules to diffuse from one region to another.

Osmosis is the diffusion of water molecules

In a *solution*, the liquid in which *solute* molecules are dissolved is called the *solvent*. In an aqueous solution, water is the solvent. Water molecules can pass directly through the phospholipid bilayer. Some membranes also have protein channels known as aquaporins which allow water molecules to cross the membrane more rapidly. The inside of cells, the cytoplasm, contains water, and the external medium of cells is also watery – as cells are surrounded by extracellular tissue fluid. Water molecules also have kinetic energy and move randomly, but will spread out.

The net diffusion from a region where there are relatively more water molecules to an area where there are fewer water molecules, across a partially permeable membrane, is called **osmosis** (see Figure 1).

When solute molecules are added to water, the relative number of water molecules, in the resulting solution, is changed. If the solute molecules dissociate into charged ions, such as sodium chloride dissociating into sodium ions and chloride ions, they exert more effect on the relative number of water molecules than do larger but non-ionic molecules such as glucose. This is because, as sodium chloride molecules dissociate into sodium ions and chloride ions, the number of particles in the solution doubles.

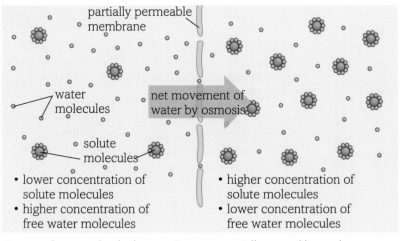

Figure 1 Passage of water molecules by osmosis across a partially permeable membrane.

DID YOU KNOW?
Osmosis and prokaryote cells
Prokaryote cells are also subject to osmosis. Water molecules can enter or leave them across the partially permeable plasma membrane. If they lose enough water, then their metabolism cannot proceed and they cannot reproduce. Adding sugar or salt to preserve food prevents spoilage as any bacteria cannot respire, grow or multiply as water is lost from the bacterial cells by osmosis.
The antibiotic penicillin prevents some types of growing bacteria from synthesising their peptidoglycan wall. This makes them vulnerable to the effects of osmosis. If they swell up as water enters, then they will burst.

Water potential

Water potential is a measure of the tendency of water molecules to diffuse from one region to another (see Figure 2).

- Pure water has the highest possible water potential.
- When solute molecules are added, they lower the water potential of the solution. The more solute molecules in the solution, the lower the water potential.
- If two aqueous solutions are separated by a partially permeable membrane, such as a plasma membrane or a cell organelle membrane, water molecules will move from the solution with the higher water potential to the solution with the lower water potential.
- If and when the water potential on both sides of the membrane becomes equal, there will be no net osmosis, although water molecules will continue to move randomly.

LEARNING TIP
We always describe the passage of water by osmosis in the context of differing water potential. Water moves via osmosis from a region of higher to lower water potential, *down* the water potential gradient.

Units for measuring water potential

- Water potential is measured in kilopascals (kPa). Pure water has the highest water potential possible, and is given the value of 0 kPa.
- As solute molecules are added, the water potential of the solution is lowered so, in numerical value, it becomes more negative. The more negative the value, the lower the water potential.

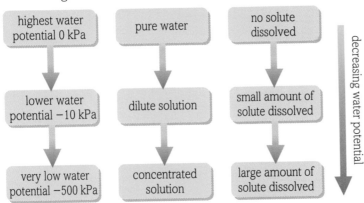

Figure 2 Ladder showing direction of osmosis as determined by differences in water potential.

Water potential of cells

- The water potential inside cells is lower than that of pure water, as there are solutes in solution, in the cytoplasm and inside the large vacuole of plant cells.
- When cells are placed in a solution of higher water potential, then water molecules move by osmosis, down the water potential gradient, across the plasma membrane, into the cell (see Figure 3(a)).

INVESTIGATION
Determining the water potential of potato-tuber cells
If cells are immersed in a solution that has the same water potential as they do, then there is no net osmosis and the cells will neither swell nor shrink. They will not gain or lose any mass.

- In animal cells, if a lot of water molecules enter, the cell will swell and burst as the plasma membrane breaks. This is called **cytolysis**.

DID YOU KNOW?
The term cytolysis literally means splitting (lysis) of cells. This phenomenon was first observed in erythrocytes, and so it is often also called haemolysis, but scientists have since found that it occurs for all animal cells placed in a solution of higher water potential.

- In plant cells, the rigid and strong cellulose cell wall will prevent bursting. The cell will swell up to a certain size when its contents push against the cell wall, which will resist any further swelling. This swollen cell is described as **turgid**. Turgidity of plant cells helps support plants, especially those that are not woody.
- When cells are placed in a solution of lower water potential, water leaves the cells by osmosis, across the partially permeable plasma membrane (see Figure 3(b)).
- Animal cells shrivel and are described as **crenated**.

INVESTIGATION
Observing crenation in animal cells
You can carry out this practical using your own cheek cells. Wear eye protection.

- The cytoplasm of plant cells shrinks and the membrane pulls away from the cellulose cell wall. The cells are described as **plasmolysed**. Plant *tissue* with plasmolysed cells is described as **flaccid**.
- Cells that are plasmolysed suffer a degree of dehydration and their metabolism cannot proceed, as enzyme-catalysed reactions need to be in solution.

(a) pure water

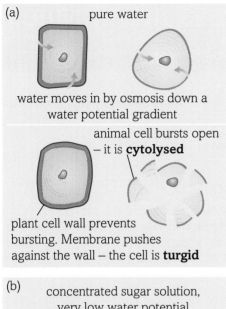

water moves in by osmosis down a water potential gradient

animal cell bursts open – it is **cytolysed**

plant cell wall prevents bursting. Membrane pushes against the wall – the cell is **turgid**

(b) concentrated sugar solution, very low water potential

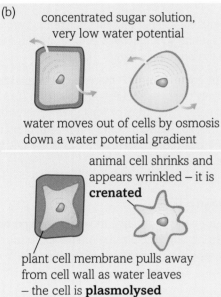

water moves out of cells by osmosis down a water potential gradient

animal cell shrinks and appears wrinkled – it is **crenated**

plant cell membrane pulls away from cell wall as water leaves – the cell is **plasmolysed**

Figure 3 (a) Plant and animal cells in solutions of high water potential – before and after. (b) Plant and animal cells in solutions of low water potential – before and after.

Observing plasmolysis in plant cells

Wear eye protection.

Use epidermis cells from red onions, as the cytoplasm will be more easily observed, without any staining, under the optical microscope.

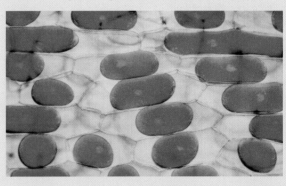

Figure 4 Epidermal cells of the red onion, *Allium cepa,* showing plasmolysis (×400).

Questions

1 If plant cells are placed in concentrated sucrose (sugar) solution, then they become plasmolysed and resemble the cells in Figure 4. What is in the space between the cellulose cell wall and the cell surface membrane?

2 Think about the factors that affect the rate of diffusion. Suggest how each of these factors will affect the rate of osmosis.

3 The water potential of blood plasma has to be regulated. Suggest why. (*Think about the effect on blood cells if the plasma were to have too high or too low a water potential.*)

4 What happens to the water potential inside a cell if many ions, such as potassium ions or sodium ions, enter the cell?

5 What will be the consequence of the change in water potential outlined in question 4?

6 Explain why the onion epidermis cells in distilled water, that you observed, swelled but did not burst.

7 Explain the following terms: plasmolysis, crenation, haemolysis, turgid, flaccid.

8 What happens to the water potential inside a cell as water enters the cell by osmosis?

(4) How substances cross membranes using active processes

By the end of this topic, you should be able to demonstrate and apply your knowledge and understanding of:

* * the movement of molecules across membranes

> **KEY DEFINITIONS**
>
> **active transport:** the movement of substances against their concentration gradient (from low to high concentration of that substance) across a cell membrane, using ATP and protein carriers.
> **endocytosis:** bulk transport of molecules, too large to pass through a cell membrane even via channel or carrier proteins, into a cell.
> **exocytosis:** bulk transport of molecules, too large to pass through a cell membrane even via channel or carrier proteins, out of a cell.

Active transport – moving against the concentration gradient

Sometimes cells need to move certain substances in or out, across their plasma membranes, *against* each substance's concentration gradient.

- This is like swimming against the tide and needs more energy than the kinetic energy of the molecules.
- This energy is provided by the hydrolysis of ATP. ATP is often described as the universal energy currency, as all cells make use of it to supply their energy needs.
- Cells or organelles may need to accumulate more of a particular ion than they could do by simple or facilitated diffusion alone. For example, root hair cells use active transport to absorb ions from soil.

Carrier proteins

These membrane proteins have specific regions, or sites, that combine reversibly with only certain solute molecules or ions. They also have a region that binds to and allows the hydrolysis of a molecule of ATP, to release energy, and in this way they act as enzymes.

The energy helps the carrier protein change its *conformation* (shape), and in doing so it carries the ion from one side of the cell membrane to the other (see Figure 1).

For example, in guard cells ATP made by chloroplasts provides energy to actively transport potassium ions from surrounding cells into the guard cells. This influx of ions lowers the water potential in the guard cells, so that water enters from surrounding cells, by osmosis. As the guard cells swell, their tips bulge and this opens the stoma between them.

Bulk transport

Some cells need to transport large molecules and particles that are too large to pass through the plasma membrane, in or out. They do this by bulk transport, a process that requires energy from ATP.

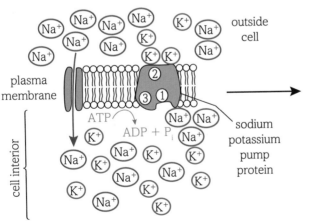

sodium ions diffuse into the cell

3 sodium ions bind to a specific site, ①, on the sodium potassium pump protein ATP binds to its site, ③ and is hydrolysed to ADP + P$_i$ releasing energy 2 potassium ions bind to their specific site, ②

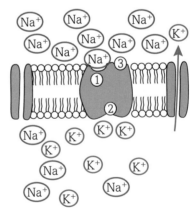

The energy released from the hydrolysis of ATP enables the sodium–potassium pump protein to change its shape so that the three sodium ions are now on the outside of the cell and the two potassium ions are inside the cell.

Figure 1 How sodium ions and potassium ions are actively transported using a special protein embedded in the plasma (cell surface) membrane. This type of transport protein is also called an antiport as it carries two different types of ions in opposite directions.

Endocytosis

This is how large particles may be brought into a cell. They do not pass through the plasma membrane. Instead, a segment of the plasma membrane surrounds and encloses the particle and brings it into the cell, enclosed in a vesicle.

- The type of **endocytosis** shown in Figure 2 is also called **phagocytosis**. This means 'eating by cells' and refers to this type of intake of solid matter.

- If cells ingest liquids by endocytosis, this is called **pino(endo)cytosis**.

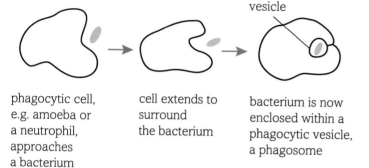

| phagocytic cell, e.g. amoeba or a neutrophil, approaches a bacterium | cell extends to surround the bacterium | bacterium is now enclosed within a phagocytic vesicle, a phagosome |

Figure 2 Endocytosis: a phagocytic cell ingests a bacterium.

ATP is needed to provide energy to form the vesicles and move them, using molecular motor proteins, along cytoskeleton threads into the cell interior.

Exocytosis

This is how large molecules may be exported out of cells. They do not pass through the plasma membrane. Instead, a vesicle containing them is moved towards and then fuses with the plasma membrane.

Another example is seen at synapses (gaps between neurones), where chemicals in vesicles are moved, by motor proteins moving along cytoskeleton threads, to the presynaptic membrane. Here, the vesicle membranes and plasma membranes fuse and the neurotransmitter chemicals are released into the synaptic cleft.

In all cases, ATP is needed to fuse the membranes together as well as for moving the vesicles. A molecule of ATP is hydrolysed for every step that a motor protein takes along the cytoskeleton thread, as it drags its cargo – the vesicle.

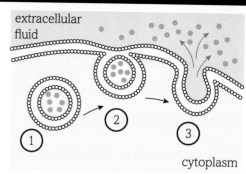

① A membrane-bound vesicle, containing the substance to be secreted, is moved towards the cell surface membrane.

② The cell surface membrane and the membrane of the vesicle fuse together.

③ The fused site opens, releasing the contents of the secretory vesicle.

Figure 3 Exocytosis. The substance to be exported out of the cell is moved towards and then fuses with the plasma membrane. The fused site opens releasing the contents of the vesicle.

DID YOU KNOW?

Macrophages ingest certain fats that are in your blood. They 'eat themselves to death' and become foam cells, sinking into the lining of the blood vessels and forming fatty plaques.

At a wound site, you may have noticed white pus. This consists mainly of dead phagocytic white blood cells that have ingested invading bacteria and also eaten themselves to death.

Questions

1 In root-hair cells, nitrate ions are actively transported in from soil. What is the effect of this influx of nitrate ions on the water potential of the root-hair cell? How will this affect osmosis into or out of this cell?

2 Explain the role of the cytoskeleton in endocytosis and exocytosis.

3 Mast cells are present in many tissues, and when stimulated by allergens they release chemicals, such as histamine, from vesicles. (a) By what process are these chemicals released? (b) Does this process use ATP?

4 Some types of unicellular protoctist, *Amoeba* spp., that do not have a cell wall, live in fresh water. Water enters their cells by osmosis and collects into a large vacuole that is emptied regularly by exo(pino)cytosis. If the amoebae are exposed to a chemical that reduces their rate of respiration, the vacuoles are not emptied as often. (a) What can you deduce about how the vacuole of water is emptied by the amoebae? (b) Predict what would happen to the amoebae if the vacuole did not empty at all.

5 Compare and contrast active transport and facilitated diffusion.

(5) Factors affecting membrane structure and permeability

By the end of this topic, you should be able to demonstrate and apply your knowledge and understanding of:

* factors affecting membrane structure and permeability

* practical investigations into factors affecting membrane structure and permeability

Temperature and kinetic energy

Increasing the temperature gives all molecules more kinetic energy, and as a result these molecules move faster.

Decreasing the temperature lowers the kinetic energy of the molecules, causing them to move more slowly (topic 2.4.4).

Phospholipids and changing temperature

Remember that many organisms do not generate heat to maintain their body temperature and so their temperature varies with their environment.

When temperature drops

* Saturated fatty acids become compressed.
* However, there are many **unsaturated fatty acids** making up the cell membrane phospholipid bilayer, and as they become compressed the kinks in their tails push adjacent phospholipid molecules away. This maintains the membrane fluidity.
* Therefore, the proportions of unsaturated and saturated fatty acids within a cell membrane determine the membrane's fluidity at cold temperatures.
* Cholesterol in the membrane also buffers the effect of lowered temperature, to prevent a reduction in the membrane's fluidity. It does this by preventing the phospholipid molecules from packing together too closely, because cholesterol molecules are in between groups of phospholipid molecules. (Look back to the diagram of a cell surface membrane in topic 2.5.1.)

Some organisms, such as fish and microorganisms (see Figure 1), can change the composition of the fatty acids in their cell membranes in response to lowered temperatures. Some plants can do this too.

Figure 1 The Atlantic cod, *Gadus morhua*, lives in cold waters and deep-sea regions of the North Atlantic.

When temperature increases

* The phospholipids acquire more kinetic energy and move around more, in a random way. This increases the membrane's fluidity.
* Permeability increases.
* It also affects the way membrane-embedded proteins are positioned and may function. If some of the proteins that act as enzymes in a membrane drift sideways, this could alter the rate of the reactions they catalyse.
* An increase in membrane fluidity may affect the infolding of the plasma membrane during phagocytosis.
* An increase in membrane fluidity may also change the ability of cells to signal to other cells by releasing chemicals, often by exocytosis.
* The presence of cholesterol molecules buffers, to some extent, the effects of increasing heat as it reduces the increase in membrane fluidity.

Proteins and temperature

Whereas changing temperature can alter the movement of phospholipids, it does not drastically alter their integral molecular structure. However, proteins are not as stable as lipids.

* High temperatures cause the atoms within their large molecules to vibrate, and this breaks the hydrogen bonds and ionic bonds that hold their structure together – they unfold.
* Their tertiary structure (shape) changes and cannot change back again when they cool – they are **denatured**. See Chapter 2.2 for more about protein structure.

Just underneath the plasma membrane are cytoskeleton threads, made of protein (see topic 2.1.5). If both the membrane-embedded proteins and the cytoskeleton threads become denatured, then the plasma membrane will begin to fall apart. It will become more permeable as holes will appear in it.

Membrane-embedded enzymes will cease to function if they become denatured. If the shape of their active site changes slightly or the enzymes move within the membrane, the rate of the reactions that they catalyse will be slowed. (Look back at Chapter 2.4 for more on enzymes.)

The effect of temperature on beetroot cell membranes

Wear eye protection.

Inside beetroot cells, within the large vacuole that is bound by a tonoplast membrane, are nitrogenous, water-soluble pigments called betacyanins, a type of betalain.

If you heat pieces of beetroot tissue, the plasma membrane and tonoplast membrane will be disrupted and the pigment will leak out. The amount of leakage of the red pigment is proportional to the degree of damage to the beetroot plasma and tonoplast membranes, and can be measured using a colorimeter, by measuring the absorbance of green light (wavelength range 530–550 nm).

Effect of solvents on phospholipids

Organic solvents such as acetone and ethanol will damage cell membranes as they dissolve lipids.

The effect of chemicals on beetroot cell membranes

Betalain pigments do not change as pH changes.

Wear eye protection.

1. Design your investigation – you may wish to use the solvents at different concentrations. This may involve doing a serial dilution of a stock solution. Write a list of the equipment that you will need.
2. Make a prediction that you can test.
3. State the independent variable, dependent variable and control variables.
4. Carry out the investigation. Record your data and suggest an explanation for your results.
5. You may also investigate the effect of detergents, dilute acids, such as hydrochloric acid, and alkalis, such as sodium hydroxide, on the structure and permeability of beetroot cell membranes.

Heat breaks some of the bonds that hold a protein's tertiary structure in place. The molecules vibrate as they acquire more kinetic energy, breaking the hydrogen bonds. Disulfide bonds are not broken by heat, but they are broken by strong reducing agents. Vibration of the protein molecules does not break the peptide bonds between amino acids.

The heat-shock pathway

In some plants, the plasma membrane acts as a heat sensor. When proteins in the membrane begin to denature, this triggers the expression of genes to make heat-shock proteins. These are chaperone proteins that bind to misshapen proteins, preventing further misfolding. The heat-signalling pathway in plant cells is triggered by an influx of calcium ions, caused by the opening of calcium ion channels in plasma membranes when plant cells are subjected to mild heat.

In bacteria, calcium ion influx can increase membrane fluidity. Scientists make use of this when genetically modifying bacteria. They give the bacteria more calcium ions and subject them to heat shock (exposure to 0 °C followed by exposure to 40 °C). This increases the influx of calcium ions and the chance of bacteria taking up DNA.

The betacyanin pigments, e.g. betanin, in beetroot help to lower your blood pressure. This is because they are nitrogenous compounds and stimulate the linings of your arteries to produce nitric oxide, which dilates arteries and reduces blood pressure.

Questions

1. Explain how cholesterol in cell membranes can reduce the effects of changing temperature on their stability.

2. Explain why having a greater proportion of unsaturated fatty acids in the cell membranes maintains their fluidity at low temperatures.

3. Cell-membrane fluidity can be altered in some organisms in response to environmental changes, such as temperature change. How would you expect the ratio of saturated : unsaturated fatty acids to change in the plasma membranes of such organisms, as the environmental temperature (a) increases from 10 °C to 25 °C and (b) decreases from 10 °C to 2 °C?

4. Describe and explain how increasing the temperature can affect the rate of reactions catalysed by enzymes embedded in cell membranes, either at the cell surface or within the cell in organelles.

5. How would you expect the feeding ability of a phagocytic cell, such as *Amoeba*, to change as its pond-water temperature decreased to 3 °C during the winter?

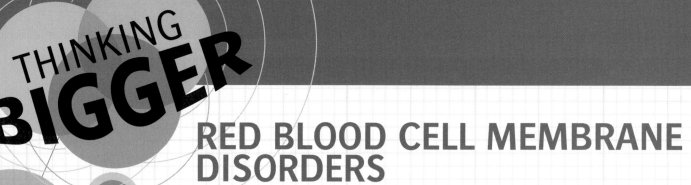

THINKING BIGGER

RED BLOOD CELL MEMBRANE DISORDERS

The plasma membrane of erythrocytes is responsible for the mechanical properties of the cell, and for some of its physiological functions. Disorders can result if there are mutations in genes that code for its plasma-membrane proteins. This article is adapted from an article published in the *British Journal of Haematology*.

RED BLOOD CELL MEMBRANE DISORDERS

The red blood cell plasma membrane comprises a lipid bilayer, integral membrane proteins and a membrane skeleton. This skeleton is a multi-protein complex formed by structural proteins including actin, ankyrin, spectrins, tropomyosin and protein 4.1. The membrane skeleton proteins interact with transmembrane proteins and the lipid bilayer to give the red cell membrane its strength and flexibility. They form scaffolding on the inner side of the lipid bilayer.

The genes of the major red cell membrane proteins have been cloned and their amino acid sequences deduced. One important protein forms a transmembrane channel for chloride–hydrogencarbonate exchange.

Disruption of vertical interactions between the membrane skeleton and the lipid bilayer, or disruption of horizontal interactions between the components of the membrane skeleton network may cause loss of structural and functional integrity of the membrane. Defects in horizontal interactions lead to hereditary **elliptocytosis**, and defects in the vertical interactions lead to hereditary **spherocytosis**.

Hereditary spherocytosis (HS) is the most common cause of non-immune **haemolytic anaemia** among people of northern European ancestry, with an incidence of approximately 1 in 2000. There are various mutations in the gene coding for ankyrin protein, including a point mutation at position 204, a nonsense mutation and frameshift mutations.

Some mutations therefore lead to defects in the structure of red blood cell membranes, and this leads to their shape being overly rounded or elongated. The membranes lose mechanical strength and become more fragile, so the cells break up into oddly shaped fragments, with **haemolysis** and sometimes life-threatening anaemia.

One form of hereditary elliptocytosis, called South East Asian **ovalocytosis** (SAO) is widely found in Malaysia, Indonesia, Papua New Guinea and the Philippines. When sufferers' blood is observed under a microscope, 20–50% of their erythrocytes are oval in shape. These distinctive red cells are not seen in any other condition. Most individuals are **asymptomatic**, but a few have some degree of haemolysis. **Homozygosity** is probably lethal. SAO is associated with increased rigidity and decreased deformability of the membrane. The geographical distribution of SAO corresponds with malaria **endemicity**, suggesting that the condition conveys protection against malaria.

Source

● Tse, W.T. and Lux, S.E. (2005) Red blood cell membrane disorders. *British Journal of Haematology*, Vol. 104, Issue 1. Published online on 8 February 2005. http://onlinelibrary.wiley.com/doi/10.1111/j.1365-2141.1999.01130.x/abstract (last accessed February 2015).

KEY DEFINITIONS

asymptomatic: not having any symptoms.
elliptocytosis: cells being more elliptical in shape than they usually are.
endemicity: refers to degree of a condition being endemic – always present in an area/community.
haemolysis: lysis of animal cells, in this case it is referring to lysis of red blood cells.
haemolytic anaemia: anaemia with chronic premature destruction of red blood cells.
homozygosity: having two copies of an allele at a specific gene locus.
ovalocytosis: cells being more oval in shape than they usually are.
spherocytosis: cells being more spherical in shape than they usually are.

Where else will I encounter these themes?

1.1   2.1   2.2   2.3   2.4   2.5   YOU ARE HERE

Consider the style of writing.

1. Who do you think this article is aimed at? What features of the article lead you to your conclusion?
2. Is the 'Key definitions' box useful to you? (This was not part of the original article, but has been added here to help you understand the content of the article.)

Now let's look at the content and apply what we have learned about cells and membranes to decipher the information in this article. You may need to refer to the topics on membranes and the cell diversity topic that describes features of red blood cells. You may also need to recall what you learnt at GCSE about inherited diseases and re-read chapter 2.3 on nucleic acids.

3. What do you understand by the term 'transmembrane proteins'?
4. What do you think the following sentence means – 'The genes of the major red cell membrane proteins have been cloned and their amino acid sequences deduced.'
5. Explain why haemolysis of red blood cells in circulation leads to anaemia.
6. Suggest why severe anaemia is life-threatening.
7. SAO is associated with increased rigidity and decreased deformability of the red blood cell plasma membrane. Explain why this hinders the red cells from performing their function.
8. Suggest why there is a high incidence of people with SAO in South East Asia.

DID YOU KNOW?

In the lipid bilayer of the plasma membrane there are different types of lipids. Cells have special enzymes that are dependent on ATP (which means they use energy) to mix these lipids around and maintain asymmetry – unequal distribution of them in the two layers of lipid. One of the enzymes is called *scramblase* and another is called *floppase*.

Activity

Red blood cells have many other proteins embedded in their plasma membrane. Some of these are markers or antigens and can be used to classify people according to their blood group.

Research, using GCSE texts and the Internet, and find out more about these membrane proteins or antigens.

Now complete one of the following activities:

1. Make a poster for use in a blood-donation unit to inform people about blood groups and explaining which types of blood can be donated or received.
2. Make a poster for display in an antenatal clinic, to inform prospective parents about Rhesus factor and haemolytic disease of the newborn.

When designing and producing your poster, keep the information short and sweet, whilst informative and eye-catching and easily understood.

1. Which of the following is not a feature of plasma membranes? [1]
   A. containing enzymes
   B. making antigens
   C. releasing signalling chemicals
   D. containing receptors

2. How would you describe a plant cell that has lost much water by osmosis? [1]
   A. crenated
   B. flaccid
   C. plasmolysed
   D. turgid

3. Read the following statements.
   (i) Diffusion of molecules is dependent on the molecules' kinetic energy.
   (ii) Facilitated diffusion needs cellular energy to open the membrane channels through which ions move into or out of cells.
   (iii) The rate of diffusion decreases as temperature increases.
   Which of these statements is/are true? [1]
   A. (i), (ii) and (iii)
   B. Only (i) and (ii)
   C. Only (ii) and (iii)
   D. Only (i)

4. Which of the following processes are dependent on ATP? [1]
   (i) Movement of organelles within a cell along cytoskeleton threads
   (ii) Ingestion of a bacterium by a phagocytic white blood cell
   (iii) Influx of sodium ions into a neurone down a concentration gradient
   A. (i), (ii) and (iii)
   B. Only (i) and (ii)
   C. Only (ii) and (iii)
   D. Only (i)

5. Read the following statements.
   (i) Some organisms can change the composition of the plasma membranes in response to a change in environmental temperature.
   (ii) When the temperature drops the saturated fatty acids in plasma membranes become compressed.
   (iii) Cholesterol molecules in plasma membranes keep plasma membranes fluid when temperatures are lowered.
   Which of these statements is/are true? [1]
   A. (i), (ii) and (iii)
   B. Only (i) and (ii)
   C. Only (ii) and (iii)
   D. Only (i)

   [Total: 5]

In questions 6–8, you may need to use your knowledge of other topics, such as biological molecules, nucleic acids, enzymes, cell structure and transport in plants.

6. Figure 1 shows the concentrations in mmol dm^{-3} of potassium and chloride ions inside and around a cell of *Nitella*, a freshwater alga.

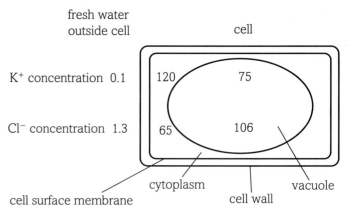

Figure 1

An investigation was carried out to determine the relationship between the uptake of potassium ions and glucose consumption by a cell of *Nitella* at different oxygen concentrations. The results are shown in Figure 2.

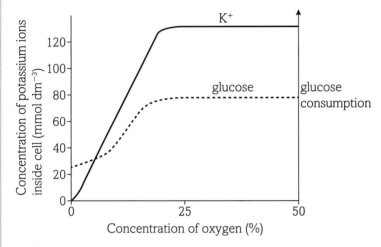

Figure 2

(a) By which process do potassium ions enter the *Nitella* cytoplasm from the surrounding water? Give two reasons for your answer. [2]

(b) Compare the ways in which potassium ions and chloride ions might pass from the cytoplasm into the vacuole. [6]

(c) Describe, in terms of water potential, what would happen if the *Nitella* cell was placed in a strong sugar solution. [5]

(d) Suggest how other species of algae are able to survive in strong salt solutions, such as sea water. [2]

[Total: 15]

7. Figure 3 is a diagrammatic representation of a plasma (cell surface) membrane.

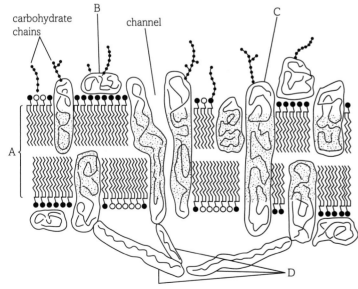

Figure 3

(a) How is this model of the plasma membrane structure usually described? State two reasons why the plasma membrane structure is described in this way. [3]

(b) Name the structures labelled A, B, C and D. [4]

(c) What is the function of the channel shown in the diagram? [1]

(d) State three features of a molecule that influence its ability to pass through a plasma membrane. [3]

(e) Give four reasons why transport across membranes is vital to a living cell. [4]

(f) Describe two functions of the structures labelled D. [2]

(g) Suggest a possible function of the carbohydrate chains attached to the outer surface of the plasma membrane. [1]

(h) Describe two features of membranes that affect diffusion across them. [2]

[Total: 20]

8. Figure 4 shows what happened in an investigation when a species of amoeba that normally lives in the sea was placed in fresh water. Water entering the cell is enclosed in vesicles that then coalesce into a contractile vacuole, which is moved towards the plasma membrane and the excess water is then released out of the cell.

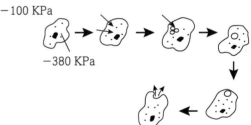

Figure 4

(a) Explain, in terms of water potential, why water enters this amoeba when it is placed in fresh water. [3]

(b) Describe the mechanism that moves the contractile vacuole, within the amoeba cell, to the plasma membrane. [2]

(c) Name the process by which the excess water leaves the cell. [1]

(d) If these cells are exposed to a chemical that inhibits aerobic respiration, the contractile vacuole is not moved to the cell surface membrane.

(i) Suggest how this would affect the entry of water into the cell. [3]

(ii) What would you expect to happen to the cell as a result of the contractile vacuole not emptying its contents to the surrounding environment? [1]

[Total: 10]

Foundations in biology

CELL DIVISION, CELL DIVERSITY AND CELL DIFFERENTIATION

Introduction

Since Schleiden and Schwann developed cell theory in the 19th century, scientists have carried out a lot more research, not just into the ultrastructure of cells but also into how cells divide, differentiate and give rise to the many different cell types found in eukaryotic organisms.

Scientists have also found how cells are organised into tissues and tissues organised into organs. This unit will often focus on structure and function as specialised cells, tissues and organs are well adapted to carry out their specific functions and this thread is part of the unifying theme of biology, namely how those best adapted structures enable organisms to survive and reproduce, passing on those advantageous characteristics.

Research into how the cell cycle is regulated is key for helping scientists to understand the underlying mechanisms that give rise to tumours, and to help develop treatments for cancer.

It is more than likely that within your lifetime scientists will be able to take stem cells from you and use them to grow replacement organs or tissues if you have a degenerative disease or have lost tissues because of an injury.

Differentiated cells can be reprogrammed to 'turn back the clock' and go back to being undifferentiated so they can then be made into a specific cell type. The research going on now is about finding out how best to grow and nurture these cells and make sure they are directed to grow into particular cell types.

All the maths you need

To unlock the puzzles in this section you need the following maths:

- Units of measurement
- Powers of 10
- Division and multiplication

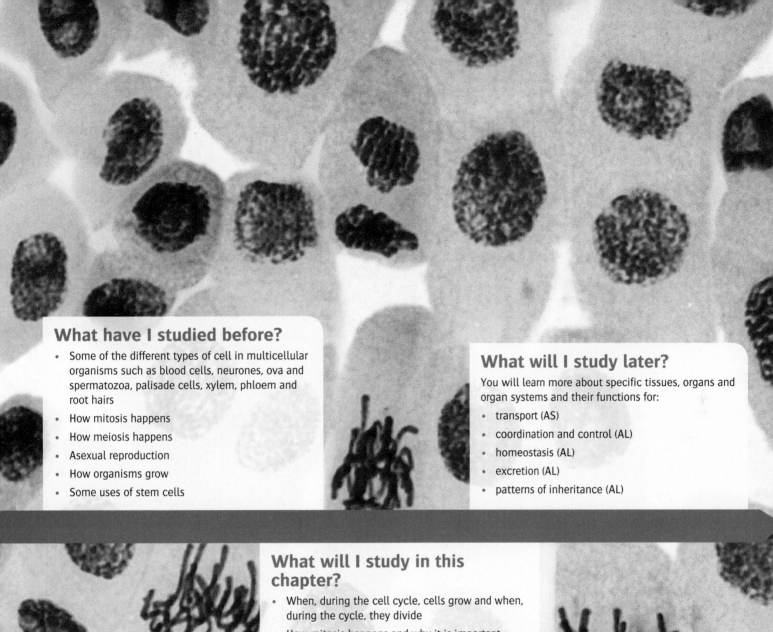

What have I studied before?

- Some of the different types of cell in multicellular organisms such as blood cells, neurones, ova and spermatozoa, palisade cells, xylem, phloem and root hairs
- How mitosis happens
- How meiosis happens
- Asexual reproduction
- How organisms grow
- Some uses of stem cells

What will I study later?

You will learn more about specific tissues, organs and organ systems and their functions for:

- transport (AS)
- coordination and control (AL)
- homeostasis (AL)
- excretion (AL)
- patterns of inheritance (AL)

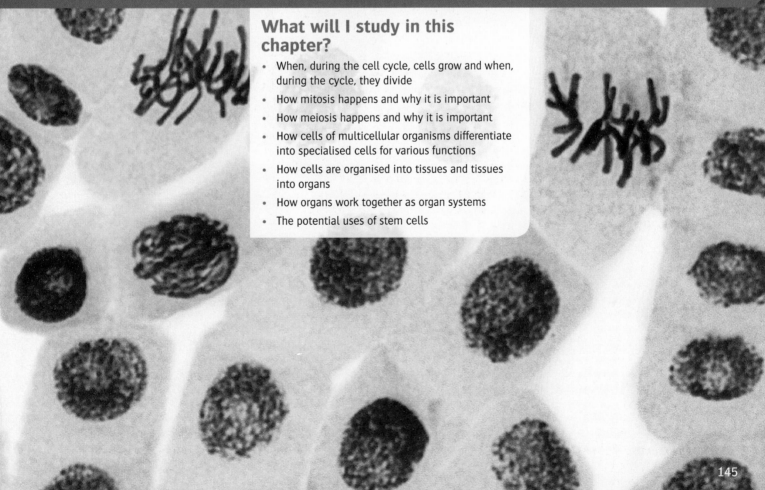

What will I study in this chapter?

- When, during the cell cycle, cells grow and when, during the cycle, they divide
- How mitosis happens and why it is important
- How meiosis happens and why it is important
- How cells of multicellular organisms differentiate into specialised cells for various functions
- How cells are organised into tissues and tissues into organs
- How organs work together as organ systems
- The potential uses of stem cells

(1) The cell cycle and its regulation

By the end of this topic, you should be able to demonstrate and apply your knowledge and understanding of:

* the cell cycle
* how the cell cycle is regulated

KEY DEFINITIONS

cytokinesis: cytoplasmic division following nuclear division, resulting in two new daughter cells.
interphase: phase of cell cycle where the cell is not dividing; it is subdivided into growth and synthesis phases.
mitosis: type of nuclear division that produces daughter cells genetically identical to each other and to the parent cell.

The eukaryotic cell cycle

Cells reproduce by duplicating their contents and then splitting into two daughter cells.

Early researchers observing cell division under the microscope could easily see the behaviour of chromosomes during **mitosis**, which is nuclear division, followed by **cytokinesis** or cytoplasmic division, resulting in two daughter cells (see topic 2.6.2).

However, nuclear and cytoplasmic division, called the M phase, occupy only a small part of the cell cycle (see Figure 1). Between each M phase is an **interphase**. Interphase, when studied under the microscope, appears to be uneventful. However, more sophisticated techniques have enabled scientists to learn that during interphase there are elaborate preparations being made for cell division, in a carefully ordered and controlled sequence, with *checkpoints*.

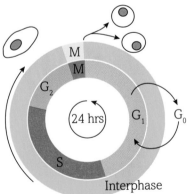

Figure 1 The cell cycle in eukaryotic cells. M indicates the division phase and interphase is divided into G_1, S and G_2. Cells may also enter a G_0 phase, where they may undergo differentiation, or **apoptosis** (programmed cell death) or enter senescence – where they can no longer divide.

Regulation of the eukaryotic cell cycle

Cell-cycle checkpoints

There are two main checkpoints, the G_1/S checkpoint, also called the *restriction point*, and the G_2/M checkpoint (see Table 1). There

are other checkpoints, for example there is one halfway through mitosis and one in early G_1. The purpose of the checkpoints is:

* to prevent uncontrolled division that would lead to tumours (cancer)
* to detect and repair damage to DNA (for example, damage caused by UV light).

Because the molecular events that control the cell cycle happen in a specific sequence, they also ensure that:

* the cycle cannot be reversed
* the DNA is only duplicated *once* during each cell cycle.

DID YOU KNOW?

Cells should normally only undergo a certain number of cycles or divisions. The number is about 50 and is known as the Hayflick constant. If cell division becomes uncontrolled, then a tumour can form which may become malignant or cancerous. There are proto-oncogenes that help regulate cell division by coding for proteins that help regulate cell growth and differentiation. If these proto-oncogenes mutate, then they may become oncogenes and can cause cells to fail to undergo apoptosis and instead to keep on dividing, leading to a tumour.

LEARNING TIP

Remember that tumours form because cells divide *uncontrollably*. The division phase, mitosis, of each cancerous cell takes the same length of time as in a non-cancerous cell. Because the cell division is uncontrolled, there is a greater proportion of cells dividing within a tumour than within normal tissue.

DID YOU KNOW?

The *p53* gene is important as it triggers the two main checkpoints in the regulation of the cell cycle. Hence it is known as a tumour suppressor gene. Other regulatory chemicals are proteins called cyclins (the scientist who discovered them actually named them for his love of cycling, but the title is apt) and CDKs (cyclin-dependent kinases – you will learn more about kinase proteins in the second year of your A level course). Cyclins are synthesised in response to cell-signalling molecules such as growth factors.

DID YOU KNOW?

The prokaryotic cell cycle
This occurs by a process called *binary fission*. The cell grows to its limit of size and then splits into two. Before the cell divides, its DNA is replicated. The two new loops of DNA are pulled to opposite ends of the cell and a cell wall forms which begins to separate the bacterial cell. Each new cell also contains replicated plasmids and synthesised ribosomes. Mitochondria and chloroplasts, within eukaryotic cells, also divide by binary fission.

Phase of cell cycle and checkpoints	Events within the cell
M phase • A checkpoint chemical triggers condensation of chromatin. • Halfway through the cycle, the metaphase checkpoint ensures that the cell is ready to complete mitosis.	• Cell growth stops. • Nuclear division (mitosis) consisting of stages: prophase, metaphase, anaphase and telophase (see topic 2.6.2 for more on mitosis). • Cytokinesis (cytoplasmic division).
G_0 (gap 0) phase • A resting phase triggered during early G_1 at the restriction point (see below), by a checkpoint chemical. • Some cells, e.g. epithelial cells lining the gut, do not have this phase.	• In this phase, cells may undergo apoptosis (programmed cell death), differentiation or senescence. • Some types of cells (e.g. neurones) remain in this phase for a very long time or indefinitely.
G_1 (gap 1) phase – also called the growth phase • A G_1 checkpoint control mechanism ensures that the cell is ready to enter the S phase and begin DNA synthesis.	• Cells grow and increase in size. • Transcription of genes to make RNA occurs. • Organelles duplicate. • Biosynthesis, e.g. protein synthesis, including making the enzymes needed for DNA replication in the S phase. • The p53 (tumour suppressor) gene helps control this phase.
S (synthesis) phase of interphase • Because the chromosomes are unwound and the DNA is diffuse, every molecule of DNA is replicated. There is a specific sequence to the replication of genes: housekeeping genes – those which are active in all types of cells, are duplicated first. Genes that are normally inactive in specific types of cells are replicated last.	• Once the cell has entered this phase, it is committed to completing the cell cycle. • DNA replicates. • When all chromosomes have been duplicated, each one consists of a pair of identical sister **chromatids**. • This phase is rapid, and because the exposed DNA base pairs are more susceptible to mutagenic agents, this reduces the chances of spontaneous mutations happening.
G_2 (gap 2) phase of interphase • Special chemicals ensure that the cell is ready for mitosis by stimulating proteins that will be involved in making chromosomes condense and in formation of the spindle.	• Cells grow.

Table 1 The events that take place during the different phases of the eukaryotic cell cycle.

Questions

1. If a sample of a population of cells is fixed and stained and examined under a microscope, cells in various stages of the cell cycle will be seen. Assuming the duration of the cell cycle in this cell population is 24 hours, if $\frac{1}{12}$ cells are in mitosis, then the duration of mitosis is approximately $\frac{1}{12} \times 24$ hours = 2 hours. In the same population of 12 cells, four were in the S phase. What is the approximate duration of the S phase for these cells?

2. Epithelial cells lining the intestine divide two or three times a day. Liver cells divide about once a year. In which parts of their cell cycle will liver cells differ from intestine epithelial cells?

3. List the purpose of checkpoints to control the cell cycle.

4. During the cell cycle, the organelles in a cell are duplicated so that each new daughter cell receives roughly equal numbers of organelles. However, when DNA is replicated, each molecule is copied and each new daughter cell receives exactly the same amount of genetic material. Suggest why it is crucial that each new cell receives the same amount of DNA.

5. In which phase of the eukaryotic cell cycle do the following happen? (a) Replication of chloroplasts and mitochondria; (b) condensation of chromatin (chromosomes); (c) replication of DNA; (d) differentiation; (e) apoptosis (programmed cell death); and (f) growth of the cell.

(2) Mitosis

By the end of this topic, you should be able to demonstrate and apply your knowledge and understanding of:

* the main stages of mitosis

* sections of plant tissue showing the cell cycle and stages of mitosis

* the significance of mitosis in life cycles

The significance of mitosis in the life cycle

All living organisms need to produce genetically identical daughter cells, by **mitosis**, for the following reasons:

* **Asexual reproduction** – single-celled protoctists such as *Amoeba* and *Paramecium* divide by mitosis to produce new individuals. Some plants, e.g. strawberry, reproduce asexually by forming new plantlets on the ends of stolons (runners). Fungi, such as single-celled yeasts, can reproduce asexually by mitosis. Asexual reproduction is rarer in animals but some female sharks kept in captivity without any males have produced female offspring that are genetically identical to themselves. Aphids may sometimes produce eggs, by mitosis, that do not need fertilising.

* **Growth** – all multicellular organisms grow by producing more cells that are genetically identical to each other and to the parent cell from which they arose by mitosis.

* **Tissue repair** – wounds heal when growth factors, secreted by platelets and macrophages (white blood cells) and damaged cells of the blood-vessel walls, stimulate the proliferation of endothelial and smooth muscle cells to repair damaged blood vessels.

The main stages of mitosis

Although mitosis is a continuous process, scientists observing the process have defined four main stages of mitosis, namely prophase, metaphase, anaphase and telophase (see Figures 1–4 and Table 1).

Stage of mitosis: diagram	Stage of mitosis: photo	Events during the stage
Prophase  spindle forming nuclear envelope breaking down sister chromatids centromere	**Prophase** 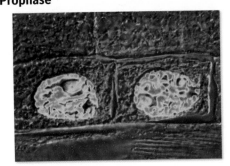 **Figure 1** Prophase of mitosis in hyacinth root-cells. The condensed chromatin is visible.	• The chromosomes that have replicated during the S phase of interphase and consist of two identical sister chromatids, now shorten and thicken as the DNA supercoils. • The nuclear envelope breaks down. • The centriole in animal cells (normally found within a region of the cell called a centrosome) divides and the two new daughter centrioles move to opposite poles (ends) of the cell. • Cytoskeleton protein (tubulin) threads form a spindle between these centrioles. The spindle has a 3D structure and is rather like lines of longitude on a virtual globe. In plant cells, the tubulin threads are formed from the cytoplasm.

continued

continued

Stage of mitosis: diagram	Stage of mitosis: photo	Events during the stage
Metaphase chromosomes attached to spindle equator centromere tubulin threads	**Metaphase** **Figure 2** Fluorescent micrograph of a cell during metaphase of mitosis. The microtubules forming the spindle are stained green, with the tubulin threads emanating from the centrioles (pink dots). The chromosomes are stained blue and the actin cytoskeleton filaments of the cell are stained red.	• The pairs of chromatids attach to the spindle threads at the equator region. • They attach by their centromeres.
Anaphase chromatids begin to be pulled apart	**Anaphase** **Figure 3** Anaphase of mitosis in a bluebell cell.	• The centromere of each pair of chromatids splits. • Motor proteins, walking along the tubulin threads, pull each sister chromatid of a pair, in opposite directions, towards opposite poles (see also topic 2.1.5). • Because their centromere goes first, the chromatids, now called chromosomes, assume a V shape.
Telophase two sets of chromosomes new nuclear membranes form	**Telophase** **Figure 4** 3D immunofluorescent light micrograph showing telophase in a rat-kangaroo kidney-cell. Chromosomes are stained blue; spindle threads of tubulin are stained green.	• The separated chromosomes reach the poles. • A new nuclear envelope forms around each set of chromosomes. • The cell now contains two nuclei each genetically identical to each other and to the parent cell from which they arose.

Table 1 Events occurring during each stage of mitosis.

Cytokinesis

Once mitosis is complete, the cell splits into two, so that each new cell contains a nucleus.

- In animal cells, the plasma membrane folds inwards and 'nips in' the cytoplasm.
- In plant cells, an end plate forms where the equator of the spindle was, and new plasma membrane and cellulose cell-wall material are laid down on either side along this end plate.

Two new daughter cells are now formed. They are genetically identical to each other and to the parent cell.

INVESTIGATION

Making a root-tip squash to examine cells in stages of mitosis

Once plant cells have divided and differentiated and have a vacuole and rigid cellulose cell wall, they cannot divide. There are meristems, such as shoot and root tips and cambium between xylem and phloem tissue, where plant cells are undifferentiated and can divide by mitosis. Cells of root tips can be stained with acetic orcein (which stains chromosomes) and observed under a microscope, to see the stages of mitosis.

Questions

1. In a cell in the root of a black nightshade plant, *Solanum nigrum*, there are 72 chromosomes. How many chromosomes are in this cell during: (a) G_2 interphase, (b) G_1 interphase, (c) prophase of mitosis, (d) telophase of mitosis, and (e) after **cytokinesis**?

2. Explain why plant palisade mesophyll cells cannot undergo mitosis.

3. Distinguish (explain their meanings) between these terms: centriole, centrosome, chromosome and centromere.

4. State four reasons why mitosis is necessary in the cell cycle of living organisms.

5. Male honeybees develop from unfertilised eggs, and each of their cells is haploid (contains only one set of chromosomes). The function of the males in the hive is for some of them to fertilise the new queen. Suggest which type of cell division is used to make the bee spermatozoa.

(3) Meiosis

By the end of this topic, you should be able to demonstrate and apply your knowledge and understanding of:

* the significance of meiosis in life cycles

* the main stages of meiosis

KEY DEFINITIONS

haploid: having only one set of chromosomes; represented by the symbol '*n*'.

homologous chromosomes: matching chromosomes, containing the same genes at the same places (loci). They may contain different alleles for some of the genes.

meiosis: type of nuclear division that results in the formation of cells containing half the number of chromosomes of the parent cell.

The significance of meiosis in life cycles

Sexual reproduction increases genetic variation because it involves the combining of genetic material from two (usually) unrelated individuals of the same species, by the process of **fertilisation**. Genetic variation within a population increases its chances of survival when the environment changes, as some individuals will have characteristics that enable them to be better adapted to the change. In many organisms, the body cells are **diploid**. For sexual reproduction to occur they must produce **haploid gametes**, so that when two gamete nuclei fuse during fertilisation, a diploid **zygote** is produced and the normal chromosome number is maintained through the generations.

Meiosis means 'reduction', and it occurs in diploid germ cells to produce haploid gametes. The diploid cells undergoing meiosis are in specialised organs called gonads – ovaries and testes. These cells have been in interphase before they enter meiosis (see topic 2.6.1).

Homologous chromosomes

In your body cells there are 46 chromosomes. Twenty-three came from your mother, in the egg nucleus, and 23 came from your father in the sperm nucleus. These can form matching pairs – one maternal and one paternal chromosome containing the same genes at the same places on the chromosome. These matching pairs are called **homologous chromosomes**. Although they have the same genes, they may contain different **alleles** (**variants**) for the genes (see Figure 1).

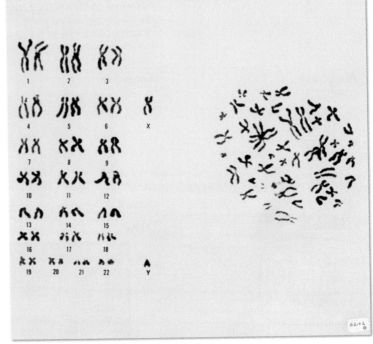

Figure 1 A karyotype (photomicrograph of chromosomes in a cell) showing the pairs of homologous chromosomes. This individual is male and one pair of chromosomes, XY, is not fully homologous, but enough homology exists to allow them to pair up. The chromosomes in each of the homologous pairs consist of two chromatids, because these chromosomes were observed using a light microscope and photographed as the nucleus was dividing.

The main stages of meiosis

Before meiosis, during the S phase of interphase, each chromosome was duplicated as its DNA replicated, after which each chromosome consists of two sister chromatids. In meiosis, the chromosomes pair up in their homologous pairs.

* There are two *divisions* in meiosis, and in each division there are four stages.

* In the first meiotic division (see Table 1), the four stages are: prophase 1, metaphase 1, anaphase 1 and telophase 1.

* The cell may then enter a short interphase, before embarking on the second meiotic division that also has four stages: prophase 2, metaphase 2, anaphase 2 and telophase 2 (see Figure 2). This takes place in a plane at right angles to that of meiosis 1.

* At the end of the second division, cytokinesis may occur.

Stage of meiosis 1	Events during the stage
Prophase 1 **Figure 2** Prophase 1 may last for days, months or years, depending on the species and type of gamete (male or female) being formed. **Figure 3**	• The chromatin condenses and each chromosome supercoils. In this state, they can take up stains and can be seen with a light microscope. • The nuclear envelope breaks down, and spindle threads of tubulin protein form from the centriole in animal cells. • The chromosomes come together in their homologous pairs. • Each member of the pair consists of two chromatids. • Crossing over occurs where non-sister chromatids wrap around each other and may swap sections so that alleles are shuffled.
Metaphase 1 **Figure 4**	• The pairs of homologous chromosomes, still in their crossed over state, attach along the equator of the spindle. • Each attaches to a spindle thread by its centromere. • The homologous pairs are arranged randomly, with the members of each pair facing opposite poles of the cell. This arrangement is independent assortment. • The way that they line up in metaphase determines how they will segregate independently when pulled apart during anaphase.
Anaphase 1 **Figure 5**	• The members of each pair of homologous chromosomes are pulled apart by motor proteins that drag them along the tubulin threads of the spindle. • The centromeres do not divide, and each chromosome consists of two chromatids. • The crossed-over areas separate from each other, resulting in swapped areas of chromosome and allele shuffling.
Telophase 1 **Figure 6**	• In most animal cells, two new nuclear envelopes form around each set of chromosomes, and the cell divides by cytokinesis. There is then a short interphase when the chromosomes uncoil. • Each new nucleus contains half the original number of chromosomes, but each chromosome consist of two chromatids. • In most plant cells, the cell goes straight from anaphase 1 into prophase 2.

Table 1 The first meiotic division (meiosis 1).

Stage of meiosis 2	Events during the stage
Prophase 2 centrioles replicate and move to poles new spindle fibres form at right angles to previous spindle axis **Figure 7**	• If the nuclear envelopes have reformed, then they now break down. • The chromosomes coil and condense, each one consisting of two chromatids. • The chromatids of each chromosome are no longer identical, due to crossing over in prophase 1. • Spindles form.
Metaphase 2 chromosomes lying on the equator of the cell **Figure 8**	• The chromosomes attach, by their centromere, to the equator of the spindle. • The chromatids of each chromosome are randomly arranged. • The way that they are arranged will determine how the chromatids separate during anaphase.
Anaphase 2 chromatid moving towards the pole **Figure 9**	• The centromeres divide. • The chromatids of each chromosome are pulled apart by motor proteins that drag them along the tubulin threads of the spindle, towards opposite poles. • The chromatids are therefore randomly segregated.
Telophase 2 haploid cells **Figure 10**	• Nuclear envelopes form around each of the four haploid nuclei. • In animals, the two cells now divide to give four haploid cells. • In plants, a tetrad of four haploid cells is formed. **Figure 11** Telophase 2 of meiosis in lily anther.

Table 2 The second meiotic division (meiosis 2).

How meiosis produces genetic variation

• Crossing over during prophase 1 shuffles alleles.

• Independent assortment of chromosomes in anaphase 1 leads to random distribution of maternal and paternal chromosomes of each pair.

• Independent assortment of chromatids in anaphase 2 leads to further random distribution of genetic material.

• Haploid gametes are produced, which can undergo random fusion with gametes derived from another organism of the same species.

LEARNING TIP

• Make sure that you spell meiosis correctly, so that there is no confusion with mitosis.

• Shuffling of alleles during crossing over in meiosis leads to genetic variation, but it is not an example of mutation.

DID YOU KNOW?

In the ovaries of human female fetuses, germ cells have already begun meiosis, and these cells remain suspended in prophase 1 from birth until puberty, when one each month completes meiosis 1 and enters meiosis 2. Meiosis 2 is not completed until a sperm has entered the ovum.

Questions

1 Explain why sexual reproduction involves meiosis.

2 Describe how meiosis produces genetic variation in the gametes produced.

3 Explain how fertilisation produces genetic variation in a new individual.

4 Why is genetic variation good for populations of living organisms?

5 In which stage of meiosis is the chromosome number halved?

6 What are the products of meiosis?

By the end of this topic, you should be able to demonstrate and apply your knowledge and understanding of:

* how cells of multicellular organisms are specialised for particular functions

* the features and differentiation of stem cells

* the production of erythrocytes and neutrophils derived from stem cells in bone marrow

KEY DEFINITIONS

differentiation: process by which stem cells become specialised into different types of cell.
epithelial cells: cells that constitute lining tissue.
erythrocyte: red blood cell.
neutrophil: type of white blood cell that is phagocytic (can ingest microbes and small particles).
stem cell: unspecialised cell able to express all of its genes and divide by mitosis.

The need for cell differentiation and specialisation

Within a single-celled organism, such as an amoeba, the division of labour is determined by the organelles, each of which has a specific function. Single-celled organisms are small and have a large surface area to volume (SA/V) ratio (see topic 3.1.1 for more on SA/V ratio) so that oxygen can diffuse across their plasma membrane, and waste products can diffuse out via the same membrane.

Multicellular organisms are larger and therefore have a smaller SA/V ratio, which means that most of their cells are not in direct contact with the external environment. They need specialised cells to carry out particular functions.

Differentiation

Multicellular eukaryotic organisms start life as a single undifferentiated cell, called a zygote. A zygote results when an ovum (egg cell) is fertilised by a spermatozoon and the two haploid nuclei fuse to give a cell with a diploid nucleus. The zygote is not specialised, and all the genes in its **genome** are able to be expressed. It is also able to divide by mitosis. It is a **stem cell** (see topic 2.6.9 for more about stem cells). After several mitotic divisions, an embryo forms, containing many undifferentiated embryonic stem cells.

LEARNING TIP

'Genome' refers to the genetic material within an individual. 'Gene pool' refers to all the genetic material within a population. So individuals have a genome, and populations have a gene pool.

These embryonic cells **differentiate** (become different) as certain genes are switched off and other genes may be expressed more, so that:

* the proportions of the different organelles differs from those of other cells

* the shape of the cell changes

* some of the contents of the cell change.

Owing to this differentiation, each cell type is *specialised* for a particular function.

Some specialised animal cells

Erythrocytes and neutrophils

In mammals:

* **erythrocytes** carry oxygen from the lungs to respiring cells.

* **neutrophils** ingest invading pathogens.

These cells are very different from each other, but both derive from stem cells in the bone marrow.

Erythrocytes are adapted to carry out their function in several ways:

* They are very small, about 7.5 μm in diameter, so have a large SA/V ratio. This means that oxygen can diffuse across their membranes and easily reach all regions inside the cell. Their biconcave shape also increases their SA/V ratio.

* They are flexible. A well developed cytoskeleton allows the erythrocytes to change shape so that they can twist and turn, as they travel through very narrow capillaries.

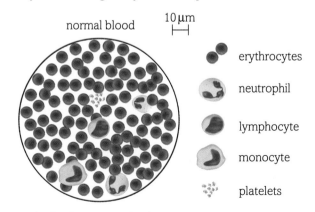

Figure 1 Blood cells as seen with a light microscope.

- Most of their organelles are lost at differentiation, so they have no nucleus, mitochondria or endoplasmic reticulum, and very little cytoplasm. This provides more space for the many haemoglobin molecules housed inside them. Haemoglobin is synthesised within immature erythrocytes, whilst they still have their nucleus, ribosomes and rough endoplasmic reticulum.

Neutrophils make up about 50% of the white blood cells in your body.

- They are about twice the size of erythrocytes, and each neutrophil contains a multilobed nucleus.
- They are attracted to and travel towards infection sites by chemotaxis.
- Their function is to ingest bacteria and some fungi by phagocytosis (see also topics 2.5.4 and 4.1.5 for more about neutrophils and phagocytosis).

Spermatozoa

Spermatozoa (sperm cells) are specialised in a number of ways (see Figure 3).

- The many mitochondria carry out aerobic respiration. The ATP provides energy for the undulipodium (tail) to move and propel the cell towards the ovum.
- Because spermatozoa are small but long and thin, they can move easily.
- Once the spermatozoon reaches an ovum, enzymes are released from the acrosome (a specialised lysosome). The enzymes digest the outer protective covering of the ovum, allowing the sperm head to enter the ovum.
- The head of the sperm contains the haploid male gamete nucleus and very little cytoplasm.

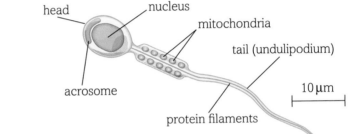

Figure 2 A human spermatozoon.

Epithelial cells

Epithelium is lining tissue. It is found on the outside of your body and on the inside – for example, making up the walls of the alveoli and capillaries (see topic 3.1.1 on exchange surfaces), and lining your intestines.

- Squamous **epithelial cells** are flattened in shape.
- Many of the cells in epithelium have cilia (see also topic 2.1.4 on cell ultrastructure).

Topic 2.6.6 has more information about these specialised cells within epithelial tissues.

INVESTIGATION

Examining prepared slides of some animal cells

You can examine prepared slides of some animal cells. In a blood smear you will see the features of erythrocytes and neutrophils. You may also measure the diameters of these cells.

Questions

1. What are stem cells?

2. Describe what happens to cells as they differentiate.

3. State the functions of erythrocytes, and explain how their structure adapts them to carry out their function.

4. Describe the function of neutrophils, and explain how their specialised structure enables them to carry it out.

5. Describe and explain how spermatozoa are adapted to carry out their function of travelling towards, and then fertilising, an ovum.

6. Suggest how the absence of mitochondria in erythrocytes enables them to carry oxygen efficiently.

7. List three metabolic activities that mature erythrocytes cannot carry out, and explain why.

⑤ Cell diversity in plants

By the end of this topic, you should be able to demonstrate and apply your knowledge and understanding of:

✳ how cells of multicellular organisms are specialised for particular functions

KEY DEFINITIONS

guard cells: in leaf epidermis, cells that surround stomata.
palisade cells: closely-packed photosynthetic cells within leaves.
root-hair cells: epidermal cells of young roots with long hair-like projections.

In topic 2.6.4, the need for cells in multicellular organisms to differentiate and specialise is explained, and some examples of specialised animal cells are given. Here we look at some examples of specialised cells in plants, each well adapted to carry out its function (see Figure 1 (a) and (b)).

Some specialised plant cells

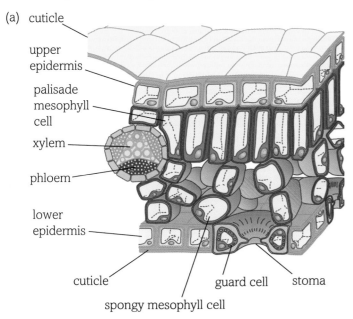

(a) cuticle
upper epidermis
palisade mesophyll cell
xylem
phloem
lower epidermis
cuticle
guard cell
stoma
spongy mesophyll cell

(b)
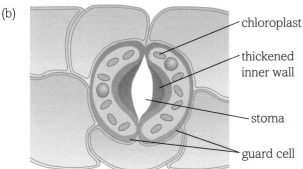
chloroplast
thickened inner wall
stoma
guard cell

Figure 1 (a) Part of a transverse section through a dicotyledonous leaf, showing the distribution of different types of cells. (b) Diagram of a stoma surrounded by a pair of guard cells.

Palisade cells

Palisade cells within leaves are well adapted for photosynthesis because:

* they are long and cylindrical, so that they pack together quite closely but with a little space between them for air to circulate; carbon dioxide in these air spaces diffuses into the cells

* they have a large vacuole so that the chloroplasts are positioned nearer to the periphery of the cell, reducing the diffusion distance for carbon dioxide

* they contain many chloroplasts – the organelles that carry out photosynthesis (see topic 2.1.4)

* they contain cytoskeleton threads and motor proteins (see topic 2.1.5) to move the chloroplasts – nearer to the upper surface of the leaf when sunlight intensity is low, but further down when it is high.

Guard cells

There are pairs of specialised cells, known as **guard cells** (see Figure 1 (b) and Figure 2), within the lower epidermis, that do contain chloroplasts. However, they cannot carry out photosynthesis, as they do not have the enzymes needed for the second stage of the process.

* Light energy is used to produce ATP.

* The ATP actively transports potassium ions from surrounding epidermal cells into the guard cells, lowering their water potential.

* Water now enters the guard cells from neighbouring epidermal cells, by osmosis (see topic 2.5.3).

* The guard cells swell, but at the tips the cellulose cell wall is more flexible, and it is more rigid where it is thicker (see Figure 1 (b)). The tips bulge, and the gap between them, the stoma (from a Greek word meaning 'mouth') enlarges.

* As these stomata open, air can enter the spaces within the layer of cells beneath the palisade cells.

* Gaseous exchange can occur, and carbon dioxide will diffuse into the palisade cells. As they use it for photosynthesis, this will then maintain a steep concentration gradient.

* Oxygen produced during photosynthesis can diffuse out of the palisade cells into the air spaces and out through the open stomata.

When the stomata are open, water vapour also exits from them, and you will learn more about this process (transpiration) in Chapter 3.3.

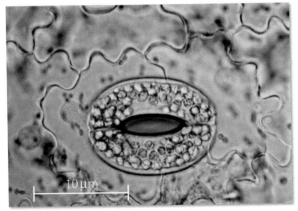

Figure 2 Photomicrograph of guard cells and stoma on the underside of a leaf of a wood fern, *Dryopteris*.

LEARNING TIP

Make sure you know how to spell 'stoma' correctly and do not confuse it with 'stroma', which is the fluid-filled matrix inside chloroplasts.

Root hair cells

Root hair cells are epidermal cells on the outer layer of young plant roots (see Figures 3 and 4).

- The hair-like projection greatly increases the surface area for absorption of water and mineral ions, such as nitrates, from the soil into which it projects.

- Mineral ions are actively transported into the root hair cells, lowering the water potential within them and causing water to follow by osmosis, down the water-potential gradient.

- The root hair cells have special carrier proteins in the plasma membranes in order to actively transport the mineral ions in.

- These cells will also produce ATP, as this is needed for active transport (see topics 2.5.4 and 3.3.5).

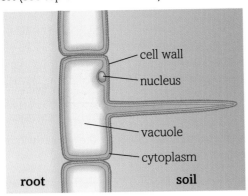

Figure 3 Root hair cells.

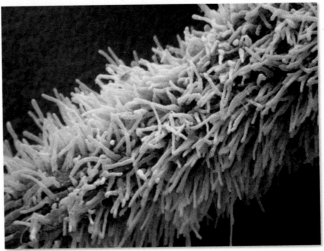

Figure 4 False-colour SEM of a root of marjoram, *Origanum vulgare*, showing many root hairs, which greatly increase the surface area for absorption (×133).

Xylem and phloem

Xylem and phloem form the vascular tissue of plants and are described in topic 2.6.7, with more information given in chapter 3.3 on transport in plants. Xylem vessels and phloem sieve tubes are present in vascular bundles, and their location can be seen in Figure 1 (a) in this topic.

DID YOU KNOW?

Palisade cells are so named because they resemble a fence made from a row of posts – a palisade.

INVESTIGATION

Examining prepared slides of plant cells
Wear eye protection.
You can examine prepared slides of leaf sections to identify and note the features of palisade cells, vascular bundles and guard cells.
If you use forceps to strip a piece of lower epidermis from a geranium leaf and mount it in water, on a microscope slide, you can observe guard cells and stomata.

Questions

1. Suggest why plants die if their roots are in waterlogged soil for several days.

2. Outline the function of guard cells and explain how they are adapted to carry out their function.

3. Make a prediction, with an explanation, about the number of mitochondria in root hair cells.

4. Suggest how you could estimate the number of stomata on the lower surface of a leaf.

(6) Animal tissues

By the end of this topic, you should be able to demonstrate and apply your knowledge and understanding of:

* the organisation of cells into tissues

You have seen that cells are the basic building blocks of living organisms and how, within multicellular organisms, cells become specialised for different functions. A group of similar cells working together to perform a certain function is called a **tissue**.

Your body has four main tissue types:

* **epithelial** or lining tissue
* **connective tissues** – these hold structures together and provide support, e.g. blood, bone and **cartilage**
* **muscle tissue** – made of cells that are specialised to contract and cause movement
* **nervous tissue** – made of cells specialised to conduct electrical impulses.

Epithelial tissue

This covers and lines free surfaces in the body such as the skin, cavities of the digestive and respiratory systems (gut and airways), blood vessels, heart chambers and walls of organs. The characteristics of epithelial tissue are as follows:

* Epithelial tissue is made up almost entirely of cells.
* These cells are very close to each other and form continuous sheets. Adjacent cells are bound together by lateral contacts, such as tight junctions and desmosomes.
* There are no blood vessels within epithelial tissue; cells receive nutrients by diffusion from tissue fluid in the underlying connective tissue.
* Some epithelial cells have smooth surfaces, but some have projections, either cilia or microvilli.
* Epithelial cells have short cell cycles and divide up to two or three times a day to replace worn or damaged tissue.
* Epithelial tissue is specialised to carry out its functions of protection, absorption, filtration, excretion and secretion.

DID YOU KNOW?

Every four days your whole intestinal lining will be replaced with new epithelial cells.

LEARNING TIPS

Do not confuse microvilli with cilia.

* Microvilli are extensions of the plasma membrane to increase its surface area.
* Cilia are hair-like organelles (see Figure 2), some of which beat and propel substances along the epithelial surfaces. In some cases, a single cilium on a cell acts as an antenna. It has receptors on it to receive chemical signals from its surroundings.

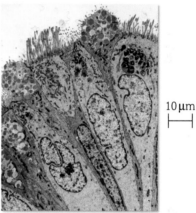

10 µm

Figure 1 False-colour SEM of ciliated cells lining the oviduct. The cilia beat and move the ovum along the oviduct, from ovary to uterus.

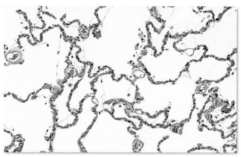

Figure 2 Light micrograph showing walls of alveoli and some nearby capillaries, the walls of which are also made of squamous epithelial cells (×40). The walls of alveoli (air sacs) in your lungs are one cell thick. They are made of a single layer of flattened squamous epithelial cells. This provides a short diffusion distance for gaseous exchange.

Connective tissue

Connective tissue is widely distributed in the body. It consists of a non-living extracellular matrix containing proteins (collagen and elastin) and polysaccharides (such as hyaluronic acid, which traps water). This matrix separates the living cells within the tissue and enables it to withstand forces such as weight.

Blood, bone, cartilage, tendons and ligaments are examples of connective tissue. Skin also contains connective tissue.

Cartilage

Immature cells in cartilage are called *chondroblasts*. They can divide by mitosis and secrete the extracellular matrix. Once the matrix has been synthesised, the chondroblasts become mature, less active *chondrocytes*, which maintain the matrix. There are three types of cartilage – hyaline, fibrous and elastic.

- Hyaline cartilage forms the embryonic skeleton, covers the ends of long bones in adults, joins ribs to the sternum, and is found in the nose, trachea (forming the C-shaped rings of cartilage that keep the trachea open) (see Figure 3) and larynx (voice box).

- Fibrous cartilage occurs in discs between vertebrae in the backbone (spine) and in the knee joint.

- Elastic cartilage makes up the outer ear (pinna) and the epiglottis (flap that closes over the larynx when you swallow).

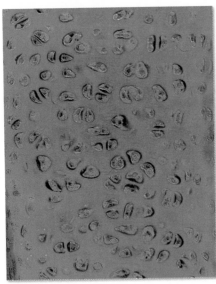

Figure 3 Photomicrograph of stained hyaline cartilage from a human trachea. Chondrocytes are within the matrix that was secreted by chondroblasts (×83).

Muscle tissue

Muscle tissue is well vascularised (has many blood vessels). Muscle cells are called fibres; they are elongated and contain special organelles called myofilaments made of the proteins actin and myosin. These myofilaments allow the muscle tissue to contract.

Functions of muscle

Muscle tissues allow movement. There are three types:

- Skeletal muscles (see Figure 4 (a)), packaged by connective tissue sheets, joined to bones by tendons; these muscles, when they contract, cause bones to move.

- Cardiac muscle (see Figure 4 (b)) makes up the walls of the heart and allows the heart to beat and pump blood.

- Smooth muscle (see Figure 4 (c)) occurs in the walls of intestine, blood vessels, uterus and urinary tracts, and it propels substances along these tracts.

You will learn more about muscle tissue during the second year of your course.

(a)

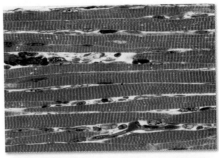

(b)

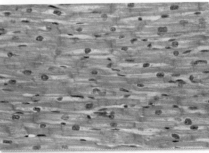

(c)

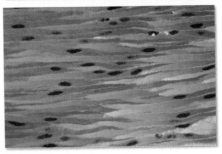

Figure 4 (a) Light micrograph of longitudinal section through skeletal muscle fibres (×300). (b) Light micrograph of human cardiac muscle (×160). (c) Light micrograph of human smooth muscle cells (×157).

Questions

1. Name one part of the human body where you would find (a) elastic cartilage, (b) fibrous cartilage, and (c) hyaline cartilage.

2. State the four main tissue types found in humans.

3. What is the significance of the fact that the walls of the alveoli (see Figure 2) consist of squamous epithelial cells?

4. Distinguish between chondroblasts and chondrocytes.

5. Cartilage tissue is not vascularised (supplied with blood vessels). Explain how the cells in this tissue receive nutrients.

6. (a) What is the function of muscle tissue? (b) Name the three types of mammalian muscle tissue.

By the end of this topic, you should be able to demonstrate and apply your knowledge and understanding of:

* the organisation of cells into tissues, organs and organ systems

* the production of xylem vessels and phloem sieve tubes from meristems

KEY DEFINITIONS

meristem: area of unspecialised cells within a plant that can divide and differentiate into other cell types.
organ: collection of tissues working together to perform a function/ related functions.
phloem: tissue that carries products of photosynthesis, in solution, within plants.
xylem: tissue that carries water and mineral ions from the roots to all parts of the plant.

Plant tissues

You learned in topic 2.6.6 that, in animals, groups of similar cells work together to perform a certain function. Plants also contain tissues, made up of specialised cells, and the main types are described here.

Epidermal tissue

This is equivalent to epithelial tissue in animals. It consists of flattened cells that, apart from the guard cells, lack chloroplasts and form a protective covering over leaves, stems and roots. Some epidermal cells also have walls impregnated with a waxy substance, forming a cuticle. This is particularly important to plants that live in dry places, as the cuticle reduces water loss.

Vascular tissue

Vascular tissue is concerned with transport. There are two sorts – **xylem** (see Figure 1) and **phloem** (see Figure 2), both present in vascular bundles.

* Xylem vessels carry water and minerals from roots to all parts of the plant.

* Phloem sieve tubes transfer the products of photosynthesis (mainly sucrose sugar), in solution, from leaves to parts of the plant that do not photosynthesise, such as roots, flowers and growing shoots.

See Chapter 3.3 for more on plant transport, and details of the structure and function of xylem and phloem.

DID YOU KNOW?

Aphids (greenfly) insert their needle-like mouthparts into the phloem tissue of some plants. Early investigations to analyse the composition of phloem made use of this and decapitated 'plugged-in' aphids, leaving their inserted mouthparts intact, through which phloem sap could be collected.

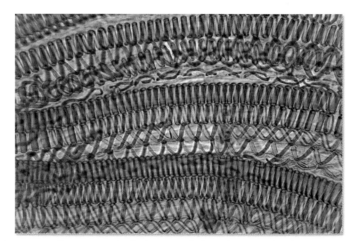

Figure 1 Light micrograph of xylem vessels from the stem of a ribwort plantain, *Plantago lanceolata*. The walls of the cells have lignin deposited in them, for strength, and here they are stained red. You can see that the lignin sometimes occurs in rings, sometimes in coils, and sometimes as spirals (×117).

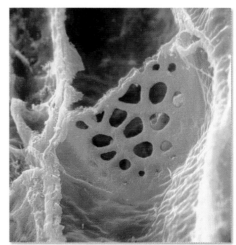

Figure 2 Scanning electron micrograph showing a sieve plate at the end of a phloem sieve tube. Phloem sap flows through the pores in the sieve plate (×4800).

Meristematic tissue

Meristematic tissue contains stem cells. It is from this tissue that all other plant tissues are derived by cell differentiation. It is found at root and shoot tips, and in the cambium of vascular bundles. These areas are called **meristems**.

The cells in meristems:

* have thin walls containing very little cellulose

* do not have chloroplasts

- do not have a large vacuole
- can divide by mitosis and differentiate into other types of cells.

The onion-root tips that you may have examined when studying mitosis (topic 2.6.2) are meristems, hence many cells there were dividing.

How xylem and phloem derive from meristems

As most plant cells mature, they develop a large vacuole and rigid cellulose cell wall. These prevent the cell from dividing. However, plants need to grow and produce new cells. New cells arise at the meristems, by mitosis.

Some cambium cells differentiate into xylem vessels.

- Lignin (a woody substance) is deposited in their cell walls to reinforce and waterproof them; however, this also kills the cells.
- The ends of the cells break down so that the xylem forms continuous columns with wide lumens to carry water and dissolved minerals.

Other cambium cells differentiate into phloem sieve tubes or companion cells.

- Sieve tubes lose most of their organelles, and sieve plates develop between them.
- Companion cells retain their organelles and continue metabolic functions to provide ATP for active loading of sugars into the sieve tubes.

LEARNING TIPS
Plant vascular bundles (see Figure 3) contain xylem tissue, cambium tissue and phloem tissue.
Xylem carries **w**ater and minerals. Remember this by thinking that X and W are very close together in the alphabet.
Phloem carries dissolved products of **ph**otosynthesis – remember this by thinking phloem and **f**ood or **p**hotosynthesis – all have 'ph' sounds.
Cambium cells can divide by mitosis and make new cells that may differentiate into either xylem or phloem.

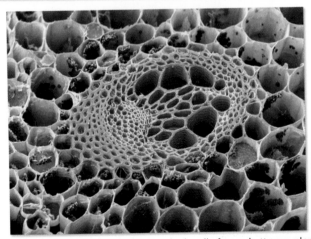

Figure 3 False-coloured SEM of a vascular bundle from a buttercup plant. The vascular bundle is surrounded by parenchyma cells (cream). Outside of the parenchyma are chlorenchyma cells with chloroplasts. Phloem tissue is coloured deep yellow, and the large xylem vessels appear orangey-red (×180).

DID YOU KNOW?
Parenchyma, sclerenchyma and collenchyma
Parenchyma is packing tissue and fills spaces between other tissues. In roots, parenchyma cells may store starch. In leaves, some parenchyma cells (called chlorenchyma) have chloroplasts and can photosynthesise. In aquatic plants, aerenchyma tissue is parenchyma with air spaces to keep the plant buoyant.
Collenchyma cells have thick cellulose walls, and they strengthen the vascular bundles and outer parts of stems, whilst also allowing some flexibility in these regions.
Sclerenchyma cells have lignified walls, and these cells strengthen stems and leaf midribs.

Plant organs

A collection of tissues working together to perform the same function is called an **organ** (see Table 1).

Plant organ	Main functions
Leaf	• Photosynthesis
Root	• Anchorage in soil • Absorption of mineral ions and water • Storage, e.g. carrot, parsnip, dahlia and swede roots store carbohydrates
Stem	• Support • Holds leaves up so that they are exposed to more sunlight • Transportation of water and minerals • Transportation of products of photosynthesis • Storage of products of photosynthesis, e.g. potato tubers store starch; rhubarb stems store sugars and polysaccharides
Flower	• Sexual reproduction

Table 1 Examples of plant organs.

DID YOU KNOW?
Flowers could be regarded as an organ system, as they include different organs within them – sepals, petals, ovaries and stamens.

Questions

1. List the organs and tissues of a plant that are involved in photosynthesis, and explain how each one is important. Think about the needs of the plant – how do the water and carbon dioxide enter the plant and arrive at the site of photosynthesis? Where does photosynthesis take place, and how are the products of photosynthesis transported to other parts of the plant?

2. Plant epidermal tissue can be considered to be equivalent to epithelial tissues in animals. Which animal tissue do you think parenchyma plant tissue is equivalent to?

3. Suggest one way in which the function of plant vascular tissue differs from the function of animal vascular tissue.

4. State the functions of (a) roots, and (b) flowers.

5. Which of the following have lignin in their cell walls? Phloem sieve tubes, xylem vessels, collenchyma cells, parenchyma cells, epidermal cells, sclerenchyma cells.

8 Organs and organ systems in animals

By the end of this topic, you should be able to demonstrate how to apply your knowledge and understanding of:

* the organisation of cells into tissues, organs and organ systems

Animal organs

In the previous topic about plants, you learned that a collection of tissues working together to perform the same function is called an **organ**.

Examples of animal organs include the heart, kidney, liver, brain, optic nerve, biceps muscle, lungs and the eye. You will learn more about the heart in Chapter 3.2 on transport in animals, and more about the lungs in Chapter 3.1 on gaseous exchange. In the second year of your course, you will learn about the structure and functions of the kidneys.

KEY DEFINITION

organ system: a number of organs working together to carry out an overall life function.

Organ systems

A number of organs working together to carry out an overall life function is called an **organ system**.

System	Organs and tissues involved	Examples of life processes carried out
digestive system	oesophagus; stomach, intestines plus associated glands; the liver and pancreas	nutrition to provide ATP and materials for growth and repair
circulatory system	heart and blood vessels	transport to and from cells
respiratory system	airways and lungs, plus diaphragm and intercostal muscles	breathing and gaseous exchange excretion
urinary system	kidneys, ureters and bladder	excretion and osmoregulation
integumentary system	skin, hair and nails	waterproofing, protection, temperature regulation
musculo-skeletal system	skeleton and skeletal muscles	support, protection and movement
immune system	bone marrow, thymus gland, skin, stomach acid, blood	protection against pathogens
nervous system	brain, spinal cord and nerves	communication, control and coordination
endocrine system	glands that make hormones, e.g. thyroid, ovaries, testes, adrenals	communication, control and coordination
reproductive system	testes, penis, ovaries, uterus, vagina	reproduction
lymph system	lymph nodes and vessels	transports fluid back to the circulatory system and is also important in resisting infections

Table 1 Organ systems in your body.

LEARNING TIP

The skin is an organ – in fact it is your largest organ. Within it are other smaller organs, such as hair follicles.

Questions

1 List the following in order of size, beginning with the smallest:

organ   organ system   cell   tissue   organelle

2 Describe the roles of your organ systems that are involved when you cook and eat a meal.

3 Describe the roles of your organ systems that are involved when you take part in a named sporting activity.

⑨ Stem cells and their potential uses

By the end of this topic, you should be able to demonstrate and apply your knowledge and understanding of:

* the features and differentiation of stem cells

* the potential uses of stem cells in research and medicine

> **KEY DEFINITIONS**
>
> **stem cell:** unspecialised cell able to express all of its genes and divide by mitosis.

Stem cells:

- are undifferentiated cells, capable of becoming any type of cell in the organism.

- are described as **pluripotent**.

- are able to express all their genes.

- can divide by mitosis and provide more cells that can then differentiate into specialised cells, for growth and tissue repair.

These characteristics of stem cells have made them potentially very important in research for medical use.

Sources of stem cells

There are different types of stem cells that can be obtained from different sources:

- embryonic stem cells – these are present in an early embryo formed when the zygote begins to divide (see topic 2.6.4 on differentiation)

- stem cells in umbilical-cord blood

- adult stem cells (also found in infants and children) are found in developed tissues, such as blood, brain, muscle, bone, adipose (fat storage) tissue and skin, amongst the differentiated cells; they act like a repair system because they are a renewing source of undifferentiated cells

- induced pluripotent stem cells (iPS cells) developed in laboratories by reprogramming differentiated cells to switch on certain key genes and become undifferentiated.

Potential uses in research and medicine

Bone-marrow transplants

Stem cells from bone marrow are already extensively used in bone-marrow transplants to treat diseases of the blood (such as sickle-cell anaemia and leukaemia) and immune system (e.g. severe combined immunodeficiency or SCID). They are also used to restore the patient's blood system after treatment for specific types of cancer, where the patient's bone-marrow cells can be obtained before treatment, stored, and then put back inside the patient after treatment.

Drug research

If stem cells can be made to develop into particular types of human tissue, then new drugs can be tested first on these tissues, rather than on animal tissue.

Developmental biology

Scientists can make use of stem cells, in many ways, to research developmental biology and enable a better understanding of how multicellular organisms develop, grow and mature.

- They can study how these cells develop to make particular cell types (e.g. blood, bone, muscle and skin) and can learn how each cell type functions and see what goes wrong when they are diseased.

- They are trying to find out if they can extend the capacity that embryos have for growth and tissue repair, into later life.

Repair of damaged tissues or replacement of lost tissues

It is quite difficult to culture stem cells in a lab, so research into this is ongoing. Also, it is necessary to find out which cytokine cell-signalling molecules are needed to direct the differentiation of stem cells into particular cell types.

- Stem cells have been used to treat mice with type 1 diabetes by programming iPS cells to become pancreatic beta cells. Research is under way to develop such treatment for type 1 diabetes in humans.

- Bone-marrow stem cells can be made to develop into liver cells (hepatocytes) and could be used to treat liver disease.

- Stem cells directed to become nerve tissue could be used to treat Alzheimer and Parkinson diseases or to repair spinal-cord injuries.

- Stem cells may be used to populate a bioscaffold of an organ, and then directed to develop and grow into specific organs for transplanting. This is called **regenerative medicine**. If the patient's cells are obtained, reprogrammed to become iPS cells, and then used to make such an organ, there will be no need for immunosuppressant drugs.

- Stem cells may eventually be used to treat many conditions, including arthritis, strokes, burns, vision and hearing loss, Duchenne muscular dystrophy and heart disease.

Pluripotent literally means able to make many cell types – in the case of iPS cells very very many, but there is also a special word for a cell which can make not only all the cells of the body, but placental cells too: totipotent.

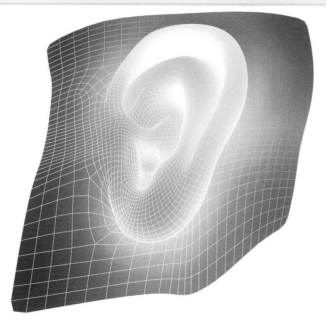

Figure 1 Computer artwork, showing the topography of a human ear. A bioscaffold of collagen could be populated with a patient's stem cells and grown into a new ear to be transplanted. This is an example of regenerative medicine or tissue engineering.

In 2013 a two-year-old girl, who was born without a trachea, received an artificial trachea grown from stem cells from her bone marrow. These were removed using a needle, and seeded, in a lab, on to a collagen scaffold, where they multiplied and made a new windpipe. The doctors think that, after the trachea was implanted, the stem cells directed the body of the recipient to send other cells to the trachea, which then spread out and sent the correct tissues to grow on the inside and outside of the tube. The recipient does not need to take immunosuppressant drugs, as the new trachea consists of her own cells, recognised as 'self' by her immune system and therefore not rejected. The scaffolds for organs made in this way can also be made of a type of plastic, using 3D printing.

Questions

1. Describe three characteristics of stem cells.

2. Discuss the problems that arose with using embryonic stem cells for research.

3. Suggest why using human tissue derived from iPS cells for testing new drugs is better than using animal tissue.

4. iPS cells may be used to treat vision loss caused by loss or damage to rods and cones in the retina. One form of blindness is caused by mutation in mitochondrial genes. Suggest why using patient-derived iPS cells may not be useful for treating mitochondrial-disease blindness.

5. Suggest why a patient given an organ transplant, where the organ is produced by populating a bioscaffold with iPS cells derived from the patient, would not need immunosuppressant drugs.

THE EASTLACK SKELETON

A biological exhibit in the Mütter Museum in Philadelphia, USA, raises interesting questions about cell differentiation and the origin of bone and muscle tissues. This information comes from an article in The Society of Biology magazine: *The Biologist*.

THE EASTLACK SKELETON

Harry Eastlack Jr suffered from a rare and poorly understood disease, fibrodysplasia ossificans progressiva (FOP). This condition causes damaged muscle, tendons and ligaments to grow back as bone, hence it is sometimes called 'stone man syndrome'. As plates and bridges of bone grow throughout the body, joints are permanently locked in place.

Harry Eastlack was born in 1933, and at the age of five he broke his leg. Uncontrolled bone growth followed as his leg healed, and this growth continued throughout his life, spreading to other parts of his body. He endured several operations to remove masses of **heterotropic** bone that had formed in his muscles. However, the surgical interventions exacerbated the condition. By 1946, his body was permanently angled, his spine was fused, and the muscles in his upper back were **ossified**. By the time that he died in 1973, he could only move his lips.

Eastlack gave permission for his skeleton to be preserved for scientific research, and it was donated to the Mütter Museum in Philadelphia, USA. The skeleton has become a valuable asset to scientists studying this extremely rare disease, which is believed to affect 1 in every 2 million people.

The condition usually arises from a spontaneous mutation, in a **gamete**, in a gene that encodes a key protein involved in the growth and development of bones and muscles. The gene is called *ACVR1*, and the mutation is a change to one base triplet, so that at position 206 the amino acid histidine is substituted for arginine. Bone, cartilage and ligaments are all connective tissue; they and muscle tissue both derive from the **mesoderm** layer of the embryo. Normally, the gene *ACVR1* is switched off in the fetus once its bones have formed. In FOP, the gene keeps working. This leads to an overgrowth of bone and the fusion of joints. Sufferers are imprisoned by their own skeletons. The excessive bone growth starts at the neck and works downwards as ossification does in the developing fetus. The only muscles in the body that do not ossify are the tongue, the diaphragm, the heart, the muscles that move the eyes, and smooth muscle.

Source

- Anon. Museum piece – biological exhibits from around the world. *The Biologist*, Vol. 61, no. 4, Aug/Sept, p. 46.

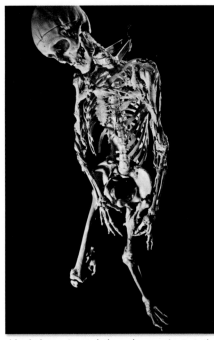

Figure 1 Eastlack's skeleton. It needed no glue or wire to articulate it, as it was almost fused into one piece at the time of his death.

KEY DEFINITION

gamete: sex cell, e.g. ovum/spermatozoon.
heterotropic ossification: overgrowth of bone, often in the wrong place, e.g. muscle tissue.
mesenchyme: connective tissue.
mesoderm: the middle of the three layers in the early embryo; gives rise to connective tissue, muscles and part of the gonads (ovaries and testes).
ossification: process of changing cartilage to bone by depositing calcium phosphate.

Where else will I encounter these themes?

1.1   2.1   2.2   2.3   2.4   2.5

Let's think about the style of writing.

1. What features of this article indicate that it is aimed at readers with a knowledge of biology?

2. The key definitions were not included in the original article; they have been added here to help you. Does this glossary help you to understand the article? Did you have to look up the meanings of any other words?

Here are some questions relating to the content of the article:

3. During what process within a gamete is the spontaneous mutation of the *ACVR1* gene likely to happen?

4. Most people who suffer from FOP are unable to bear children. This genetic inherited disease is caused by a dominant allele, as in Huntington disease. Why do you think the incidence of FOP is much rarer than the incidence of Huntington disease?

5. Why do you think FOP is called a disease of the connective tissue?

6. Explain the difference between 'tissues' and 'cells'.

7. Because this disease is rare, cases may be misdiagnosed by doctors who think the lump of extra bone is a tumour. How could a biopsy, to find out if there is a tumour, make things worse?

8. What is the name given to the process of the mesenchymal stem cells becoming bone cells?

9. Why do you think damaged muscle tissue becomes bone tissue, rather than skin tissue?

10. What effect do you think the extra bone around the rib cage has on the patient's breathing?

11. How do you think interference RNA could be developed into a treatment for FOP?

DID YOU KNOW?

Since 1800 there have been about 2500 recorded cases of people turning to stone, 700 of which have been confirmed as being due to FOP.

Surgery to remove the extra bone does more harm than good as, in the damaged tissue, cells transform to **mesenchymal** stem cells, which then become bone cells.

Activity

Use the information in this article and from any other research you may carry out using the Internet, to produce a poster that could be used in a History of Medicine Museum to inform members of the public about this condition. Be aware of the level of biological knowledge of your target audience, and simplify the language accordingly but keep the information accurate.

Practice questions

1. Which row correctly shows events that occur in the main stages of the eukaryotic cell cycle? [1]

	M	G$_1$	S	G$_2$
A	division	transcription	DNA replication	growth
B	division	growth	protein synthesis	organelles replicate
C	division	differentiation	DNA replication	apoptosis
D	division	senescence	DNA replication	growth

2. Which row shows the correct sequence of stages during mitosis? [1]

	Stage 1	Stage 2	Stage 3	Stage 4
A	interphase	prophase	metaphase	anaphase
B	prophase	anaphase	metaphase	cytokinesis
C	prophase	metaphase	anaphase	telophase
D	anaphase	metaphase	telophase	prophase

3. At which stage of meiosis do the chromatids of each chromosome separate? [1]

A. anaphase 1

B. metaphase 1

C. metaphase 2

D. anaphase 2

4. Read the following statements.
   (i) Some epithelial tissues have ciliated cells.
   (ii) Some epithelial cells are also glands.
   (iii) Blood, bone and cartilage are all connective tissue.

   Which of these statements is/are true? [1]

A. (i), (ii) and (iii)

B. Only (i) and (ii)

C. Only (ii) and (iii)

D. Only (i)

5. Which of the following is **not** a feature of plant meristematic cells? [1]

A. They have thick cellulose cell walls.

B. They do not have chloroplasts.

C. They do not have a large vacuole.

D. They can divide by mitosis.

[Total: 5]

In questions 6–9, you may need to use your knowledge of other topics, such as biological molecules, nucleic acids, enzymes, cell structure and transport in plants.

Bear in mind that when you sit your exam at the end of your course, you will have covered and learned about all the topics in the specification and should be able to make links between them.

6. S-cyclin proteins are thought to activate cyclin dependent kinase (CDK) enzymes that then promote DNA replication. In an investigation, soya bean tissue cultures were grown on media containing different concentrations of S-cyclin.

   The DNA content was measured after six days.

   The results are shown in the table below.

Concentration of S-cyclin in medium (mol dm^{-3})	DNA content of tissue culture (mg per million cells)
0	0.75
10^{-6}	1.20
10^{-4}	2.15
10^{-2}	4.03

(a) Do the data in the table support the hypothesis that S-cyclin stimulates DNA replication? Explain your answer. [2]

(b) The activation of CDK also requires phosphorylation (addition of a phosphate group). Suggest which type of molecule in cells undergoes hydrolysis to provide the phosphate. [1]

(c) The cyclin-CDK complex activates enzymes needed for DNA replication. Name two enzymes involved in catalysing DNA replication. [2]

(d) Scientists setting up plant tissue cultures may use root or shoot tips as a source of cells. The small pieces of tissue are obtained, using aseptic technique, and grown on sterile agar that contains other growth factors.
   (i) Suggest why tissue for culturing in this investigation is obtained from root and shoot tips. [2]
   (ii) Suggest why aseptic technique and sterile agar are used. [2]

(e) During which phase of the cell cycle do chloroplasts divide? [1]

(f) By what process do chloroplasts divide? [1]

[Total: 11]

7. Figure 1 is a diagram representing the quantity of DNA present in an animal cell during the different phases of the cell cycle.

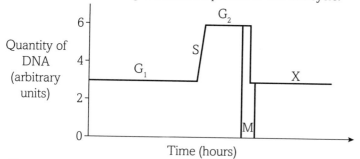

Figure 1

(a) Describe how the quantity of DNA is increased during the cell cycle, before the cell divides. You may use annotated diagrams if you wish. [6]

(b) (i) What is the quantity of DNA, in arbitrary units, in a cell after cytokinesis? [1]

(ii) Describe how the quantity of DNA is returned to this level. [4]

(iii) What type of division is this cell undergoing? Give a reason for your answer. [2]

(c) Name the stage of the cell cycle indicated by the letter X on the diagram. [1]

(d) Name one metabolic process that cannot take place during mitosis. [1]

[Total: 15]

8. (a) The table below refers to some of the events during mitosis, the first division of meiosis and the second division of meiosis.

Statement	Mitosis	Meiosis 1	Meiosis 2
Chromosomes pair up			
Chromosomes, consisting of two chromatids, attach to the equator of the spindle			
Independent assortment of chromosomes occurs			
Independent assortment of chromatids occurs			

If a statement is true, place a tick in the appropriate box. If a statement is false, place a cross in the appropriate box. [4]

(b) (i) When during the cell cycle is the spindle made? [1]

(ii) Name the protein present in the spindle threads. [1]

(c) Describe the products of:

(i) mitosis [2]

(ii) meiosis. [2]

(d) Explain how meiosis causes genetic variation. [4]

(e) Honey bees are social insects. A newly hatched queen takes a nuptial flight and mates with up to 20 drones (males), some from her hive and many from nearby hives. She stores the sperm and uses it throughout her life. Some of the eggs she lays are fertilised and these diploid individuals develop into females, the worker bees. Some eggs are unfertilised, and these develop into drones, which are all haploid.

(i) Which type of cell division is used in the testes of drones to produce sperm? Give reasons for your answer. [3]

(ii) Suggest why it is advantageous for a queen to mate with several drones, many of which are from other hives. [1]

[Total: 18]

9. In each human skin cell there are 46 chromosomes.

(a) How many chromosomes are there in each of the following human cells?

(i) liver cell [1]

(ii) brain cell [1]

(iii) red blood cell in circulation [1]

(iv) a spermatozoon [1]

(b) Describe the significance of mitosis in the life cycle of eukaryotic organisms. [4]

(c) Describe the process of mitosis in a bone marrow stem cell. [8]

[Total: 16]

10. Describe how the structure of each of the following tissues allows them to carry out their functions.

You may use annotated diagrams.

(a) xylem [5]

(b) phloem [4]

(c) cartilage. [6]

[Total: 15]

Exchange and transport

EXCHANGE SURFACES AND BREATHING

Introduction

All living things need to take up substances from their environment and remove wastes.

In order to respire and grow, living things need to take up:
- oxygen
- water
- nutrients such as mineral ions, vitamins, sugars, fats and amino acids
- carbon dioxide (in plants)

Living things also need to remove by-products and wastes such as:
- carbon dioxide
- urea
- oxygen (in plants)

In this module you will learn about these exchange surfaces and how substances are moved across them. The gas exchange surface in the lungs is used to exemplify the properties and functions of exchange surfaces in living things. You will also learn about the importance of ventilating gas exchange surfaces to ensure that exchange remains effective.

All the maths you need

To unlock the puzzles in this section you need the following maths:
- The significance of surface area to volume ratios
- Interpreting graphs
- Calculating the gradient of a line

What have I studied before?

- The role of the plasma membrane as a selectively permeable barrier
- How certain substances can diffuse across the plasma membrane
- The role of a concentration gradient that causes diffusion

What will I study later?

- Transport of substances around plants (AS)
- Transport of substances around animals (AS)

What will I study in this chapter?

- The properties of a good exchange surface
- Exchange surfaces in plants, insects, fish and mammals
- The cells and tissues involved in an exchange surface

Exchange surfaces

By the end of this topic, you should be able to demonstrate and apply your knowledge and understanding of:

* the need for specialised exchange surfaces

* the features of an efficient exchange surface

Size matters

All living cells need a supply of oxygen and nutrients to survive. They also need to remove waste products so that these do not build up and become toxic. In very small organisms, this exchange can take place over the surface of the body. They do not need a specialised exchange system. However, in larger organisms with more than two layers of cells, the body surface is no longer sufficient. Therefore larger organisms need a specialised surface for exchange of substances with their environment.

There are three main factors that affect the need for an exchange system:

* size
* surface area to volume ratio
* level of activity.

Size

In very small organisms, such as single-celled organisms, all the cytoplasm is very close to the environment in which they live. Diffusion will supply enough oxygen and nutrients to keep the cells alive and active. However, multicellular organisms may have several layers of cells. Here, any oxygen or nutrients diffusing in from the outside have a longer diffusion pathway. Diffusion is too slow to enable a sufficient supply to the innermost cells.

Surface area to volume ratio

Small organisms have a small surface area, but they also have a small volume. Their surface area is relatively large compared with their volume. We say that they have a large surface area to volume ratio. This means that their surface area is large enough to supply all their cells with sufficient oxygen (see Table 1).

Larger organisms have a larger surface area, but they also have a larger volume. As size increases, the volume rises more quickly than the surface area. Therefore, their surface area is relatively small compared with their volume. We say that they have a small surface area to volume ratio.

Some organisms increase their surface area by adopting a different shape. An animal such as a flatworm has a very thin, flat body. This gives it a larger surface area to volume ratio ($SA:V$). But such a body form limits the overall size that the animal can reach. Most large organisms need a range of tissues to give the body support and strength. Their volume increases as their body gets thicker, but the surface area does not increase as much. Therefore, the surface area to volume ratio of a large organism is relatively small.

Radius (r) (mm)	Surface area (mm^2) ($SA = 4\pi r^2$)	Volume (mm^3) ($V = 4/3\pi r^3$)	$SA:V$
1	12.568	4.189	3.00 : 1.00
2	50.272	33.515	1.50 : 1.00
5	314.200	523.667	0.60 : 1.00

Table 1 Effect of increasing radius (r) on surface area to volume ratio of an organism with a spherical shape.

Level of activity

Some organisms are more active than others. Metabolic activity uses energy from food and requires oxygen to release the energy in aerobic respiration. The cells of an active organism need good supplies of nutrients and oxygen to supply the energy for movement. This need for energy is increased in those animals, such as mammals, that keep themselves warm.

The features of a good exchange surface

All good exchange surfaces have certain features in common:

- A large surface area to provide more space for molecules to pass through. This is often achieved by folding the walls and membranes involved. A good example is the root hairs in plants (see Figure 1).

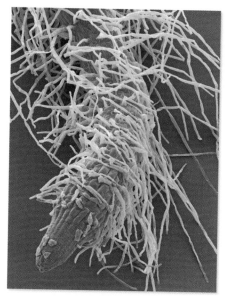

Figure 1 Root hairs at the tip of a root increase the surface area (×66).

- A thin barrier to reduce the diffusion distance – and that barrier must be permeable to the substances being exchanged. This is shown well in the alveoli of the lungs (see Figure 2). You will learn more about the lungs later in this chapter, in topic 3.1.2.

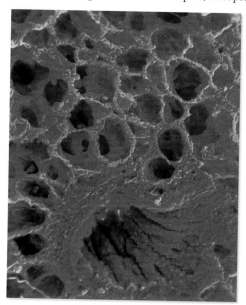

Figure 2 The walls of alveoli are one cell thick. The barrier between the blood and the air is just 0.3 micrometres (×150).

- A good blood supply. This can bring fresh supplies of molecules to one side (supply side), keeping the concentration high, or it may remove molecules from the demand side to keep the concentration low. This is important to maintain a steep concentration gradient so that diffusion can occur rapidly. The gills in fish are a good example (Figure 3).

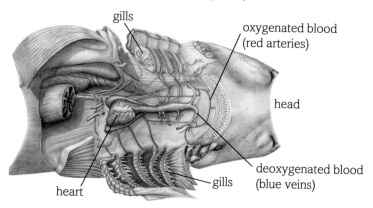

Figure 3 The heart and gill anatomy of a shark. The heart pumps blood forward to the gills. Each gill has a good blood supply to absorb oxygen from the water.

Questions

1. Calculate the surface area to volume ratio of a sphere with a radius of 10 mm.

2. Calculate the surface area to volume ratio of a human, given that the typical skin surface area is 1.8 m² and the volume is 0.07 m³

3. Calculate the ratio of lung surface area to body volume of a human using lungs with a surface area of 70 m².

4. Comment on the need for lungs in a human.

5. Explain why plants need a large surface area for absorption in their roots.

6. Explain the importance of maintaining a steep concentration gradient for exchange.

7. Why do active organisms need a specialised exchange surface?

8. Explain why an elephant has a smaller surface area to volume ratio than a single-celled organism.

(2) Mammalian gaseous exchange system

By the end of this topic, you should be able to demonstrate and apply your knowledge and understanding of:

* the structures and functions of the components of the mammalian gaseous exchange system
* the mechanism of ventilation in mammals

The gaseous exchange system in mammals consists of the lungs and associated airways that carry air into and out of the lungs (see Figure 1). The lungs are a pair of inflatable sacs lying in the chest cavity. Air can pass into the lungs through the nose and along the **trachea** (windpipe), **bronchi** and **bronchioles**. Finally, it reaches tiny air-filled sacs called **alveoli**. These are the surfaces where the exchange of gases takes place.

The lungs are protected by the ribcage. The ribs are held together by the **intercostal muscles**. The action of these muscles and the **diaphragm** (a layer of muscular tissue beneath the lungs) helps to produce breathing movements (**ventilation**).

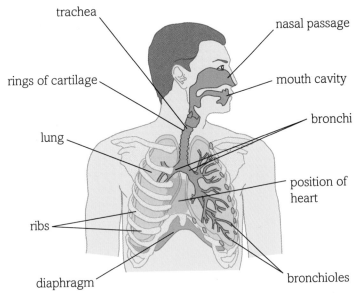

Figure 1 The lungs and associated structures.

trachea
nasal passage
rings of cartilage
mouth cavity
bronchi
lung
position of heart
ribs
diaphragm
bronchioles

Gaseous exchange in the lungs

Gases pass by diffusion through the thin walls of the alveoli. Oxygen passes from the air in the alveoli to the blood in the capillaries. Carbon dioxide passes from the blood to the air in the alveoli. The lungs must maintain a steep concentration gradient in each direction in order to ensure that diffusion can continue.

In the last topic, we learned about the features of a good exchange surface. We will consider these in relation to the lungs.

Large surface area to provide more space for molecules to pass through

The individual alveoli are very small – about 100–300 μm across. However, they are so numerous that the total surface area of the lungs is much larger than that of our skin. It has been calculated that the total surface area of the exchange surface in humans is about 70 m^2, or about half the size of a tennis court.

The alveoli are lined by a thin layer of moisture, which evaporates and is lost as we breathe out. The lungs must produce a surfactant that coats the internal surface of the alveoli to reduce the cohesive forces between the water molecules, as these forces tend to make the alveoli collapse.

The barrier to exchange is permeable to oxygen and carbon dioxide

The barrier to exchange is comprised of the wall of the alveolus and the wall of the blood capillary. The cells and their plasma (cell-surface) membranes readily allow the diffusion of oxygen and carbon dioxide, as the molecules are small and non-polar.

Thin barrier to reduce the diffusion distance

There are a number of adaptations to reduce the distance the gases have to diffuse:

* The alveolus wall is one cell thick.
* The capillary wall is one cell thick.
* Both walls consist of squamous cells – this means flattened or very thin.
* The capillaries are in close contact with the alveolus walls.
* The capillaries are so narrow that the red blood cells are squeezed against the capillary wall – making them closer to the air in the alveoli and reducing their rate of flow.

So, the total barrier to diffusion is only two flattened cells, and is less than 1 μm thick.

A good blood supply

The blood supply helps to maintain a steep concentration gradient, so that the gases continue to diffuse.

- The blood system transports carbon dioxide from the tissues to the lungs. This ensures that the concentration of carbon dioxide in the blood is higher than that in the air of the alveoli. Therefore carbon dioxide diffuses into the alveoli (see Figure 2).

- The blood also transports oxygen away from the lungs. This ensures that the concentration of oxygen in the blood is kept lower than that in the alveoli – so that oxygen diffuses into the blood.

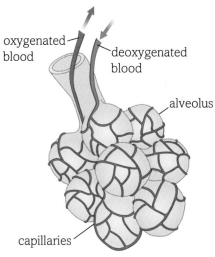

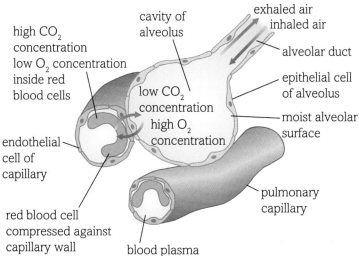

Figure 2 Capillary network over the alveoli and gaseous exchange.

Ventilation

The breathing movements ventilate the lungs. This replaces the used air with fresh air, bringing in more oxygen and removing carbon dioxide. Ventilation (see Table 1) ensures that:

- the concentration of oxygen in the air of the alveolus remains higher than that in the blood

- the concentration of carbon dioxide in the alveoli remains lower than that in the blood.

Therefore, the concentration gradient necessary for diffusion is maintained.

Inspiration (inhaling)	Expiration (exhaling)
the diaphragm contracts to move down and become flatter – this displaces the digestive organs downwards	the diaphragm relaxes and is pushed up by the displaced organs underneath
the external intercostal muscles contract to raise the ribs	the external intercostal muscles relax and the ribs fall; the internal intercostal muscles can contract to help push air out more forcefully – this usually only happens during exercise or coughing and sneezing
the volume of the chest cavity is increased	the volume of the chest cavity is decreased
the pressure in the chest cavity drops below the atmospheric pressure	the pressure in the lungs increases and rises above the pressure in the surrounding atmosphere
air is moved into the lungs	air is moved out of the lungs

Table 1 Ventilation in mammals.

Questions

1. Explain why it is essential to have a steep oxygen concentration gradient across the alveolus wall.

2. Describe how the steep gradient in the oxygen concentration across the alveolus wall is maintained.

3. Describe how the steep gradient in carbon dioxide concentration across the alveolus wall is maintained.

4. The surface area to volume ratio of a human is 25.7 : 1.0. Using the lungs as the surface area, we have a surface area to volume ratio of 1000 : 1.0. Calculate the percentage increase in the $SA : V$ ratio provided by the lungs.

5. Describe the role of the blood in gaseous exchange.

6. When exercising, the volume of air breathed in and out of the lungs changes. Suggest what physiological changes occur, and how these enable exercise to occur.

7. Describe the changes in the mechanism of ventilation that enable more air to enter and leave the lungs than normal.

8. When we breathe deeply, the alveoli stretch. How does this benefit gaseous exchange?

③ Tissues in the gaseous exchange system

By the end of this topic, you should be able to demonstrate and apply your knowledge and understanding of:

* the structures and functions of the components of the mammalian gaseous exchange system

* the examination of microscope slides to show the histology of exchange surfaces

Lung tissue

You learned in the previous topic that the lungs consist of large numbers of tiny air-filled sacs called **alveoli**. These are comprised of squamous epithelium (see Figure 1) and are surrounded by blood capillaries, so that the distance that gases must diffuse is very short. The alveolus walls contain elastic fibres that stretch during inspiration but then **recoil** to help push air out during expiration. The alveolus walls are so thin that it may not be possible to distinguish separate cells under a light microscope.

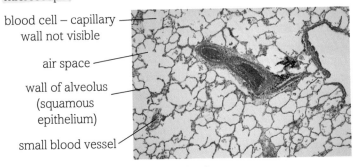

blood cell – capillary wall not visible

air space

wall of alveolus (squamous epithelium)

small blood vessel

Figure 1 Squamous epithelium in the alveoli.

The airways

You have learnt that the **trachea**, **bronchi** and **bronchioles** allow the passage of air into the lungs and out again. To be effective, these airways must meet certain requirements:

* be large enough to allow sufficient air to flow without obstruction

* be supported to prevent collapse when the air pressure inside is low during inspiration

* be flexible in order to allow movement.

The airways are lined by **ciliated epithelium**, which contributes to keeping the lungs healthy. Goblet cells in the epithelium release mucus, which traps pathogens. The cilia then move the mucus up to the top of the airway, where it is swallowed. The glandular tissue in the loose tissue also produces mucus.

The trachea and bronchi

The trachea and bronchus walls have a similar structure (see Figure 2). However, the bronchi are narower than the trachea.

These airways are supported by rings of **cartilage** which prevent collapse during inspiration. The rings of cartilage in the trachea are C-shaped rather than a complete ring which allows flexibility and space for food to pass down the oesophagus.

The bronchioles

The bronchioles are much narrower than the bronchi (see Figure 3). The larger bronchioles may have some cartilage, but smaller ones have no cartilage. The wall is comprised mostly of **smooth muscle** and elastic fibres. The smallest bronchioles end in clusters of alveoli.

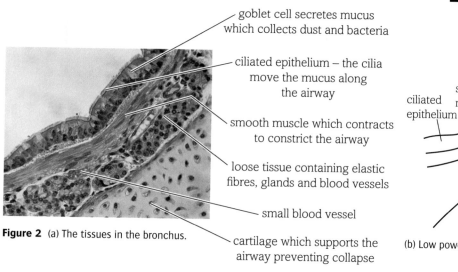

goblet cell secretes mucus which collects dust and bacteria

ciliated epithelium – the cilia move the mucus along the airway

smooth muscle which contracts to constrict the airway

loose tissue containing elastic fibres, glands and blood vessels

small blood vessel

cartilage which supports the airway preventing collapse

Figure 2 (a) The tissues in the bronchus.

ciliated epithelium

smooth muscle

loose tissue

cartilage

(b) Low power tissue plan (×40).

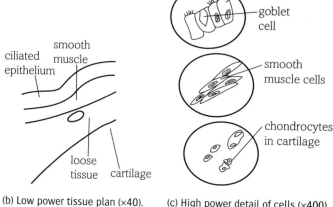

ciliated cell

goblet cell

smooth muscle cells

chondrocytes in cartilage

(c) High power detail of cells (×400).

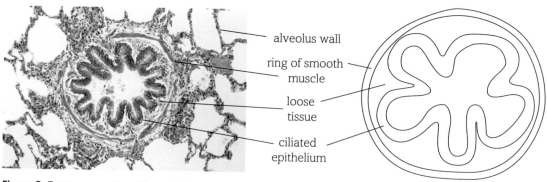

alveolus wall

ring of smooth muscle

loose tissue

ciliated epithelium

Figure 3 Transverse section of bronchiole (×68).

Smooth muscle and elastic tissue

The smooth muscle can contract. The action of the smooth muscle will constrict the airway. This makes the lumen of the airway narrower. Constriction of the lumen can restrict the flow of air to and from the alveoli. Controlling the flow of air to the alveoli might be important if there are harmful substances in the air. The contraction of the smooth muscle and control of airflow is not a voluntary act and may occur as a result of an allergic reaction. Once the smooth muscle has contracted, it cannot reverse this effect on its own. The smooth muscle is elongated again by the elastic fibres. When the muscle contracts, it deforms the elastic fibres. As the muscles relax, the elastic fibres recoil to their original size and shape. This acts to dilate the airway.

Questions

1. What is the meaning of the term 'tissue'?

2. Describe the differences between squamous epithelium and ciliated epithelium.

3. Construct a table listing all the tissues found in the gaseous exchange system in the left-hand column and their functions in the right-hand column.

4. The cartilage supports the airways holding them open. Explain what may cause the airways to collapse without this support.

5. The cilia and mucus act together to keep the lungs free from infection. Describe how they work together.

6. Explain why smoking (which paralyses the cilia) leads to increased infections.

7. Lung tissue contains many small blood vessels. Explain why these blood vessels are important.

8. Use the tissues in the lungs as examples to explain how tissues work together to produce a functioning organ.

INVESTIGATION

Drawing slides of lung tissue to show tissue histology

When viewing slides of tissue it is important to follow certain guidelines.

Always wear eye protection when using microscopes.

View at low power (×40) or medium power (×100) to select the best part of the slide.

Draw a tissue plan – this will contain no individual cells.

Select each tissue in turn and view at high power (×400).

Draw a few cells, showing as much detail of cell shape and cell contents as possible.

Each of the photomicrographs shown in this topic has been interpreted with both a tissue plan and a few cells drawn at high power.

DID YOU KNOW?

Some people overreact to certain substances in the air, and their bronchioles constrict unnecessarily. This is the cause of asthma.

LEARNING TIP

Remember that elastic fibres recoil after deformation – they do not contract. The elastic fibres in the airways recoil to dilate the airways. The elastic fibres in the alveoli recoil to reduce the size of the alveoli during expiration.

(4) Measuring lung volumes

By the end of this topic, you should be able to demonstrate and apply your knowledge and understanding of:

* the relationship between vital capacity, tidal volume, breathing rate and oxygen uptake

Using a spirometer

Lung volumes can be measured using a **spirometer** (see Figure 1). This is a device that measures the movement of air in and out of the lungs as a person breathes. A float-chamber spirometer consists of a chamber of air or medical-grade oxygen floating on a tank of water. During inspiration, air is drawn from the chamber so that the lid moves down. During expiration, the air returns to the chamber, raising the lid. These movements may be recorded on a datalogger.

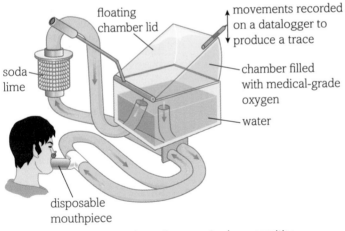

Figure 1 The spirometer – a device for measuring lung capacities.

The carbon dioxide-rich air exhaled is passed through a chamber of soda lime, which absorbs the carbon dioxide. This allows the measurement of oxygen consumption.

Precautions that must be taken when using a spirometer

* The subject should be healthy and, in particular, free from asthma.
* The soda lime should be fresh and functioning.
* There should be no air leaks in the apparatus, as this would give invalid or inaccurate results.
* The mouthpiece should be sterilised.
* The water chamber must not be overfilled (or water may enter the air tubes).

Modern spirometers may be small and simple hand-held devices. These record the movements of air in and out of the lungs. However, many cannot measure the rate of oxygen consumption.

Lung volumes

The total lung volume consists of the *vital capacity*, which can be measured, and the *residual volume*, which cannot be measured using the spirometer (see Figure 2).

* **Vital capacity** is the maximum volume of air that can be moved by the lungs in one breath. This is measured by taking a deep breath and expiring all the air possible from the lungs. Vital capacity depends upon a number of factors such as:

  o the size of a person (particularly their height)

  o their age and gender

  o their level of regular exercise.

Vital capacity is usually in the region of 2.5–5.0 dm^3, but may rise above this in trained athletes.

* The **residual volume** is the volume of air that remains in the lungs even after forced expiration. This air remains in the airways and alveoli. This is approximately 1.5 dm^3.

* **Tidal volume** is the volume of air moved in and out with each breath. It is normally measured at rest. A typical tidal volume at rest might be 0.5 dm^3. This is usually sufficient to supply all the oxygen required in the body at rest.

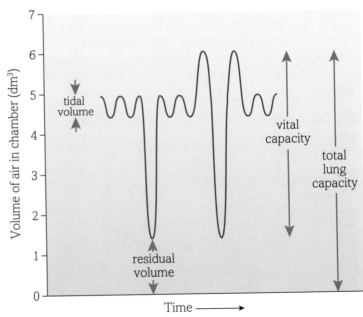

Figure 2 A typical trace from a spirometer, showing the volumes that can be measured.

Oxygen uptake

Breathing supplies oxygen for respiration and removes carbon dioxide produced in respiration. As a person breathes from the spirometer, oxygen is absorbed by the blood and replaced by carbon dioxide. This carbon dioxide is absorbed by the soda lime in the spirometer, so that the volume of air in the chamber decreases. This decrease can be observed and measured on the spirometer trace. We can assume that the volume of carbon dioxide released and absorbed by the soda lime equals the volume of oxygen absorbed by the blood. Therefore, measuring the gradient of the decrease in volume enables us to calculate the rate of oxygen uptake.

Calculating oxygen uptake from a spirometer trace

(a)

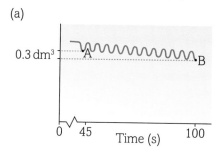

(b)

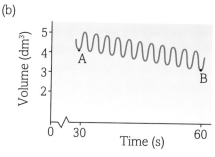

Figure 3 How to measure the oxygen uptake. Graph (a) is recorded from a person at rest and graph (b) is from the same person during exercise.

- On trace (a), draw a line from point A down to the horizontal axis, and another line from point B to the horizontal axis. Measure the length of time between these points (55 seconds).
- Measure the difference in volume between points A and B (0.3 dm³).
- Divide by the time taken for this decrease (55 s).
- The unit will be dm³ s⁻¹.
- Rate of oxygen uptake on trace (a) = 0.3/55 = 0.0055 dm³ s⁻¹.

Breathing rate and oxygen uptake

The **breathing rate** can also be measured from a spirometer trace. Simply count the number of peaks in each minute. Breathing rate at rest is usually about 12–14 breaths per minute.

Oxygen uptake will depend upon a number of factors. A higher oxygen uptake will result from increased demand, such as during exercise when the muscles are respiring more.

Increased oxygen uptake will result from:

- increased breathing rate.
- deeper breaths.

LEARNING TIP

Remember that under normal conditions you breathe in air – not oxygen. You breathe out air rich in carbon dioxide – not carbon dioxide.

Questions

1. Use trace (b) in Figure 3 in order to measure the tidal volume of this person during exercise. (Remember that you should always record at least three measurements and calculate a mean.)

2. Explain why you should always record at least three measurements and calculate a mean.

3. The tidal volume in trace (a) (at rest) is only 0.3 dm³. What might this suggest about the person from whom it was measured?

4. Explain why the tidal volume in trace (b) is so much higher than that in trace (a) taken from the same person.

5. Use graph (b) to calculate the rate of oxygen uptake for this person during exercise.

6. There is always a certain volume of air left in the lungs after exhaling – this is called the residual volume. Suggest why we are unable to exhale all the air from our lungs.

7. Collect data about the vital capacity of your class members.
   (a) Present the data in the form of a histogram.
   (b) Use the data to calculate the mean vital capacity of your class.
   (c) Use the data to determine if either height or gender affect the vital capacity.

⑤ Gas exchange in other organisms

By the end of this topic, you should be able to demonstrate and apply your knowledge and understanding of:

* the mechanisms of ventilation and gas exchange in bony fish and insects

* the dissection, examination and drawing of the gaseous exchange system of a bony fish and/or insect trachea

---

KEY DEFINITIONS

buccal cavity: the mouth.
countercurrent flow: where two fluids flow in opposite directions.
filaments: slender branches of tissue that make up the gill. They are often called primary lamellae.
lamellae (sometimes called secondary lamellae): folds of the filament to increase surface area. They are also called gill plates.
operculum: a bony flap that covers and protects the gills.
spiracle: an external opening or pore that allows air in or out of the tracheae.
tracheal fluid: the fluid found at the ends of the tracheoles in the tracheal system.
tracheal system: a system of air-filled tubes in insects.

Bony fish

Bony fish must exchange gases with the water in which they live. They use gills in order to absorb oxygen dissolved in the water and release carbon dioxide into the water. The oxygen concentration will be typically much lower than is found in air. Most bony fish have five pairs of gills (see Figure 1) which are covered by a bony plate called the **operculum**.

Each gill consists of two rows of gill **filaments** (primary lamellae) attached to a bony arch. The filaments are very thin, and their surface is folded into many secondary **lamellae** (or gill plates). This provides a very large surface area. Blood capillaries carry deoxygenated blood close to the surface of the secondary lamellae where exchange takes place.

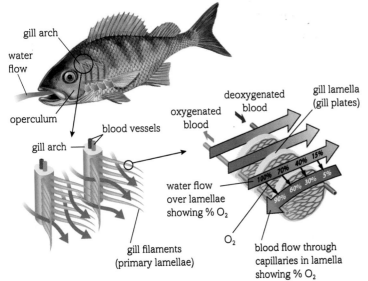

Figure 1 The gills of a bony fish.

Dissection of a bony fish would allow the gills to be seen, as shown in Figure 2.

INVESTIGATION

Dissection of fish gill

1 Find the operculum (bony covering on each side of the fish's head).
2 Lift the operculum and observe the gills. Note their colour.
3 Cut away one operculum to view the gills. Note the gill slits or spaces between the gills.
4 Carefully cut out one gill. Note the bony support (gill arch) and the soft gill filaments that make up each gill. Draw the gill.

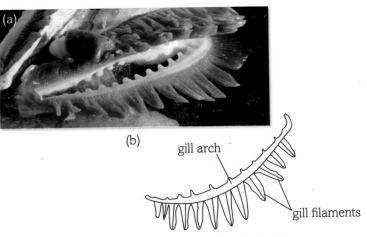

Figure 2 (a) The dissected gills of a bony fish. (b) A single gill.

Countercurrent flow

Blood flows along the gill arch and out along the filaments to the secondary lamellae. The blood then flows through capillaries in the opposite direction to the flow of water over the lamellae. This arrangement creates a **countercurrent flow** that absorbs the maximum amount of oxygen from the water.

Ventilation in bony fish

Bony fish can keep water flowing over the gills by using a buccal–opercular pump. The **buccal cavity** (mouth) can change volume. The floor of the mouth moves downwards, drawing water into the buccal cavity. The mouth closes and the floor is raised again pushing water through the gills. Movements of the operculum are coordinated with the movements of the buccal cavity. As water is pushed from the buccal cavity, the operculum moves outwards. This movement reduces the pressure in the opercular cavity (the space under the operculum), helping water to flow through the gills.

Insects

Insects do not transport oxygen in blood. Insects have an open circulatory system in which the body fluid acts as both blood and tissue fluid. Circulation is slow and can be affected by body movements.

Insects possess an air-filled **tracheal system** (see Figure 3), which supplies air directly to all the respiring tissues. Air enters the system via a pore in each segment, called a **spiracle**. The air is transported into the body through a series of tubes called tracheae (singular 'trachea'). These divide into smaller and smaller tubes, called *tracheoles*. The ends of the tracheoles are open and filled with fluid called **tracheal fluid**. Gaseous exchange occurs between the air in the tracheole and the tracheal fluid. Some exchange can also occur across the thin walls of the tracheoles.

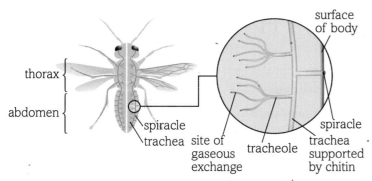

Figure 3 The tracheal system of an insect.

Many insects are very active and need a good supply of oxygen. When tissues are active, the tracheal fluid can be withdrawn (see Figure 4) into the body fluid in order to increase the surface area of the tracheole wall exposed to air. This means that more oxygen can be absorbed when the insect is active.

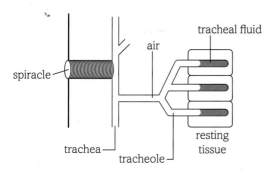

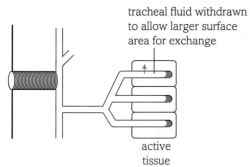

Figure 4 Withdrawal of tracheal fluid to enable larger surface area for exchange.

Ventilation in insects

Larger insects can also ventilate their tracheal system by movements of the body. This can be achieved in a number of ways.

- In many insects, sections of the tracheal system are expanded and have flexible walls. These act as air sacs which can be squeezed by the action of the flight muscles. Repetitive expansion and contraction of these sacs ventilate the tracheal system.

- In some insects, movements of the wings alter the volume of the thorax. As the thorax volume decreases, air in the tracheal system is put under pressure and is pushed out of the tracheal system. When the thorax increases in volume, the pressure inside drops and air is pushed into the tracheal system from outside.

- Some insects have developed this ventilation even further. Locusts can alter the volume of their abdomen by specialised breathing movements. These are coordinated with opening and closing valves in the spiracles. As the abdomen expands, spiracles at the front end of the body open and air enters the tracheal system. As the abdomen reduces in volume, the spiracles at the rear end of the body open and air can leave the tracheal system.

DID YOU KNOW?

The mechanism of ventilation in insects is similar to that in mammals. An increase in the volume of the body cavity reduces pressure so that air enters the tracheal system; a decrease in volume raises the pressure to push air out again.

Questions

1. Fish cannot survive out of water because their gills collapse. Explain why they cannot survive in these conditions.

2. Explain why there are many filaments and lamellae in fish gills.

3. Sharks do not have an operculum and some have lost their ability to pump water through the buccal cavity. Explain why these sharks will drown if they stop moving.

4. Why do insects have a separate tracheal system instead of relying on blood to transport oxygen around the body?

5. Explain why the tracheal fluid is withdrawn into the body when the tissue is very active.

ASTHMA

Asthma is a common condition apparently on the increase in western society today. Some parents are concerned that the drugs used to treat asthma in children may have other side-effects. Here, we explore one of the possible side-effects.

ASTHMA DRUGS STUNT GROWTH – BUT ONLY BY A CENTIMETRE

Parents of children with asthma can rest a little easier. The long-standing worry that some asthma drugs stunt children's growth looks to be overblown.

It seems that children who take inhaled steroids may be about 1 centimetre shorter on average than their peers who do not. "This is very reassuring," says Andrew Bush of Imperial College London, who was not involved in the research.

Steroid drugs have long been used to combat asthma, which arises when people's airways become inflamed and constricted. High doses of oral steroids over long periods are known to stunt children's growth – as does leaving asthma untreated. What was unclear was whether inhaled steroids, which deliver a lower but more direct dose, would also affect height.

Various trials have explored the issue over the years. Now Francine Ducharme at the University of Montreal in Canada and colleagues have pulled the results together in a systematic review of 25 randomised trials, which tested six inhaled steroids against placebo treatments. The trials lasted from 3 months to 6 years, and the children involved were followed up for various lengths of time after the trials had ended.

Most of the studies they analysed looked at children's growth rates. On average, the steroids reduced the growth rate by about 0.5 centimetres in the first year of treatment, but this effect tailed off in subsequent years, and there was some catch-up in growth if the treatment ended or was wound down.

Only one study measured the participants' adult height: it found there was a 1.2-centimetre difference between people who had used inhaled steroids as a child and those who were given a placebo. A three-year study found a 0.7-centimetre disparity at the end of the trial.

Small price to pay

"That is a small price to pay for the benefits of the drugs," says Ducharme. About 20 children die from asthma attacks each year in the UK.

A separate review by Ducharme and a different group of colleagues looked at 22 studies that tested varying the dose of steroids, and found that lower doses had less of an effect on height. One less puff of the medicine per day translated into an additional height increase of 0.25 centimetres in the first year.

As well as making sure that steroid doses are as low as possible, we should show children how to use their inhalers properly, says Bush. If they do not, the medicine may be deposited in the mouth and absorbed into the bloodstream where it can more easily affect growth, instead of reaching the lungs as intended.

"But the first priority must be to treat and control the disease," says Bush. "Undertreated asthma creates a huge burden in terms of time off school and children not able to do sports."

Source

- http://www.newscientist.com/article/dn25909-asthma-drugs-stunt-growth--but-only-by-a-centimetre.html#.VOcX_PmsXgl

Figure 1 A girl using an asthma inhaler.

Where else will I encounter these themes?

1.1   2.1   2.2   2.3   2.4   2.5

Let's start by considering the way in which science is presented in the article.

1. Another recent newspaper article based on this scientific research suggested that using certain types of inhaler reduces growth in height by over a centimetre. The article had the title 'Asthma inhalers "stunt growth"'. Comment on the use of such a title in a popular daily newspaper and discuss how the article above has been written to avoid sensationalising the issue.

2. Scientific theories change as new research is carried out and more evidence becomes available. What new evidence is presented in this article and how might this evidence modify people's views about the use of steroid inhalers to treat asthma?

Now we will look at the biology in, or connected to, this article. Don't worry if you are not ready to answer these questions yet. You may like to return to the questions once you have covered other topics later in the book. Use the timeline at the bottom of the page to help you put this work in context with what you have already learned and what is ahead in your course.

3. At the end of the article, Professor Bush states "Undertreated asthma creates a huge burden in terms of time off school and children not able to do sports." Suggest what the long-term consequences of untreated asthma may be on the children with asthma.

4. Name the tissue in the airways that causes them to constrict.

5. There has been much research that shows that using an inhaler is beneficial to people with asthma. Asthma restricts the flow of air into and out of the lungs. Suggest the mechanism by which the chemical in the inhaler may work on this tissue in the airways.

6. Research about asthma is often funded by the pharmaceutical companies that manufacture the medicines. Their reports may not reflect a truly independent view. What is truly independent research? What are the advantages and disadvantages of truly independent research into medical issues?

7. In a related article, one doctor suggested that wheezing in the very young may be a mechanism to protect the lungs and immature immune system. Suggest how the constriction of the airways that causes wheezing could offer such protection.

8. Francine Ducharme concludes that a minor limitation of growth in height is a small price to pay for the protection against potentially lethal asthma attacks, and that parents should not discourage their children from using inhalers. The article above refers to a review of 25 trials. These trials involved 6471 children and teenagers up to 18 years old. The trials lasted from three months to six years and only one study looked at adult height. Evaluate this information.

Activity

Write an information leaflet about asthma. Your leaflet should explain to a young child (10–11 years old) what causes asthma and what the symptoms may be. You should include some scientific content about the structure of the airways, including how a treatment such as an inhaler may work to relieve the symptoms. You should present a balanced view which includes some possible reasons why inhalers should not be the instant quick fix for any newly diagnosed asthmatic.

Don't forget that annotated diagrams can help to explain structures clearly. The level of detail in your diagrams should be appropriate to the age of the target audience.

Practice questions

1. What is the correct surface area to volume ratio for a cube of length 15 mm? [1]

   A. 2.5

   B. 2.5 : 1

   C. 0.4

   D. 0.4 : 1

2. Which of the following organisms is least likely to need a specialised exchange surface? [1]

   A. A large active organism.

   B. A large inactive organism.

   C. A small active organism.

   D. A small inactive organism.

3. Which row best describes the features of a good exchange surface? [1]

   A. Thin, large surface area with a good blood supply.

   B. Large surface area to volume ratio.

   C. Thin with a large surface area to volume ratio.

   D. Large volume with a good blood supply.

4. Statements i–iv describe features of an exchange surface.

   (i) A high concentration on the delivery side.

   (ii) A low concentration on the demand side.

   (iii) A partially permeable membrane.

   (iv) A thick barrier.

   Which combination of features makes a concentration gradient steeper? [1]

5. Inhalation in mammals involves the following steps:

   (i) Air moves into the lungs

   (ii) Increase in volume of the chest cavity

   (iii) Decrease in pressure inside the lungs

   (iv) Contraction of the muscular diaphragm

   (v) Expansion of the alveoli

   Describe the correct sequence of these events. [1]

   [Total: 5]

6. The tables below list the components found in the mammalian gas exchange system and their functions.

	Tissue
A	cartilage
B	ciliated epithelium
C	goblet cells
D	smooth muscle
E	squamous epithelium
F	elastic fibres

	Function
1	release mucus
2	constrict the airways
3	expel air from the alveoli
4	provides a short diffusion distance
5	waft mucus along airway
6	holds airways open

   Write the correct number next to each letter in the table below.

A	
B	
C	
D	
E	
F	

   [Total: 6]

7. Figure 1 is a photograph of a part of a gill from a bony fish, taken using an electron microscope.

   (a) Suggest what type of electron microscope was used to take this photograph and explain the reasons for your choice. [3]

   (b) Name the components labelled A and B. [2]

   (c) Label C shows a cell on the surface of the gill. Name the type of cell found on the surface of the gill. [1]

   (d) Describe how the gills in a bony fish are ventilated. [6]

   [Total: 12]

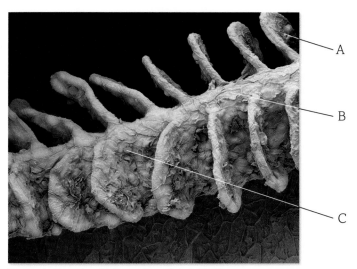

Figure 1 Part of a gill from a bony fish (×3000).

8. Figure 2 shows the trace from a spirometer recording the breathing movements of a healthy 17-year-old boy.

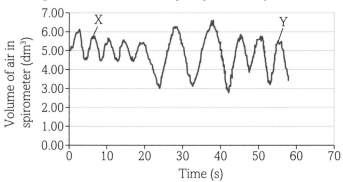

Figure 2 A trace from a spirometer recording the breathing movements of a healthy 17-year-old boy.

(a) Use the trace to calculate the mean tidal volume over the first 20 seconds. Show your working. [3]

(b) The coordinates of point **X** are: 7 seconds and 5.80 dm³. The coordinates of point **Y** are 55 seconds and 5.50 dm³. Use this information to calculate the rate of oxygen consumption. [2]

(c) Distinguish between the terms *tidal volume* and *vital capacity*. [4]

[Total: 9]

9. Figure 3 shows a part of the gaseous exchange system of an insect.

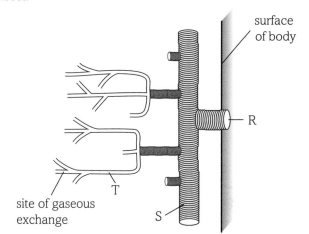

Figure 3 Part of the gaseous exchange system of an insect.

(a) Name the parts labelled R, S and T. [3]

(b) Describe how oxygen reaches the muscle of the wings. [3]

(c) Describe how the insect gaseous exchange system responds to increased demand for oxygen. [3]

(d) Explain why, in insects, oxygen is delivered to the tissues in gaseous form rather than being transported by the blood. [7]

[Total: 16]

10. In the spring of 2014 weather conditions in the UK resulted in a build-up of pollutants in the atmosphere. The health authorities advised people – especially asthmatics – not to go outdoors unless it was essential. Despite this advice 86% of asthmatics reported needing to use their inhaler more often than usual during the unusual weather.

(a) Does this suggest a link between pollution and asthma? [1]

(b) Does this show that pollution causes asthma? Justify your answer. [3]

(c) Asthma is caused by irritants that produce constriction of the airways. Name the tissue that causes constriction of the airways. [1]

(d) Suggest a mechanism by which the irritant could cause constriction of the airway. [4]

(e) Outline a plan for an investigation to show that pollutants in the atmosphere can cause asthma. [4]

[Total: 13]

Exchange and transport

TRANSPORT IN ANIMALS

Introduction

Problems with the heart and circulatory system are still one of the greatest causes of ill-health and death in humans. The importance of effective transport around the body can be highlighted by the training undertaken by sportsmen and women. Most of the training carried out by athletes is designed to improve the transport of oxygen and nutrients to the tissues. It is the supply of enough oxygen to the muscles that enables athletes to generate more power and endurance athletes to keep going longer.

All living cells require a supply of oxygen and nutrients. In small organisms these can diffuse from the surrounding environment into the cells. However, larger organisms present something of a problem – the distance that these substances must diffuse is too great. Diffusion is too slow and the supply to cells in the centre of the organism is very limited. Therefore larger organisms must develop a way of transporting these substances more quickly. It is also important to remember that removal of wastes such as carbon dioxide is just as important as supply of oxygen.

All the maths you need

To unlock the puzzles in this section you need the following maths:

- Recognise and make use of appropriate units in calculations
- Use ratios, fractions and percentages
- Estimate results
- Use an appropriate number of significant figures
- Translate information between graphical, numerical and algebraic forms
- Understand and use the symbols: $=$, $<$, $<<$, $>>$, $>$, $\sim$
- Calculate the circumferences, surface areas and volumes of regular shapes
- Solve algebraic equations
- Construct and interpret frequency tables and diagrams, bar charts and histograms

What have I studied before?

- The relationship between size and surface area to volume ratio
- How substances move by diffusion and osmosis
- How substances pass through membranes
- How oxygen and carbon dioxide are absorbed into the blood
- The need for pressure differences to make fluids flow from one place to another
- How the heart creates pressure to make blood flow
- How cells make up tissues and tissues make up organs
- The need to absorb substances from the environment and transport them around the body in a series of vessels

What will I study later?

- How the oxygen and nutrients are used in the cells (AL)
- The transport of hormones in the blood (AL)
- The role of circulation in homeostasis – particularly its importance to thermoregulation (AL)
- How blood pressure and heart rate are regulated (AL)

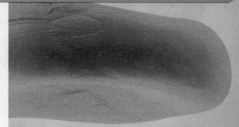

What will I study in this chapter?

- How size affects the need for a circulatory system
- How size and level of activity are affected by the type of circulatory system
- The pressure changes during a cardiac cycle
- The action of the valves to ensure a one way flow
- How the blood vessels are adapted to their position in the circulatory system
- How exchange at the capillaries is achieved and tissue fluid is formed
- How oxygen and carbon dioxide are transported

Transport in animals

By the end of this topic, you should be able to demonstrate and apply your knowledge and understanding of:

* the need for transport systems in multicellular animals

* the different types of circulatory systems

KEY DEFINITIONS

double circulatory system: one in which the blood flows through the heart twice for each circuit of the body.
single circulatory system: one in which the blood flows through the heart once for each circuit of the body.
transport: the movement of substances such as oxygen, nutrients, hormones, waste and heat around the body.

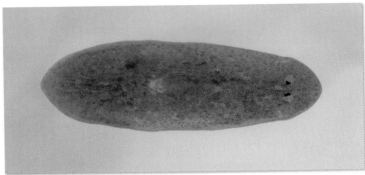

Figure 1 A planarian worm is a flatworm. It is very flat, so that the surface area to volume ratio remains large and sufficient exchange can occur across its surface. Most planarian worms range in length from approximately 0.3 to 2.5 cm.

The need for a transport system

As you learned at the start of Chapter 3.1, all living animal cells need a supply of oxygen and nutrients to grow and survive. They also need to remove waste products so that these do not build up and become toxic.

Very small animals do not need a separate transport system, because all their cells are surrounded by (or very close to) the environment in which they live. Diffusion will supply enough oxygen and nutrients to keep the cell alive. However, a larger animal with a complex anatomy will have more than two layers of cells. The diffusion distance becomes too long, and diffusion alone will be too slow to supply all the requirements.

There are three main factors that influence the need for a transport system:

* size
* surface area to volume ratio
* level of metabolic activity.

Size

The cells inside a large organism are further from its surface – the diffusion pathway is increased. The diffusion rate is reduced, and diffusion is too slow to supply all the requirements. Also, the outer layers of cells use up the supplies, so that less will reach the cells deep inside the body.

Surface area to volume ratio

As already explained in Chapter 3.1, small animals have a large surface area to volume ratio, e.g. the flatworm (see Figure 1).

This means that for each gram of tissue in their body they have a sufficient area of the body surface through which exchange can occur. However, larger animals have a smaller surface area to volume ratio. This means that each gram of tissue has a smaller area of body surface for exchange.

Level of metabolic activity

Animals need energy from food, so that they can move around. Releasing energy from food by aerobic **respiration** requires oxygen. If an animal is very active, its cells need good supplies of nutrients and oxygen to supply the energy for movement. Animals that keep themselves warm, such as mammals, need even more energy.

Features of a good transport system

An effective **transport** system will include:

* a fluid or medium to carry nutrients, oxygen and wastes around the body – this is the blood

* a pump to create pressure that will push the fluid around the body – this is the heart

* exchange surfaces (see Chapter 3.1) that enable substances to enter the blood and leave it again where they are needed – these are the capillaries.

An efficient transport system will also include:

* tubes or vessels to carry the blood by mass flow

* two circuits – one to pick up oxygen and another to deliver oxygen to the tissues.

Single and double circulatory systems

Fish have a **single circulatory system** (see Figure 2 (a)). The blood flows through the heart once for each circuit of the body. The blood takes the following route:

heart → gills → body → heart

Mammals have a **double circulatory system** (see Figure 2 (b)). The system has two separate circuits. One circuit carries blood to the lungs to pick up oxygen. This is pulmonary circulation. The other circuit carries the oxygen and nutrients around the body to the tissues. This is systemic circulation. Blood flows through the heart twice for each circuit of the body. The blood takes the following route:

heart → body → heart → lungs → heart

Advantages of a double circulation

An efficient circulatory system will deliver oxygen and nutrients quickly to the parts of the body where they are needed. The blood can be made to flow more quickly by increasing the blood pressure created by the heart. By comparing the transport systems of fish and mammals, we shall see the advantages of a double circulatory system.

In the single circulatory system of fish:

- the blood pressure drops as blood passes through the tiny capillaries of the gills

- blood has a low pressure as it flows towards the body, and will not flow very quickly

- the rate at which oxygen and nutrients are delivered to respiring tissues, and carbon dioxide and urea are removed, is limited.

Fish are not as metabolically active as mammals, as they do not maintain their body temperature. Therefore, they need less energy. Their single circulatory system delivers sufficient oxygen and nutrients for their needs.

In the double circulatory system of mammals:

- the blood pressure must not be too high in the pulmonary circulation, otherwise it may damage the delicate capillaries in the lungs.

- the heart can increase the pressure of the blood after it has passed through the lungs, so the blood is under higher pressure as it flows to the body and flows more quickly

- the systemic circulation can carry blood at a higher pressure than the pulmonary circulation.

Mammals are active animals and maintain their body temperature. Supplying the energy for activity and the heat needed to keep the body warm requires energy from food. The energy is released from food in the process of respiration. To release a lot of energy, the cells need a good supply of both nutrients and oxygen, as well as the removal of waste products.

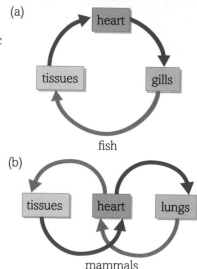

Figure 2 (a) Single and (b) double circulatory systems.

Questions

1. Explain why very small animals can survive with no transport system whilst a large animal cannot survive without a transport system.

2. Flatworms can grow to a large size. Explain how they can supply sufficient oxygen to their tissues without a transport system.

3. A good transport system has exchange surfaces at both the supply end (lungs) and the demand end (muscles). Explain why exchange surfaces are essential for rapid transport.

4. What limits the delivery of oxygen and nutrients in a single circulatory system?

5. Explain why a double circulation is essential for mammals.

6. Why don't animals simply increase the pressure of the blood so that it can flow through the capillaries of the lungs and still have enough pressure to flow quickly around the body?

7. Birds are active and regulate their body temperature using heat from respiration. What type of circulatory system do birds have?

② Blood vessels

By the end of this topic, you should be able to demonstrate and apply your knowledge and understanding of:

* the different types of circulatory systems

* the structure and functions of arteries, arterioles, capillaries, venules and veins

Open circulatory systems

Many animals, including insects, have an **open circulatory system** (see Figure 1). This means that the blood is not always held within blood vessels. Instead, the blood fluid circulates through the body cavity, so that the tissues and cells are bathed directly in blood.

In some animals, movements of the body help to circulate the blood – and without movement the blood stops moving, so that transport of oxygen and nutrients stops.

In other animals, such as insects, there is a muscular pumping organ much like a heart. This is a long, muscular tube that lies just under the dorsal (upper) surface of the body. Blood from the body enters the heart through pores called **ostia**. The heart then pumps the blood towards the head by **peristalsis**. At the forward end of the heart (nearest the head), the blood simply pours out into the body cavity. This circulation can continue when the insect is at rest, but body movements may still affect circulation.

Some larger and more active insects, such as locusts, have open-ended tubes attached to the heart. These direct the blood towards active parts of the body, such as the leg and wing muscles.

Open circulatory systems have certain disadvantages:

* Blood pressure is low and blood flow is slow.

* Circulation of blood may be affected by body movements or lack of body movements.

Closed circulatory systems

In larger animals the blood stays entirely inside vessels – this is a **closed circulatory system** (see Figure 1). A separate fluid, called tissue fluid, bathes the tissues and cells. This has certain advantages over the open system.

* Higher pressure, so that blood flows more quickly.

* More rapid delivery of oxygen and nutrients.

* More rapid removal of carbon dioxide and other wastes.

* Transport is independent of body movements.

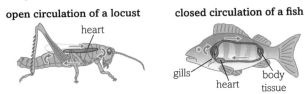

Figure 1 Open and closed circulatory systems.

Blood vessels

Blood flows through a series of vessels. Each is adapted to its particular role in relation to its distance from the heart. All types of blood vessel have an inner layer or lining, made of a single layer of cells, called the **endothelium**. This is a thin layer that is particularly smooth in order to reduce friction with the flowing blood.

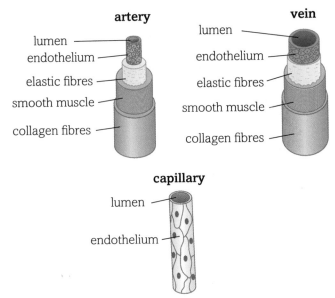

Figure 2 Structure of artery, vein and capillary walls.

Arteries

Please refer to Figure 2 on the structures of blood-vessel walls. **Arteries** carry blood away from the heart. The blood is at high pressure, so the artery wall must be thick in order to withstand that pressure. The lumen is relatively small in order to maintain high pressure, and the inner wall is folded to allow the lumen to expand as blood flow increases.

The wall consists of three layers:

- Inner layer (tunica intima) consists of a thin layer of elastic tissue which allows the wall to stretch and then recoil to help maintain blood pressure.

- Middle layer (tunica media) consists of a thick layer of smooth muscle.

- Outer layer (tunica adventitia) is a relatively thick layer of collagen and elastic tissue. This provides strength to withstand the high pressure, and recoil to maintain the pressure.

DID YOU KNOW?

Elastic and muscular arteries
Arteries near the heart have more elastic tissue in the wall, in order to allow stretch and recoil, which helps to even out the fluctuations in blood pressure created by the heart. Further from the heart, the walls contain more muscle tissue.

Arterioles

Arterioles are small blood vessels that distribute the blood from an artery to the capillaries. Arteriole walls contain a layer of smooth muscle. Contraction of this muscle will constrict the diameter of the arteriole. This increases resistance to flow and reduces the rate of flow of blood. Constriction of the arteriole walls can be used to divert the flow of blood to regions of the body that are demanding more oxygen.

Capillaries

DID YOU KNOW?

There are about 100 000 miles of capillaries in the human body.

Capillaries have very thin walls. They allow exchange of materials between the blood and tissue fluid – see Chapter 3.1.

- The lumen is very narrow – its diameter is about the same as that of a red blood cell (7 μm). The red blood cells may be squeezed against the walls of the capillary as they pass along the capillary; this helps transfer of oxygen, as it reduces the diffusion path to the tissues. It also increases resistance and reduces rate of flow.

- The walls consist of a single layer of flattened endothelial cells. This reduces the diffusion distance for the materials being exchanged.

- The walls are leaky. They allow blood plasma and dissolved substances to leave the blood.

You will learn more about the role of capillaries in the next topic.

Venules

From the capillaries blood flows into small vessels called **venules**. These collect the blood from the capillary bed and lead into the

veins. The venule wall consists of thin layers of muscle and elastic tissue outside the endothelium, and a thin outer layer of collagen.

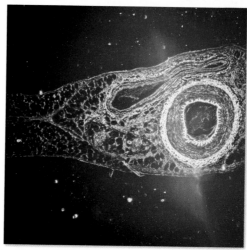

Figure 3 A transverse section of an artery (right) and vein (left and above).

Veins

Veins carry blood back to the heart. The blood is at low pressure and the walls do not need to be thick.

- The lumen is relatively large, in order to ease the flow of blood.

- The walls have thinner layers of collagen, smooth muscle and elastic tissue than in artery walls. They do not need to stretch and recoil, and are not actively constricted in order to reduce blood flow.

- The main feature of veins is that they contain valves to help the blood flow back to the heart and to prevent it flowing in the opposite direction. As the walls are thin, the vein can be flattened by the action of the surrounding skeletal muscle (as shown in Figure 3). Contraction of the surrounding skeletal muscle applies pressure to the blood, forcing the blood to move along in a direction determined by the valves.

Questions

1. What is the difference between open circulatory systems and closed circulatory systems?

2. Explain why mammals need a closed circulatory system in order to remain active.

3. Why do the arteries near the heart need a lot of elastic tissue in their walls?

4. The tissues in the artery walls help to reduce the fluctuations in blood pressure. Describe how this is achieved.

5. Explain how constricting the arterioles in the digestive system can cause blood to be diverted to the muscles.

6. Draw an annotated diagram to explain how blood at low pressure can return through the veins to the heart.

7. Draw a table comparing the structures of an artery, arteriole, capillary and vein.

③ Exchange at the capillaries

By the end of this topic, you should be able to demonstrate and apply your knowledge and understanding of:

* the formation of tissue fluid from plasma

KEY DEFINITIONS

blood: the fluid used to transport materials around the body.
hydrostatic pressure: the pressure that a fluid exerts when pushing against the sides of a vessel or container.
lymph: the fluid held in the lymphatic system, which is a system of tubes that returns excess tissue fluid to the blood system.
oncotic pressure: the pressure created by the osmotic effects of the solutes.
plasma: the fluid portion of the blood.
tissue fluid: the fluid surrounding the cells and tissues.

Blood plasma and tissue fluid

Blood is the fluid held in our blood vessels. It consists of a liquid called plasma, containing many blood cells. The plasma contains many dissolved substances, including oxygen, carbon dioxide, minerals, glucose, amino acids, hormones and plasma proteins. The cells include the red blood cells (erythrocytes), various white blood cells (leucocytes) and fragments called platelets.

Tissue fluid is similar to blood plasma. But it does not contain most of the cells found in blood, and neither does it contain plasma proteins. Tissue fluid is formed by plasma leaking from the capillaries. It surrounds the cells in the tissue, and supplies them with the oxygen and nutrients they require. As blood plasma leaks from the capillary, it carries all the dissolved substances into the tissue fluid. This movement is mass flow rather than diffusion. Waste products from cell metabolism will be carried back into the capillary as some of the tissue fluid returns to the capillary (see Figure 1).

Formation of tissue fluid

- When an artery reaches the tissues, it branches into smaller arterioles, and then into a network of capillaries. These eventually link up with venules to carry blood back to the veins. Therefore blood flowing into an organ or tissue is contained in the capillaries.

- At the arterial end of a capillary, the blood is at relatively high (hydrostatic) pressure. This pressure tends to push the blood fluid out of the capillaries through the capillary wall. The fluid can leave through the tiny gaps between the cells in the capillary wall.

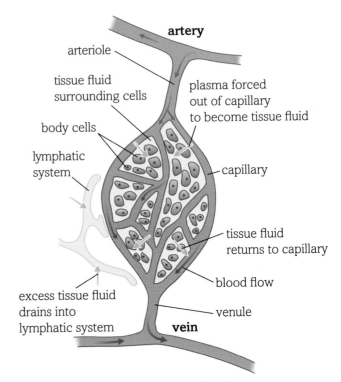

Figure 1 A capillary bed linking an artery and a vein.

- The fluid that leaves the blood consists of plasma with dissolved nutrients and oxygen. All the red blood cells, platelets and most of the white blood cells remain in the blood, as do the plasma proteins. These are too large to be pushed out through the gaps in the capillary wall.

- This tissue fluid surrounds the body cells, so exchange of gases and nutrients can occur across the plasma membranes. The exchange occurs by diffusion, facilitated diffusion and active uptake. Oxygen and nutrients enter the cells; carbon dioxide and other wastes leave the cells.

Returning to the blood

The blood pressure at the venous end of the capillary is much lower. This allows some of the tissue fluid to return to the capillary carrying carbon dioxide and other waste substances into the blood.

Not all the tissue fluid re-enters the blood.

- Some tissue fluid is directed into another tubular system called the lymph system or lymphatic system.

- This drains excess tissue fluid out of the tissues and returns it to the blood system in the subclavian vein in the chest.

- The fluid in the lymphatic system is called **lymph** and is similar in composition to the tissue fluid. It will contain more lymphocytes, as these are produced in the **lymph nodes**.

Lymph nodes are swellings found at intervals along the lymphatic system, which have an important part to play in the **immune response**.

> **DID YOU KNOW?**
> If a tissue is infected, then the capillaries become more leaky and more fluid is directed into the lymph system – this helps direct bacteria towards the lymph nodes.

Feature	Blood plasma	Tissue fluid	Lymph
hydrostatic pressure	high	low	low
oncotic pressure	more negative	less negative	less negative
cells	red blood cells, neutrophils, lymphocytes	some neutrophils, especially in infected areas	lymphocytes
proteins	plasma proteins	few proteins	few proteins
fats	transported in lipoproteins	few fats	more fats, especially near the digestive system

Table 1 The differences between plasma, tissue fluid and lymph.

Movement of fluids

The hydrostatic pressure of the blood is not the only influence on the movement of fluid into and out of the capillary. The tissue fluid has its own hydrostatic pressure, and the **oncotic pressure** of the solutes (dissolved substances) also has an influence.

Figure 2 shows the influence of these forces on movement of the fluids.

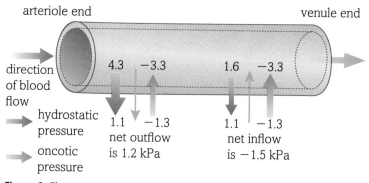

Figure 2 The pressures (in kPa) influencing movement of fluids during exchange at the capillaries.

- The hydrostatic pressure of the blood tends to push fluid out into the tissues.
- The hydrostatic pressure of the tissue fluid tends to push fluid into the capillaries.
- The oncotic pressure of the blood tends to pull water back into the blood (it has a negative figure).
- The oncotic pressure of the tissue fluid pulls water into the tissue fluid.

The net result of these forces creates a pressure to push fluid out of the capillary at the arterial end and into the capillary at the venule end.

	Arterial end	Venous end
net hydrostatic pressure (kPa)	4.3 – 1.1 = 3.2	1.6 – 1.1 = 0.5
net oncotic pressure (kPa)	−3.3 – (−1.3) = −2.0	−3.3 – (−1.3) = −2.0
net pressure (kPa)	3.2 + (−2.0) = 1.20	0.5 + (−2.0) = −1.5
movement of fluid	out of capillary	into capillary

Table 2 The net movement of fluid into and out of a capillary.

> **Questions**
>
> 1 Where is the hydrostatic pressure of the blood created?
>
> 2 What substances might contribute to the oncotic pressure of blood?
>
> 3 Blood contains many proteins called plasma proteins, which do not leave the blood. Explain why they cannot leave the blood.
>
> 4 Explain why neutrophils can enter the tissue fluid but erythrocytes cannot.
>
> 5 Certain capillaries have walls with fenestrations – small pores through the cells that allow larger molecules to pass through the capillary wall. These occur in the kidneys, pancreas, intestines and endocrine organs. Suggest why these organs have capillaries with fenestrations.
>
> 6 It is not accurate to say that oxygen diffuses from the red blood cells to the tissues. Explain why this is not accurate.

4 The structure of the heart

By the end of this topic, you should be able to demonstrate and apply your knowledge and understanding of:

* the external and internal structure of the mammalian heart

* the dissection, examination and drawing of the external and internal structure of the mammalian heart

---

KEY DEFINITIONS

atrio-ventricular valves: valves between the atria and the ventricles, which ensure that blood flows in the correct direction.
cardiac muscle: specialised muscle found in the walls of the heart chambers.
semilunar valves: valves that prevent blood re-entering the heart from the arteries.

DID YOU KNOW?

A human heart will beat over three billion times during a lifetime.

At the top of the heart are a number of tubular blood vessels. These are the veins that carry blood into the atria and the arteries that carry blood away from the heart.

The mammalian heart is a muscular pump. It is divided into two sides. The right side pumps the deoxygenated blood to the lungs to be oxygenated. The left side pumps oxygenated blood to the rest of the body. On both sides, the heart squeezes the blood, putting it under pressure. This pressure forces the blood along the arteries and through the circulatory system.

External features of the heart

In humans, the heart lies just off-centre towards the left of the chest cavity. The main part of the heart consists of firm, dark-red muscle called cardiac muscle (see Figure 1). There are two main pumping chambers – the **ventricles**. Above the ventricles are two thin-walled chambers – the **atria**. These are much smaller than the ventricles and are easy to overlook. (When you buy a heart from a butcher, the atria have often been removed.)

Lying over the surface of the heart are **coronary arteries** that supply oxygenated blood to the heart muscle. As the heart is a hard-working organ, these arteries are very important. If they become constricted, it can have severe consequences for the health of the heart. Restricted blood flow to the heart muscle reduces the delivery of oxygen and nutrients such as fatty acids and glucose. This may cause **angina** or a heart attack (**myocardial infarction**).

Internal features of the mammalian heart

The heart is divided into four chambers (see Figure 2). The two upper chambers are atria. These receive blood from the major veins. Deoxygenated blood from the body flows through the **vena cava** into the right atrium. Oxygenated blood from the lungs flows through the **pulmonary vein** into the left atrium.

From the atria, blood flows down through the **atrio-ventricular valves** into the ventricles. The action of the valves is described in the next topic. Attached to the valves are obvious **tendinous cords** (see Figure 3), which prevent the valves from turning inside out when the ventricle walls contract.

A wall of muscle called the **septum** separates the ventricles from each other. This ensures that the oxygenated blood in the left side of the heart and the deoxygenated blood in the right side are kept separate.

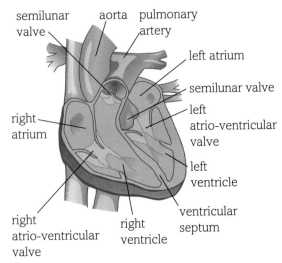

Figure 2 The internal structure of the heart.

Deoxygenated blood leaving the right ventricle flows into the **pulmonary artery** leading to the lungs, where it is oxygenated. Oxygenated blood leaving the left ventricle flows into the **aorta**. This carries blood to a number of arteries that supply all parts of

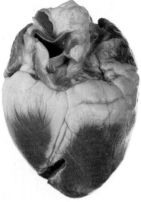

Figure 1 The external features of the heart.

the body. At the base of the major arteries, where they exit the heart, are the **semilunar valves**. These prevent blood returning to the heart as the ventricles relax.

Dissection of the heart

1 Observe a whole mammalian heart. Ensure you can identify the external features described above including the aorta and pulmonary artery.
2 Carefully cut into each ventricle noting the thickness of the wall in each case. Using scissors is safer than using scalpels.
3 Cut upwards towards the atria to expose the atrio-ventricular valves. Note the action of the valves and the role of the tendinous cords.

Blood pressure

The cardiac muscle in the wall of each chamber contracts to create pressure in the blood. The higher the pressure created in the heart, the further it will push the blood.

- **Atria** – the muscle of the atrial walls is very thin. This is because these chambers do not need to create much pressure. Their function is to receive blood from the veins and push it into the ventricles.

- **Right ventricle** – the walls of the right ventricle are thicker than the walls of the atria. This enables the right ventricle to pump blood out of the heart. The right ventricle pumps deoxygenated blood to the lungs. The lungs are in the chest cavity beside the heart, so that the blood does not need to travel very far. Also, the alveoli in the lungs are very delicate and could be damaged by very high blood pressure.

- **Left ventricle** – the walls of the left ventricle can be two or three times thicker than those of the right ventricle. The blood from the left ventricle is pumped out through the aorta and needs sufficient pressure to overcome the resistance of the **systemic circulation**.

Figure 3 A dissected heart.

Cardiac muscle structure

Figure 4 A coloured transmission electron micrograph of cardiac muscle (×1650).

Cardiac muscle (see Figure 4) consists of fibres that branch, producing cross-bridges – shown in the centre of the micrograph. These help to spread the stimulus around the heart, and also ensure that the muscle can produce a squeezing action rather than a simple reduction in length. There are numerous mitochondria between the muscle fibrils (**myofibrils**) to supply energy for contraction. The muscle cells are separated by **intercalated discs** (thick wavy blue line), which facilitate synchronised contraction. Each cell has a nucleus and is divided into contractile units called **sarcomeres** (marked by the thin blue lines).

Questions

1 Why are the walls of the atria very thin?

2 Explain why the left ventricle walls need to be so much thicker than those on the right.

3 Why must the blood in the left side of the heart be kept separate from the blood in the right side?

4 Why is it important that the blood pressure is higher in the heart than elsewhere in the circulatory system?

5 What is the role of the tendinous cords in the ventricles?

6 Cardiac muscle is unusual in having fibres that branch to produce cross-bridges. Explain why these cross-bridges are important in cardiac muscle.

⑤ The cardiac cycle

By the end of this topic, you should be able to demonstrate and apply your knowledge and understanding of:

* the cardiac cycle

cardiac cycle: the sequence of events in one full beat of the heart.

The role of the heart is to create pressure that pushes blood around the blood vessels. The muscular walls of the four chambers must all contract in a coordinated sequence, which allows the heart to fill with blood before pumping it away. This coordinated sequence is known as the **cardiac cycle** (see Figure 1).

Both right and left ventricles pump together. Contraction starts at the apex (base) of the heart so that blood is pushed upwards towards the arteries.

The muscular walls of all four chambers relax. Elastic recoil causes the chambers to increase in volume allowing blood to flow in from the veins.

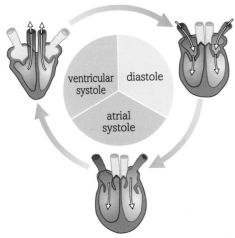

Both right and left atria contract together. The muscle in the walls is thin so only a small increase in pressure is created by this contraction. This helps to push blood into the ventricles stretching their walls and ensuring they are full of blood.

Figure 1 The cardiac cycle.

The action of the valves

The valves ensure that blood flows in the correct direction. They are opened and closed by changes in the blood pressure in the various chambers of the heart.

Atrio-ventricular valves

After **systole**, the ventricular walls relax and recoil (see Figure 2).

* The pressure in the ventricles rapidly drops below the pressure in the atria.
* Blood in the atria pushes the **atrio-ventricular valves** open.
* Blood entering the heart flows straight through the atria and into the ventricles.

* The pressure in the atria and the ventricles rises slowly as they fill with blood.
* The valves remain open while the atria contract, but close when the atria begin to relax.
* This closure is caused by a swirling action in the blood around the valves when the ventricle is full.
* As the ventricles begin to contract (systole), the pressure of the blood in the ventricles rises.
* When the pressure rises above that in the atria, the blood starts to move upwards.
* This movement fills the valve pockets and keeps them closed.
* The tendinous cords attached to the valves prevent them from turning inside out.
* This prevents the blood flowing back into the atria.

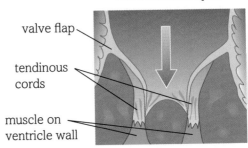

(a) valve open

higher blood pressure above valve forces it open

valve flap

tendinous cords

muscle on ventricle wall

lower blood pressure beneath valve

(b) valve closed

lower blood pressure cannot open valve

valve flaps fit together

high pressure pushes valve closed

tendinous cords stop valve inverting

Figure 2 The action of the atrio-ventricular valves.

Semilunar valves

- Before ventricular contraction, the pressure in the major arteries is higher than the pressure in the ventricles.
- This means that the **semilunar valves** are closed.
- Ventricular systole raises the blood pressure in the ventricles very quickly.
- Once the pressure in the ventricles rises above the pressure in the major arteries, the semilunar valves are pushed open.
- The blood is under very high pressure, so it is forced out of the ventricles in a powerful spurt.
- Once the ventricle walls have finished contracting, the heart muscle starts to relax (**diastole**).
- Elastic tissue in the walls of the ventricles recoils.
- This stretches the muscle out again and returns the ventricle to its original size.
- This causes the pressure in the ventricles to drop quickly.
- As it drops below the pressure in the major arteries, the blood starts to flow back towards the ventricles.
- The semilunar valves are pushed closed by the blood collecting in the pockets of the valves.
- This prevents blood returning to the ventricles.
- The pressure wave created when the left semilunar valve closes is the 'pulse' that we can easily feel at the wrist or neck.

Pressure changes in the heart chambers

The pressure changes in the heart chambers during one cardiac cycle can be seen in Figure 3. The pressure changes cause the opening and closing of the valves.

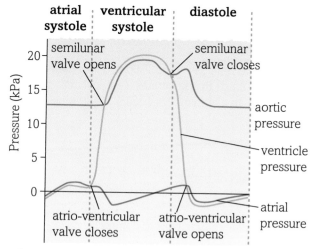

Figure 3 Pressure changes in the heart chambers and aorta during one cardiac cycle.

Pressure in the blood vessels

Blood enters the aorta and pulmonary artery in a rapid spurt, but the tissues require blood to be delivered in an even flow (see Figure 4). The structure of the artery walls plays a large part in creating a more even flow:

- The artery walls close to the heart have a lot of elastic tissue.
- When blood leaves the heart, these walls stretch.
- As blood moves on and out of the aorta, the pressure in the aorta starts to drop.
- The elastic recoil of the walls helps to maintain the blood pressure in the aorta.
- The further the blood flows along the arteries, the more the pressure drops and the fluctuations become less obvious.
- It is important to maintain the pressure gradient between the aorta and the arterioles, as this is what keeps the blood flowing towards the tissues.

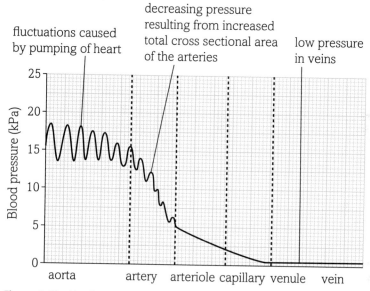

Figure 4 The blood pressure at different parts of the blood system.

Questions

1. Explain why the atria must contract before the ventricles.

2. Draw a flow chart to describe the cardiac cycle, including the action of the valves.

3. Explain why the tendinous cords are essential to the action of the atrio-ventricular valves.

4. Explain why the ventricles contract from the apex (bottom) upwards.

5. What causes the semilunar valves to close?

7. What causes the fluctuations in pressure in the aorta and arteries?

8. What is a pulse?

(6) Coordination of the cardiac cycle

By the end of this topic, you should be able to demonstrate and apply your knowledge and understanding of:

* how heart action is initiated and coordinated

* the use and interpretation of electrocardiogram (ECG) traces

---

KEY DEFINITIONS

bradycardia: a slow heart rhythm.
ectopic heartbeat: an extra beat or an early beat of the ventricles.
electrocardiogram: a trace that records the electrical activity of the heart.
fibrillation: uncoordinated contraction of the atria and ventricles.
myogenic muscle: muscle that can initiate its own contraction.
Purkyne tissue: consists of specially adapted muscle fibres that conduct the wave of excitation from the AVN down the septum to the ventricles.
sino-atrial node (SAN): the heart's pacemaker. It is a small patch of tissue that sends out waves of electrical excitation at regular intervals in order to initiate contractions.
tachycardia: a rapid heart rhythm.

The need for coordination

Heart muscle or cardiac muscle is unusual in that it can initiate its own contraction (see Figure 1). Because of this property, the heart muscle is described as **myogenic**. The muscle will contract and relax rhythmically even if it is not connected to the body. The muscles from the atria and the muscles from the ventricles each have their own natural frequency of contraction. The atrial muscle tends to contract at a higher frequency than the ventricular muscle.

This property of the muscle could cause inefficient pumping if the contractions of the chambers are not synchronised – a condition known as **fibrillation**. So the heart needs a mechanism that can coordinate the contractions of all four chambers.

Initiation and control of the heartbeat

At the top of the right atrium, near the point where the vena cava empties blood into the atrium, is the **sino-atrial node** (SAN). This is a small patch of tissue that generates electrical activity. The SAN initiates a wave of excitation at regular intervals. In a human, this occurs 55–80 times a minute. The SAN is also known as the pacemaker.

Contraction of the atria

* The wave of excitation quickly spreads over the walls of both atria.

* It travels along the membranes of the muscle tissue.

* As the wave of excitation passes, it causes the cardiac muscle cells to contract.

* This is an atrial systole (see topic 3.2.5).

The tissue at the base of the atria is unable to conduct the wave of excitation, and so it cannot spread directly down to the ventricle walls. At the top of the interventricular septum (the septum separating the two ventricles) is another node – the **atrio-ventricular node** (AVN). This is the only route that can conduct the wave of excitation through to the ventricles. The wave of excitation is delayed in the node. This allows time for the atria to finish contracting and for the blood to flow down into the ventricles before they begin to contract.

Contraction of the ventricles

- After this short delay, the wave of excitation is carried away from the AVN and down specialised conducting tissue called the **Purkyne tissue**.
- This runs down the interventricular septum.
- At the base of the septum, the wave of excitation spreads out over the walls of the ventricles.
- As the excitation spreads upwards from the base (apex) of the ventricles, it causes the muscles to contract.
- This means that the ventricles contract from the base upwards.
- This pushes the blood up towards the major arteries at the top of the heart.

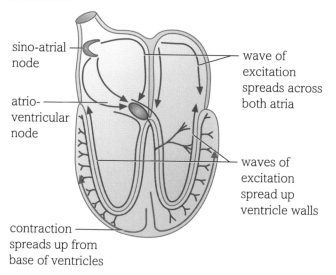

Figure 1 Coordination of the cardiac cycle.

Electrocardiograms

We can monitor the electrical activity of the heart using an **electrocardiogram** (ECG) (see Figure 2). This involves attaching a number of sensors to the skin. Some of the electrical activity generated by the heart spreads through the tissues next to the heart and outwards to the skin. The sensors on the skin pick up the electrical excitation created by the heart and convert this into a trace.

The trace of a healthy person has a particular shape. It consists of a series of waves that are labelled P, Q, R, S and T.

- Wave P shows the excitation of the atria.
- QRS indicates the excitation of the ventricles.
- T shows diastole.

(see topic 3.2.5).

QRS complex – shows ventricular stimulation

P wave – shows atrial stimulation   T wave – shows diastole

Figure 2 A normal ECG.

The shape of the ECG trace can sometimes indicate when part of the heart muscle is not healthy. The diagrams in Figure 3 show four abnormal ECG traces compared with a normal trace. Such traces can be used by medical professionals to diagnose heart problems.

sinus rhythm (normal)

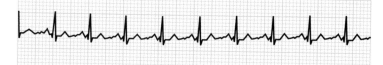

bradycardia (slow heart rate)

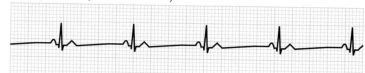

tachycardia (fast heart rate)

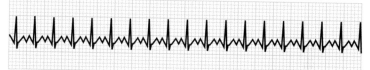

atrial fibrillation
(atria beating more frequently than ventricles – no clear P waves seen)

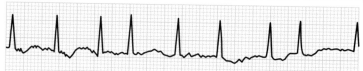

ectopic heartbeat (the third beat here is an early [ectopic] ventricular beat)
The patient often feels as if a heartbeat has been missed.

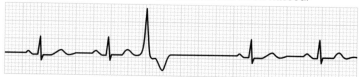

Figure 3 Abnormal ECGs, compared with a normal trace.

Questions

1. Why is the SAN known as the pacemaker?

2. Why is it essential that the AVN delays the electrical stimulation of the ventricles?

3. Suggest why the QRS complex has a larger peak than the P wave.

4. Explain why atrial fibrillation decreases the efficiency of the heart.

5. Superventricular tachycardia is caused when the electrical stimulation of the atria passes directly on to the ventricles. This causes the ventricle walls to contract from the top down, immediately followed by the normal contraction from the apex upwards. Suggest what the patient's heart rate may feel like, and explain why the patient will feel tired and faint.

6. A person with an ectopic heartbeat often feels that a heartbeat has been missed. Use the ECG trace above to explain this feeling.

(7) Transport of oxygen

By the end of this topic, you should be able to demonstrate and apply your knowledge and understanding of:

* the role of haemoglobin in transporting oxygen and carbon dioxide
* the oxygen dissociation curve for fetal and adult human haemoglobin

---

KEY DEFINITIONS

affinity: a strong attraction.
dissociation: means releasing the oxygen from the oxyhaemoglobin.
fetal haemoglobin: the type of haemoglobin usually found only in the fetus.
haemoglobin: the red pigment used to transport oxygen in the blood.

Haemoglobin

Oxygen is transported in the red blood cells (**erythrocytes**). These cells contain the protein haemoglobin. When haemoglobin takes up oxygen, it becomes **oxyhaemoglobin**.

haemoglobin + oxygen → oxyhaemoglobin

Haemoglobin is a complex protein with four subunits. Each subunit consists of a polypeptide chain and a haem (non-protein) group. The haem group contains a single iron ion in the form of Fe^{2+}. This iron ion can attract and hold an oxygen molecule. The haem group is said to have a high affinity (attraction) for oxygen. As each haem group can hold one oxygen molecule, each haemoglobin molecule can carry four oxygen molecules. It has been calculated that there are about 280 million molecules of haemoglobin in each red blood cell – so a red blood cell can carry over a billion oxygen molecules.

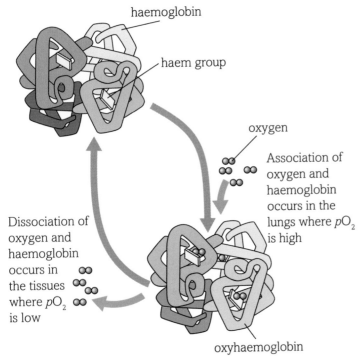

haemoglobin

haem group

oxygen

Association of oxygen and haemoglobin occurs in the lungs where pO_2 is high

Dissociation of oxygen and haemoglobin occurs in the tissues where pO_2 is low

oxyhaemoglobin

Figure 1 Reversible binding of oxygen by haemoglobin.

LEARNING TIP

Remember to say that haemoglobin has a *high* affinity for oxygen – not just an affinity. Also remember that it is haemoglobin that has a high affinity for oxygen, not oxygen that has a high affinity for haemoglobin.

Transport of oxygen

Oxygen is absorbed into the blood as it passes the alveoli in the lungs. Oxygen molecules diffusing into the blood plasma enter the red blood cells. Here they become associated with the haemoglobin. This means that the oxygen binds reversibly to the haemoglobin (see Figure 1). This takes the oxygen molecules out of solution and so maintains a steep concentration gradient, allowing more oxygen to enter the blood from the lungs and diffuse into the cells.

The blood carries the oxygen from the lungs back to the heart, before travelling around the body to supply the tissues. In the body tissues, cells need oxygen for aerobic respiration. Therefore the oxyhaemoglobin must be able to release the oxygen. This is called **dissociation**.

LEARNING TIP

Avoid using non-scientific terms such as haemoglobin 'takes up' or 'picks up' oxygen. It is better to use the correct terms such as 'associates with' or 'binds with' oxygen.

Haemoglobin and oxygen transport

The ability of haemoglobin to associate with and release oxygen depends on the concentration of oxygen in the surrounding tissues. The concentration of oxygen is measured by the relative pressure that it contributes to a mixture of gases. This is called the **partial pressure** of oxygen or pO_2. It is also called the **oxygen tension**, and is measured in units of pressure (kPa).

With a normal liquid, you might expect the concentration of oxygen absorbed into the liquid to be directly proportional to the oxygen tension in the surrounding air. A graph of percentage saturation plotted against oxygen tension would be a straight line (linear). This is not the case with blood containing haemoglobin.

Haemoglobin can associate with oxygen in a way that produces an S-shaped curve. This is called the haemoglobin dissociation curve. At low oxygen tension, the haemoglobin does not readily associate with oxygen molecules. This is because the haem groups that attract the oxygen are in the centre of the haemoglobin molecule. This makes it difficult for the oxygen molecule to reach the haem group and associate with it. This difficulty in combining with the first oxygen molecule accounts for the low saturation level of haemoglobin at low oxygen tensions.

As the oxygen tension rises, the diffusion gradient into the haemoglobin molecule increases. Eventually, one oxygen molecule enters the haemoglobin molecule and associates with one of the haem groups. This causes a slight change in the shape of the haemoglobin molecule, known as a **conformational change**. It allows more oxygen molecules to enter the haemoglobin molecule and associate with the other haem groups relatively easily. This accounts for the steepness of the curve as the oxygen tension rises.

As the haemoglobin approaches 100% saturation, the curve levels off, creating the S-shaped curve seen in Figure 2 (a). Mammalian haemoglobin is well adapted to transporting oxygen to the tissues of a mammal. The oxygen tension found in the lungs is sufficient to produce close to 100% saturation. The oxygen tension in respiring body tissues is sufficiently low to cause oxygen to dissociate readily from the oxyhaemoglobin.

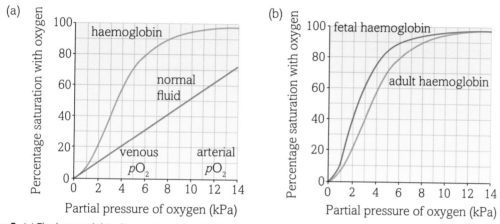

Figure 2 (a) The haemoglobin dissociation curve; (b) fetal vs adult haemoglobin dissociation curves.

Fetal haemoglobin

Fetal haemoglobin is slightly different from adult haemoglobin (see Figure 2 (b)). Fetal haemoglobin has a higher affinity for oxygen than adult haemoglobin. Therefore, the haemoglobin dissociation curve for fetal haemoglobin is to the left of the curve for adult haemoglobin. This is because fetal haemoglobin must be able to associate with oxygen in an environment where the oxygen tension is low enough to make adult haemoglobin release oxygen.

In the placenta, where the oxygen tension is low, fetal haemoglobin will absorb oxygen from the surrounding fluid. This reduces the oxygen tension even further. As a result, oxygen diffuses from the mother's blood fluid into the placenta. This reduces the oxygen tension within the mother's blood, which, in turn, makes the maternal haemoglobin release more oxygen (dissociation).

Questions

1 What is meant by the term 'affinity'?

2 Which part of the haemoglobin molecule causes the high affinity for oxygen?

3 State precisely where in the body haemoglobin binds to oxygen and where the oxygen dissociates from the haemoglobin.

4 Draw a haemoglobin dissociation curve and annotate the curve to explain the shape.

5 The haemoglobin produced by a baby after birth is different from that produced as a fetus. Explain why the haemoglobin produced must change.

6 The haemoglobin found in mammals that live at high altitude is different from that found in those at lower altitude. Suggest how it might be different, and explain why this is necessary.

(8) Transporting carbon dioxide

By the end of this topic, you should be able to demonstrate and apply your knowledge and understanding of:

* the role of haemoglobin in transporting oxygen and carbon dioxide
* the changes in the dissociation curve at different carbon dioxide concentrations

The role of haemoglobin

Carbon dioxide is released from respiring tissues. It must be removed from these tissues and transported by the blood to the lungs for excretion. Carbon dioxide is transported in three ways:

* About 5% is dissolved directly in the plasma.
* About 10% is combined directly with **haemoglobin** to form a compound called **carbaminohaemoglobin**.
* About 85% is transported in the form of **hydrogencarbonate ions** (HCO_3^-).

Formation of hydrogencarbonate ions

Carbon dioxide in the blood plasma diffuses into the red blood cells. Here, it combines with water to form a weak acid called **carbonic acid**. This reaction is catalysed by the enzyme carbonic anhydrase (see Figure 1).

$$CO_2 + H_2O \rightarrow H_2CO_3$$

This carbonic acid dissociates to release hydrogen ions (H^+) and hydrogencarbonate ions (HCO_3^-).

$$H_2CO_3 \rightarrow HCO_3^- + H^+$$

The hydrogencarbonate ions diffuse out of the red blood cell into the plasma. The charge inside the red blood cell is maintained by the movement of chloride ions (Cl^-) from the plasma into the red blood cell. This is called the chloride shift.

The hydrogen ions building up in the red blood cell could cause the contents of the red blood cell to become very acidic. To prevent this, the hydrogen ions are taken out of solution by associating with haemoglobin to produce haemoglobinic acid (HHb). The haemoglobin is acting as a buffer (a compound that maintains a constant pH).

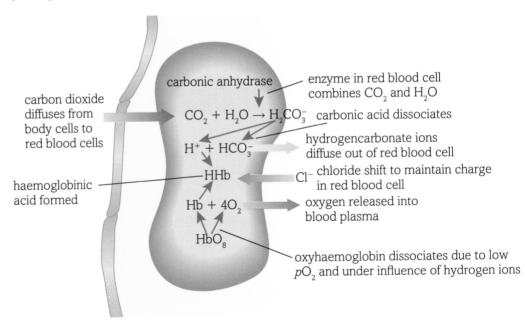

Figure 1 The formation of hydrogencarbonate ions.

The effect of increasing carbon dioxide concentration

Blood entering respiring tissues carries oxygen as oxyhaemoglobin. The partial pressure of oxygen in the respiring tissues is lower than that in the lungs, because oxygen has been used in respiration. As a result, the oxyhaemoglobin begins to dissociate and releases oxygen to the tissues. This means that the haemoglobin is available to take up the hydrogen ions, forming haemoglobinic acid. Where the tissues are very active there is more carbon dioxide released. This has a dramatic effect on the haemoglobin.

The Bohr effect

The **Bohr effect** (see Figure 2) describes the effect that an increasing concentration of carbon dioxide has on the haemoglobin.

- Carbon dioxide enters the red blood cells forming carbonic acid, which dissociates to release hydrogen ions.
- These hydrogen ions affect the pH of the cytoplasm, making it more acidic.
- As with any protein, changes in pH can affect the tertiary structure of the haemoglobin. The increased acidity alters the tertiary structure of the haemoglobin and reduces the affinity of the haemoglobin for oxygen.
- The haemoglobin is unable to hold as much oxygen, and oxygen is released from the oxyhaemoglobin to the tissues.

DID YOU KNOW?

Christian Bohr was a medical doctor in Denmark. He published his first scientific paper in 1877 and became professor of physiology at the University of Copenhagen in 1886. His particular interest was the physiology of breathing and gas exchange. He was the first person to realise that the lungs contain dead space – air that is not involved in gaseous exchange. In 1903, he described the phenomenon now known as the Bohr effect. He is the father of the famous physicist Niels Bohr, who won a Nobel prize for physics in 1922. Christian Bohr's grandson also won a Nobel prize for physics.

Where tissues (such as contracting muscles) are respiring more, there will be more carbon dioxide. As a result, there will be more hydrogen ions produced in the red blood cells. This makes the oxyhaemoglobin release more oxygen.

So, when more carbon dioxide is present, haemoglobin becomes less saturated with oxygen. This is reflected in a change to the haemoglobin dissociation curve, which shifts downwards and to the right – the **Bohr shift**.

This Bohr effect results in more oxygen being released where more carbon dioxide is produced in respiration. This is just what the muscles need for aerobic respiration to continue.

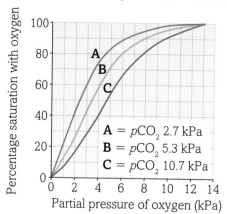

Figure 2 Dissociation curves showing the Bohr effect.

Questions

1. What effect will formation of carbaminohaemoglobin have on the amount of oxygen carried in the erythrocytes?

2. Describe how carbon dioxide is converted to hydrogencarbonate ions.

3. During conversion of carbon dioxide to hydrogencarbonate ions, chloride ions enter the red blood cell. Explain why this is necessary.

4. Explain why the presence of more carbon dioxide reduces the affinity of haemoglobin for oxygen.

5. Draw a flow chart to describe the Bohr effect.

6. When the red blood cells reach the lungs, they release carbon dioxide. Draw a flow chart to explain what happens in the red blood cells as they release carbon dioxide.

LIVING AT ALTITUDE

Mountaineers visiting high altitudes (see Figure 1) have often wondered how their local guides can carry heavy packs and still keep up a good trekking pace despite the low levels of oxygen in the air. Visitors will acclimatise to the oxygen levels by producing larger numbers of red blood cells. However, people in Tibet do not show high levels of haemoglobin in their blood. This article explores how evolution has enabled other ways of adapting to the low oxygen levels.

ADAPTATIONS TO LIVING AT ALTITUDE

A gene that controls red blood cell production evolved quickly to enable Tibetans to tolerate high altitudes, a study suggests. The finding could lead researchers to new genes controlling oxygen metabolism in the body.

An international team of researchers compared the DNA of 50 Tibetans with that of 40 Han Chinese and found 34 mutations that have become more common in Tibetans in the 2750 years since the populations split. More than half of these changes are related to oxygen metabolism.

The researchers looked at specific genes responsible for high-altitude adaptation in Tibetans. 'By identifying genes with mutations that are very common in Tibetans, but very rare in lowland populations we can identify genes that have been under natural selection in the Tibetan population,' said Professor Rasmus Nielsen of the University of California Berkeley, who took part in the study. 'We found a list of 20 genes showing evidence for selection in Tibet – but one stood out: *EPAS1*.'

The gene, which codes for a protein involved in responding to falling oxygen levels, and is associated with improved athletic performance in endurance athletes, seems to be the key to the Tibetan adaptation to life at high altitude. A mutation in the gene that is thought to affect red blood cell production was present in only 9% of the Han population, but was found in 87% of the Tibetan population.

'It is the fastest change in the frequency of a mutation described in humans,' said Professor Nielsen.

There is 40% less oxygen in the air on the 4000 m high Tibetan plateau than at sea level. Under these conditions, people accustomed to living below 2000 m – including most Han Chinese – cannot get enough oxygen to their tissues, and experience altitude sickness. They get headaches, tire easily, and have lower birth rates and higher child mortality than high-altitude populations.

Tibetans have none of these problems, despite having lower oxygen saturation in their tissues and a lower red blood cell count than the Han Chinese.

Around the world, populations have adapted to life at high altitude in different ways. One adaptation involves making more red blood cells, which transport oxygen to the body's tissues. Indigenous people in the Peruvian Andes have higher red blood cell counts than their countrymen living at sea level, for example.

But Tibetans have evolved a different method. 'Tibetans have the highest expression levels for *EPAS1* in the world,' said co-author Dr Jian Wang of the Beijing Genomics Institute in Schenzhen [*sic*], China, a research facility that collected the data. 'For Western people, after two to three weeks at altitude, the red blood cell count starts to increase. But Tibetans and Sherpas keep the same levels,' he said.

'I just summitted Everest a few weeks ago,' added Dr Wang. He said the Sherpas and Tibetans were much stronger than the Westerners or lowland Chinese on the climb. 'Their tissue oxygen concentration is almost the same as Westerners and Chinese but they are strong,' he said 'and their red blood cell count is not that high compared to people in Peru.'

'The remarkable thing about Tibetans is that they can function well in high altitudes without having to produce so much haemoglobin,' said Prof Nielsen. 'The entire mechanism is not well-understood – but is seems that the gene responsible is *EPAS1*.'

Nielsen said the gene is involved in regulating aerobic and anaerobic metabolism in the body (cell respiration with and without oxygen). 'It may be that the [mutated gene] helps balance anaerobic versus aerobic metabolism in a way that is more optimal for the low-oxygen environment of the Tibetan plateau,' he said.

Writing in *Science*, where the results are published today, the authors say: '*EPAS1* may therefore represent the strongest instance of natural selection documented in a human population, and variation at this gene appears to have had important consequences for human survival and/or reproduction in the Tibetan region.'

Dr Wang said future research will focus on comparing the levels of *EPAS1* expression in the placentas of Tibetan and Han Chinese women.

Source

- http://www.theguardian.com/science/2010/jul/02/mutation-gene-tibetans-altitude

Where else will I encounter these themes?

1.1   2.1   2.2   2.3   2.4   2.5

Let's start by considering the nature of the study described in the article.

1. Give a brief evaluation of the study described in paragraphs two and three.

You should consider the number of people studied and suggest how these people should have been selected.

Now we will look at the biology in, or connected to, this article. Don't worry if you are not ready to give answers to these questions yet. You may like to return to the questions once you have covered other topics later in the book. Use the timeline at the bottom of the page to help you put this work in context with what you have already learned and what is ahead in your course.

2. How does having a high red blood cell count help the Peruvians overcome altitude sickness?

3. What is the role of haemoglobin?

4. Explain why many elite athletes spend time training at high altitude.

5. Tibetans typically demonstrate a much higher breathing rate than other people. How might this benefit living at high altitude?

6. The article refers to selection of a gene (paragraph 3). It would be more accurate to refer to a selection of alleles – explain why.

7. The article states 'a gene that controls red blood cell production evolved quickly' (line 1). A scientist stated that this was an inaccurate statement. Explain why this may not be accurate and give a more scientifically accurate description of what has happened.

8. *EPAS1* codes for a transcription factor. It helps to regulate the expression of other genes. Suggest what effects the *EPAS1* gene may bring about that help people survive in places with low oxygen availability.

9. Suggest why researchers will focus future research on *EPAS1* expression in the placenta.

Use the terms 'mutation', 'allele', 'natural selection' and 'isolation'.

Think about availability of material for medical research and how the fetus gains oxygen.

Activity

When people climb to high altitude, the physiology of oxygen transport changes to adapt to the lower availability of oxygen. Draw a flow diagram of the path taken by oxygen from the air in our lungs to the muscles where oxygen is used in respiration. Annotate the diagram to explain the role of haemoglobin in transporting oxygen, ensuring that you use correct scientific terms as much as possible. Using a different colour, annotate your diagram with suggestions about where this pathway may be altered or adapted in people who normally live at low altitude but have climbed to an altitude of 5000 m.

Figure 1 Everest base camp.

Practice questions

1. What is the correct name of the cells that transport oxygen in the blood system of a mammal? [1]

   A. Neutrophils

   B. Haemoglobin

   C. Erythrocytes

   D. Lymphocytes

2. In what form is the majority of carbon dioxide transported in the blood of a mammal? [1]

   A. Carboxyhaemoglobin

   B. Carbaminohaemoglobin

   C. Dissolved in the plasma

   D. Hydrogencarbonate ions in the plasma

3. Figure 1 shows a diagram of the mammalian heart. Which option (A–D) gives the correct labels? [1]

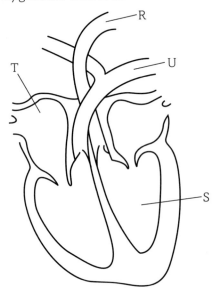

Figure 1 A diagram of a mamalian heart

	R	S	T	U
A	aorta	left ventricle	right atrium	pulmonary vein
B	aorta	left ventricle	right atrium	pulmonary artery
C	aorta	right ventricle	left atrium	pulmonary artery
D	pulmonary artery	left ventricle	right atrium	aorta

Table 1 The events that take place during the different phases of the eukaryotic cell cycle.

4. Figure 2 shows the pressures in kPa at one point in a capillary and in the surrounding tissue fluid. What is the net pressure and in what direction will the fluid move? [1]

   A. 1.2 kPa out of the capillary

   B. 0.7 kPa out of the capillary

   C. 1.2 kPa into the capillary

   D. 0.7 kPa into the capillary

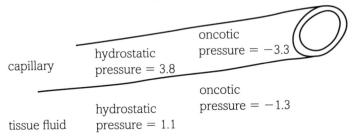

Figure 2

5. Haemoglobin has an affinity for oxygen. Some types of haemoglobin may have higher or lower affinity. Which row in the following table is correct? [1]

	Haemoglobin from a human adult	Haemoglobin from a human fetus	Haemoglobin from an animal living at high altitude
A	high	lower	higher
B	high	higher	lower
C	high	higher	higher
D	low	higher	higher

[Total: 5]

6. (a) State three reasons why some animals need a transport system. [3]

   (b) Blood systems can be open or closed.
      (i) Explain what is meant by the term 'open blood system'. [1]

      (ii) Name one type of organism that has an open blood system. [1]

   (c) Explain why a closed double circulatory system is more efficient than either an open system or a single circulatory system. [8]

   [Total: 13]

7. The table below describes the steps in the cardiac cycle.

	Step in cardiac cycle
A	The atrio-ventricular valves open
B	The atrio-ventricular valves close
C	The semilunar valves open
D	The semilunar valves close
E	The atria contract
F	The ventricles contract
G	The atria relax
H	The ventricles relax

(a) Place the steps described above in the correct sequence starting with 'The atria contract'. [7]

(b) Identify the parts labelled K and L on Figure 3. [2]

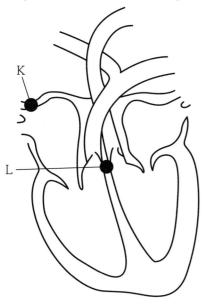

Figure 3.

(c) Describe how the contraction of the atria and ventricles is coordinated. [5]

[Total: 14]

8. Most carbon dioxide is transported in as hydrogencarbonate ions in the plasma.

(a) Describe two other ways in which carbon dioxide can be transported. [3]

(b) The series of reactions that convert carbon dioxide to hydrogencarbonate can be represented as follows:

$$CO_2 + H_2O \overset{A}{\rightarrow} H_2CO_3 \overset{B}{\rightarrow} H^+ + HCO_3^-$$

  (i) Where does this sequence of reactions occur? [1]
  (ii) Name the enzyme used at A. [1]
  (iii) What process occurs at B? [1]
  (iv) Name the compound H_2CO_3. [1]
  (v) What happens to the hydrogencarbonate ions? [1]

(c) The hydrogen ions could cause a change in the pH of the solution. Describe how this change in pH is avoided and the effect this has on the release of oxygen in tissues that have a high concentration of carbon dioxide. [6]

[Total: 14]

9. The table below compares the structure of artery walls with those of the veins.

Structural feature	Artery	Vein	Function
thickness of wall		thin	
valves		present	
smooth muscle	thick layer		
collagen	thick layer		
elastic tissue		thin layer	

(a) Complete the table. [5]

(b) Explain why arteries near the heart tend to have more elastic tissue than those further from the heart. [3]

(c) Arterioles are found between the arteries and the capillaries.
  (i) What is the role of arterioles? [2]
  (ii) Explain why you are advised not to swim shortly after eating. [3]

[Total: 13]

10. The following questions relate to transport via the blood system.

(a) List three things that are transported by the blood system. [3]

(b) Having a closed blood circulatory system enables the blood to be redistributed according to the needs of the body.
  (i) Some people go 'red in the face' when they exercise. In fact most of the skin will turn more pink. Explain why this occurs. [3]
  (ii) Describe the changes to the blood system that cause the change in skin colour. [3]

[Total: 9]

Exchange and transport

TRANSPORT IN PLANTS

Introduction

Plants play a huge part in the recycling of water. When rain falls plants protect the soil from erosion. The tree roots then take the water out of the soil and evaporation from the leaves returns the water as vapour to the atmosphere. Transporting water up the plant provides water for photosynthesis and maintains cell turgidity. In non-woody plants this water supply provides the only support that keeps the plant from wilting. The flow of water is also used to transport mineral ions up the stem from the roots to the leaves. Products of photosynthesis are transported down again in the flow of sap to the roots.

Many of the tissues in a plant are associated with transporting water and the substances dissolved in it. Substances are made by photosynthesis in the leaves but are often stored underground in the roots. In spring these substances must be transported up the plant again for use in growing new leaves.

All the maths you need

To unlock the puzzles in this section you need the following maths:

- Recognise and make use of appropriate units in calculations
- Recognise and use expressions in decimal and standard form
- Use ratios, fractions and percentages
- Estimate results
- Use an appropriate number of significant figures
- Find arithmetic means
- Construct and interpret frequency tables and diagrams, bar charts and histograms
- Understand the terms mean, median and mode
- Translate information between graphical, numerical and algebraic forms
- Plot two variables from experimental or other data
- Understand that $y = mx + c$ represents a linear relationship
- Calculate rate of change from a graph showing a linear relationship
- Draw and use the slope of a tangent to a curve as a measure of rate of change
- Identify uncertainties in measurements and use simple techniques to determine uncertainty when data are combined
- Understand and use the symbols: $=$, $<$, $<<$, $>>$, $>$, $\sim$
- Calculate the circumferences, surface areas and volumes of regular shapes

What have I studied before?

- The relationship between size and surface area to volume ratio
- How substances move by diffusion and osmosis
- How substances pass through membranes
- The structure of leaves and the role of stomata in transpiration
- The process of transpiration and the factors that affect the rate of transpiration

What will I study later?

- How communicable diseases spread through a plant by making use of the flow of water through the plant and how plants can reduce the spread of the disease (AS)
- How plants make use of the water transported in photosynthesis (AL)

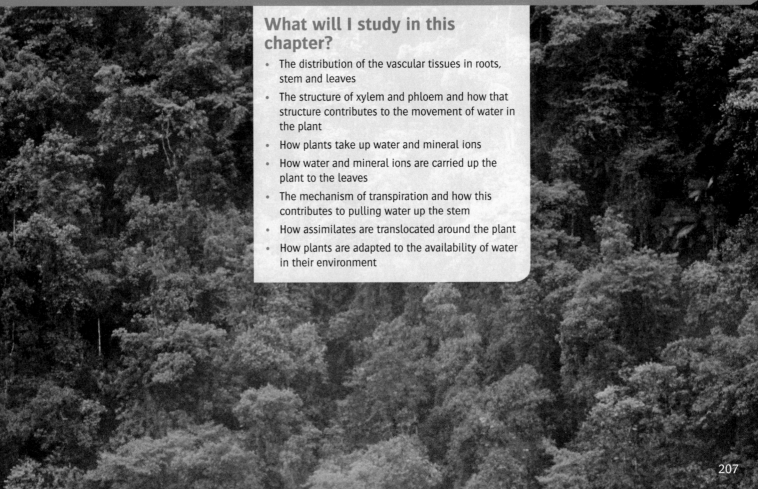

What will I study in this chapter?

- The distribution of the vascular tissues in roots, stem and leaves
- The structure of xylem and phloem and how that structure contributes to the movement of water in the plant
- How plants take up water and mineral ions
- How water and mineral ions are carried up the plant to the leaves
- The mechanism of transpiration and how this contributes to pulling water up the stem
- How assimilates are translocated around the plant
- How plants are adapted to the availability of water in their environment

(1) Transport in plants

By the end of this topic, you should be able to demonstrate and apply your knowledge and understanding of:

* the need for transport systems in multicellular plants
* the structure and function of the vascular system in the roots, stems and leaves of herbaceous dicotyledonous plants
* the examination and drawing of stained sections of plant tissue to show the distribution of xylem and phloem
* the dissection of stems, both longitudinally and transversely, and their examination to demonstrate the position and structure of xylem vessels.

KEY DEFINITIONS

dicotyledonous plants: plants with two seed leaves and a branching pattern of veins in the leaf.
meristem: a layer of dividing cells, here it is called the pericycle.
phloem: transports dissolved assimilates.
vascular tissue: consists of cells specialised for transporting fluids by mass flow.
xylem: transports water and minerals.

Why do plants need a transport system?

All living things need to take substances from, and return wastes to, their environment. As with animals (see topics 3.1.1 and 3.2.1), larger plants have a smaller surface area to volume ratio. Therefore, they need to have specialised exchange surfaces and a transport system.

Every cell of a multicellular plant needs a regular supply of oxygen, water, nutrients and minerals. Plants are not very active, and their respiration rate is low – therefore the demand for oxygen is low. This demand can be met by diffusion. However, the demand for water and sugars is still high. Plants can absorb water and minerals at the roots, but they cannot absorb sugars from the soil. The leaves can perform gaseous exchange and manufacture sugars by photosynthesis, but they cannot absorb water from the air. Therefore, plants need a transport system to move:

* water and minerals from the roots up to the leaves
* sugars from the leaves to the rest of the plant.

The vascular tissues

The transport system in plants consists of specialised **vascular tissue**:

* water and soluble mineral ions travel *upwards* in **xylem** tissue
* assimilates, such as sugars, travel *up or down* in **phloem** tissue.

Both xylem and phloem are highly specialised to carry out their transport function. But, unlike in animals, there is no pump, and respiratory gases are not carried by these tissues.

Distribution of vascular tissue

Dicotyledonous plants are those that have two seed leaves. They also have a very characteristic distribution of vascular tissue. The vascular tissue is distributed throughout the plant. The xylem and phloem are found together in vascular bundles. These bundles may also contain other types of tissue (such as **collenchyma** and **sclerenchyma**) that give the bundle some strength and help to support the plant.

Xylem and phloem in the young root

The vascular bundle is found at the centre of a young root (see Figure 1). There is a central core of xylem, often in the shape of an X. The phloem is found in between the arms of the X-shaped xylem tissue. This arrangement provides strength to withstand the pulling forces to which roots are exposed.

Around the vascular bundle is a special sheath of cells called the **endodermis**. The endodermis has a key role in getting water into the xylem vessels, which is discussed in detail in topic 3.3.5. Just inside the endodermis is a layer of **meristem** cells (cells that remain able to divide) called the **pericycle**.

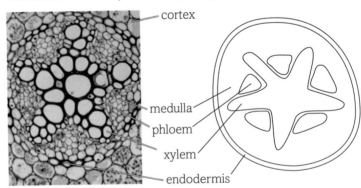

Figure 1 Transverse section of root, showing a plan view of tissues in the centre of a root observed at low power.

Xylem and phloem in the stem

The vascular bundles are found near the outer edge of the stem. In non-woody plants the bundles are separate and discrete. In woody plants the bundles are separate in young stems, but become a continuous ring in older stems. This means there is a complete ring of vascular tissue just under the bark of a tree. This arrangement provides strength and flexibility to withstand the bending forces to which stems and branches are exposed.

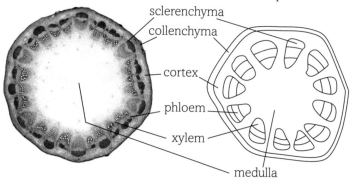

Figure 2 Transverse section of stem (×6), showing vascular bundles, with a tissue plan.

The xylem is found towards the inside of each vascular bundle and the phloem towards the outside (see Figure 2). In between the xylem and phloem is a layer of **cambium**. The cambium is a layer of meristem cells that divide to produce new xylem and phloem.

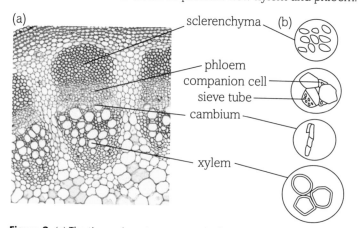

Figure 3 (a) The tissues in a stem seen under low power (×26). (b) High-power drawings of cells.

Xylem and phloem in the leaf

The vascular bundles form the midrib and veins of a leaf (see Figure 4). A dicotyledonous leaf has a branching network of veins that get smaller as they spread away from the midrib. Within each vein, the xylem is located on top of the phloem.

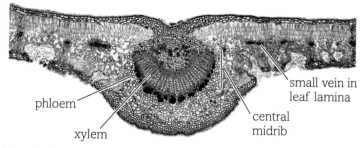

Figure 4 Transverse section of leaf, showing a tissue plan.

Dissection of plant material

The dissection of plant material to examine the distribution of vascular tissue requires staining of the tissue. This is most easily demonstrated in the leaf stalk of celery as shown in Figure 5.

However, it can also be carried out with busy lizzie (*Impatiens*) stems. Thin sections can be cut and viewed at low power. Allow the leafy stem to take up water by transpiration. The stem can then be cut longitudinally or transversely and examined with a hand lens or microscope.

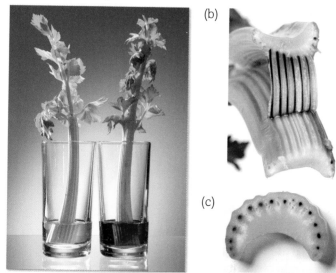

Figure 5 (a) Staining celery to show the distribution of xylem; (b) the longitudinal distribution; and (c) the transverse section.

INVESTIGATION

Wear eye protection.
(a) Observe prepared slides of plant tissue under a light microscope at low power and draw a tissue plan (see Figures 1 and 2).
(b) Observe prepared slides of plant tissue at high power and draw a few cells from each tissue type (see Figure 3).
(c) Staining plant sections to see the xylem

Both xylem and sclerenchyma contain lignin in their walls. Phloroglucinol is a stain that is specific to lignin.
1. Cut a very thin section of a plant stem using a one-sided razor blade or scalpel. (It is usually best to cut the stem straight across and then cut another very thin section at an angle so that you do not cut all the way across the stem. In this way you should achieve a section that is only one or two cells thick.)
2. Place the section on a slide.
3. Place a few drops of acidified phloroglucinol (as per CLEAPSS Recipe 93) over your section.
4. Cover with a coverslip and observe using a light microscope.

Questions

1. Explain why plants do not need to transport gases such as oxygen in their transport system.

2. Explain the difference between diffusion and mass flow, as seen in the vessels of a transport system.

3. Explain why mass flow is the best way to transport substances around the plant.

4. Describe the distribution of xylem in the root, stem and leaves.

5. Describe the distribution of phloem in the root, stem and leaves.

6. Use an annotated diagram to explain why roots are subjected to pulling forces, whilst the stem is subjected to bending forces.

By the end of this topic, you should be able to demonstrate and apply your knowledge and understanding of:

* the structure and function of the vascular system in the roots, stems and leaves of herbaceous dicotyledonous plants

Xylem

The structure and function of xylem

Xylem (see Figure 1) is a tissue used to transport water and mineral ions from the roots up to the leaves and other parts of the plant. Xylem tissue consists of:

* vessels to carry the water and dissolved mineral ions
* fibres to help support the plant
* living parenchyma cells which act as packing tissue to separate and support the vessels.

Xylem vessels

As xylem vessels develop, **lignin** impregnates the walls of the cells, making the walls waterproof. This kills the cells. The end walls and contents of the cells decay, leaving a long column of dead cells with no contents – a tube called the xylem vessel. The lignin strengthens the vessel walls and prevents the vessel from collapsing. This keeps vessels open even at times when water may be in short supply.

The lignin thickening forms patterns in the cell wall. These may be spiral, annular (rings) (see Figure 2) or reticulate (a network of broken rings). This prevents the vessel from being too rigid and allows some flexibility of the stem or branch.

In some places **lignification** is not complete, leaving gaps in the cell wall. These gaps form pits or **bordered pits**. The bordered pits in two adjacent vessels are aligned to allow water to leave one vessel and pass into the next vessel. They also allow water to leave the xylem and pass into the living parts of the plant.

Adaptations of xylem to its function

Xylem vessels can carry water and mineral ions from the roots to the very top of the plant because:

* they are made from dead cells aligned end to end to form a continuous column
* the tubes are narrow, so that the water column does not break easily and capillary action can be effective

* bordered pits in the lignified walls allow water to move sideways from one vessel to another
* lignin deposited in the walls in spiral, annular or reticulate patterns allows xylem to stretch as the plant grows, and enables the stem or branch to bend.

The flow of water is not impeded, because:

* there are no cross-walls
* there are no cell contents, nucleus or cytoplasm
* lignin thickening prevents the walls from collapsing.

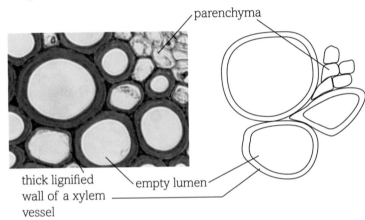

Figure 1 Transverse section of xylem, with high-power drawing.

Figure 2 Xylem in longitudinal section, showing spiral thickening.

Phloem

The structure and function of phloem

Phloem (see Figure 3) is a tissue used to transport assimilates (mainly sucrose and amino acids) around the plant. The sucrose is dissolved in water to form sap. Phloem tissue consists of sieve tubes – made up of **sieve tube elements** – and **companion cells**.

Sieve tube elements

Elongated sieve tube elements are lined up end to end to form sieve tubes. They contain no nucleus and very little cytoplasm, leaving space for mass flow of sap to occur. At the ends of the sieve tube elements are perforated cross-walls called **sieve plates**. The perforations in the sieve plate allow movement of the sap from one element to the next. The sieve tubes have very thin walls and when seen in transverse section are usually five- or six-sided.

DID YOU KNOW?

At first sight, the sieve plates appear to have no real function and actually obstruct the free flow of sap in the sieve tube. However, they may act to support the tube – keeping the lumen open. But, more importantly, they serve as a mechanism to block the sieve tube after injury or infection. The pores in the sieve plate very rapidly become blocked by deposition of **callose** (a complex carbohydrate) (see Figure 4). This prevents loss of sap and inhibits transport of pathogens around the plant.

Companion cells

In between the sieve tubes are small cells, each with a large nucleus and dense cytoplasm. These are the companion cells. They have numerous mitochondria to produce the ATP needed for active processes. The companion cells carry out the metabolic processes needed to load assimilates actively into the sieve tubes (see topic 3.3.7).

DID YOU KNOW?

The companion cells and sieve tube elements in phloem are linked by fine strands of cytoplasm, through gaps in the cell walls. This allows communication and flow of substances between the cells. See the next topic for more about the role of plasmodesmata.

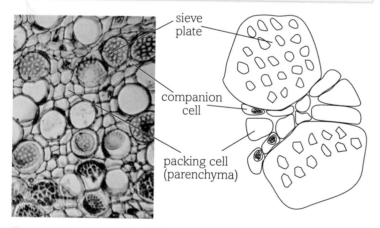

Figure 3 Transverse section of phloem, with high-power drawing.

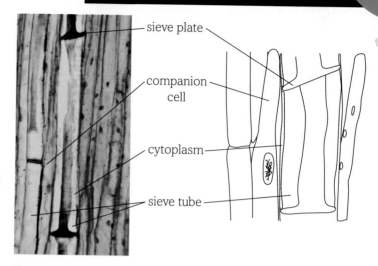

Figure 4 Longitudinal section of phloem (×90), with high-power drawing. Note that the sieve tubes may become blocked by deposition of callose, shown in red.

Questions

1. Construct a table to compare the structure of xylem tissue with that of phloem tissue.

2. Draw a diagram of a xylem vessel and annotate your diagram with the features that adapt the xylem to its function.

3. Why do xylem vessels have bordered pits?

4. Explain why sieve tube elements are not true cells.

5. Draw an annotated diagram of phloem tissue.

6. Suggest why phloem does not have walls impregnated with lignin.

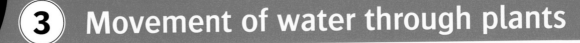

By the end of this topic, you should be able to demonstrate and apply your knowledge and understanding of:

* the transport of water into the plant, through the plant and to the air surrounding the leaves

> **KEY DEFINITION**
> **plasmodesmata:** gaps in the cell wall containing cytoplasm that connects two cells.

Pathways taken by water

The cellulose cell walls of a plant cell are fully permeable to water. Water molecules can move freely between the cellulose molecules or even in gaps between the cells. Water can also pass across the cell wall and through the partially permeable plasma membrane into the cell cytoplasm or even into the vacuole. Many plant cells are joined by special cytoplasmic bridges. These are cell junctions at which the cytoplasm of one cell is connected to that of another through a gap in their cell walls. These junctions are called **plasmodesmata** (singular: plasmodesma).

This means that there are three possible pathways taken by water through a plant. These are shown in Figure 1.

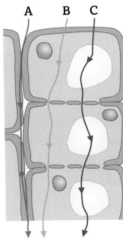

Figure 1 Three pathways taken by water through a plant (A = apoplast pathway, B = symplast pathway, C = vacuolar pathway).

The apoplast pathway

Water passes through the spaces in the cell walls and between the cells. It does not pass through any plasma membranes into the cells. This means that the water moves by mass flow rather than by osmosis. Also, dissolved mineral ions and salts can be carried with the water.

The symplast pathway

Water enters the cell cytoplasm through the plasma membrane. It can then pass through the plasmodesmata from one cell to the next.

The vacuolar pathway

This is similar to the symplast pathway, but the water is not confined to the cytoplasm of the cells. It is able to enter and pass through the vacuoles as well.

Movement from cell to cell

Water potential

Water potential (ψ) is a measure of the tendency of water molecules to move from one place to another. You will find a detailed explanation of water potential in topic 2.5.3.

Water always moves from a region of higher water potential to a region of lower water potential. The water potential of pure water is zero. In a plant cell, the cytoplasm contains mineral ions and sugars (solutes) that will reduce the water potential. This is because there are fewer 'free' water molecules available than in pure water. As a result, the water potential in plant cells is always negative.

> **LEARNING TIP**
>
> Describing water potentials as higher and lower can be confusing. It is best to describe water potentials as 'more negative' (lower) or 'less negative' (higher).

Water uptake

If you place a plant cell in pure water, it will take up water molecules by osmosis (see Figure 2 and topic 2.5.3). This is because the water potential in the cell is more negative (lower) than the water potential of the water. Water molecules will move down the water-potential gradient into the cell. But the cell will not continue to absorb water until it bursts. This is because the cell has a strong cellulose cell wall. Once the cell is full of water it is described as being **turgid**. The water inside the cell starts to exert pressure on the cell wall, called the **pressure potential** (ψ_p). As the pressure potential builds up, it reduces the influx of water.

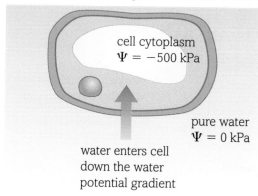

cell cytoplasm
$\Psi = -500$ kPa

pure water
$\Psi = 0$ kPa

water enters cell down the water potential gradient

Figure 2 Water entering a cell due to osmosis.

Water loss

If a plant cell is placed in a salt solution with a very negative (low) water potential, then it will lose water by osmosis. This is because the water potential of the cell is less negative (higher) than the water potential of the solution, so water moves down the water potential gradient out of the cell. As water loss continues, the cytoplasm and vacuole shrink. Eventually, the cytoplasm no longer pushes against the cell wall, and the cell is no longer turgid. If water continues to leave the cell, then the plasma membrane will lose contact with the wall – a condition known as **plasmolysis** (see Figure 3). The tissue is now **flaccid**.

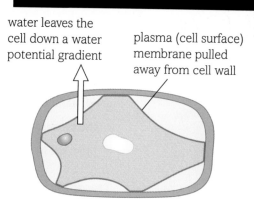

water leaves the cell down a water potential gradient

plasma (cell surface) membrane pulled away from cell wall

Figure 3 Water leaving a cell by osmosis, causing plasmolysis.

> **LEARNING TIP**
>
> Water will always move down a water potential gradient (from a less negative potential to a more negative potential).

Movement of water between cells

When plant cells are touching each other, water molecules can pass from one cell to another. The water molecules will move from the cell with the less negative (higher) water potential to the cell with the more negative (lower) water potential. This is *osmosis* (see Figure 4 and topic 2.5.3).

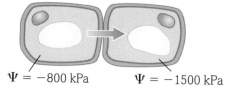

$\Psi = -800$ kPa $\Psi = -1500$ kPa

water moves from the cell with the higher water potential (-800 kPa) to the cell with the lower (more negative) water potential (-1500 kPa)

Figure 4 Water moving from one cell to another by osmosis.

Questions

1. The apoplast pathway carries water by mass flow; what additional mechanism is required to allow water molecules to enter the cells?

2. Explain what is meant by the following terms: water potential, osmosis, turgid, flaccid, plasmolysis.

3. Which component of the cell is partially permeable, allowing the movement of water molecules but not mineral ions?

4. Explain how a cell becomes plasmolysed.

5. Plant cell walls are fully permeable to water and minerals. What is found in the space between the cell wall and the plasma membrane in a plasmolysed cell?

6. Draw a diagram of three cells in a pyramid, with each cell touching the other two. Write in the water potential of each cell (cell A = -1000 kPa, cell B = -1200 kPa and cell C = -800 kPa). Now annotate the cells with arrows to show the direction of water movement, and explain why the water moves in that direction.

4 Transpiration

By the end of this topic, you should be able to demonstrate and apply your knowledge and understanding of:

* the process of transpiration and the environmental factors that affect transpiration rate

* practical investigations to estimate transpiration rates

Transpiration is the loss of water vapour from the upper parts of the plant – particularly the leaves. Some water may evaporate through the upper leaf surface, but this loss is limited by the **waxy cuticle**. Most water vapour leaves through the **stomata**, which open to allow gaseous exchange for photosynthesis. Since photosynthesis occurs only when there is sufficient light, the majority of water vapour is lost during the day.

The typical pathway taken by most water leaving the leaf is shown in Figure 1.

1. Water enters the leaf through the xylem, and moves by osmosis into the cells of the spongy mesophyll. It may also pass along the cell walls via the apoplast pathway.

2. Water evaporates from the cell walls of the spongy mesophyll.

3. Water vapour moves by diffusion out of the leaf through the open stomata. This relies on a difference in the concentration of water vapour molecules in the leaf compared with outside the leaf. This is known as the **water vapour potential gradient**. There must be a less negative (higher) water vapour potential inside the leaf than outside.

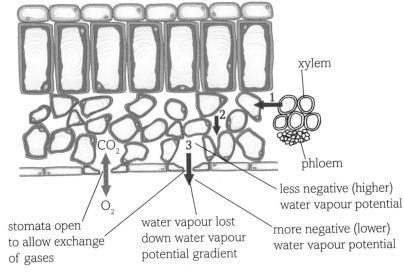

Figure 1 The route taken by most water as it leaves the leaf.

The importance of transpiration

Transpiration may be an inevitable consequence of gaseous exchange, but it is also essential for the plant to survive. As water vapour is lost from the leaf, it must be replaced from below. This draws water up the stem as a transpiration stream (see the next topic 3.3.5). This movement:

* transports useful mineral ions up the plant

* maintains cell turgidity

* supplies water for growth, cell elongation and photosynthesis

* supplies water that, as it evaporates, can keep the plant cool on a hot day.

Table 1 discusses the environmental factors that affect the transpiration rate.

Environmental factor	Explanation
Light intensity	In light, the stomata open to allow gaseous exchange for photosynthesis. Higher light intensity increases the transpiration rate.
Temperature	A higher temperature will increase the rate of transpiration in three ways. It will: • increase the rate of evaporation from the cell surfaces so that the water-vapour potential in the leaf rises • increase the rate of diffusion through the stomata because the water molecules have more kinetic energy • decrease the relative water vapour potential in the air, allowing more rapid diffusion of molecules out of the leaf.
Relative humidity	Higher relative humidity in the air will decrease the rate of water loss. This is because there will be a smaller water vapour potential gradient between the air spaces in the leaf and the air outside.
Air movement (wind)	Air moving outside the leaf will carry away water vapour that has just diffused out of the leaf. This will maintain a high water vapour potential gradient.
Water availability	If there is little water in the soil, then the plant cannot replace the water that is lost. If there is insufficient water in the soil, then the stomata close and the leaves wilt.

Table 1 The environmental factors that affect transpiration rate.

LEARNING TIP

When describing the effect of an environmental factor, remember to be clear that the environmental factor is increasing or decreasing to cause the change in transpiration rate.

INVESTIGATION

Measuring the rate of transpiration

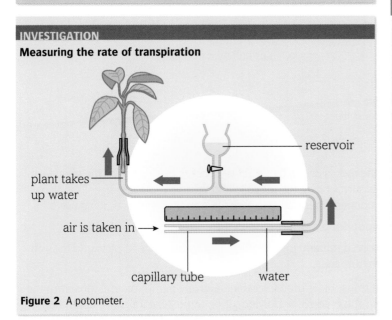

Figure 2 A potometer.

A piece of apparatus called a **potometer** (see Figure 2) can be used to estimate the rate of transpiration. What it actually measures is the rate of water uptake by a leafy shoot. Assuming that the cells are turgid, more than 95% of water taken up is lost by transpiration, so this gives a reasonable estimate of transpiration rate.

Using a potometer is straightforward once it is set up. Water vapour lost by the leaves is replaced from the water in the capillary tube. The movement of the meniscus at the end of the water column can be measured.

To study the effect of different environmental factors on the rate of transpiration, you can place the whole apparatus under different sets of conditions. Remember to vary only one factor at a time, in order to determine the effect on the transpiration rate.

When using a potometer, it is important to take certain precautions to ensure the results are valid:

1. Set it up under water to make sure there are no air bubbles inside the apparatus.
2. Ensure that the shoot is healthy.
3. Cut the stem under water to prevent air entering the xylem.
4. Cut the stem at an angle to provide a large surface area in contact with the water.
5. Dry the leaves.

INVESTIGATION

Measuring volume and transpiration rate

Measuring the volume of water taken up by the shoot involves calculating the volume of a cylinder (the length of capillary tube). The volume of a cylinder is given by the formula:

$v = \pi r^2 l$

(where r is the radius of the capillary tube and l is the length of the capillary tube).

The rate of transpiration is the volume calculated, divided by the time taken. Rate = volume/time.

LEARNING TIP

Remember that a rate must always involve time.

Questions

1. Suggest what features of a plant may affect the rate of transpiration and explain each point.

2. Explain why the water vapour potential in a leaf air space must be higher than that in the air outside.

3. Describe how you would use a potometer to investigate the effect of increasing wind speed on the rate of transpiration.

4. The distance moved by the meniscus in a potometer is 45 mm over five minutes. The radius of the capillary is 0.5 mm. Calculate the rate of transpiration.

5. The same potometer and leafy shoot are placed in brighter light. As a result, the meniscus moves 60 mm in five minutes. Calculate the percentage increase in rate of transpiration.

6. Explain the increase in the rate of transpiration caused by brighter light.

By the end of this topic, you should be able to demonstrate and apply your knowledge and understanding of:

* the transport of water into the plant, through the plant, and to the air surrounding the leaves

---

KEY DEFINITIONS

adhesion: the attraction between water molecules and the walls of the xylem vessel.
cohesion: the attraction between water molecules caused by hydrogen bonds.

The transpiration stream is the movement of water from the soil, through the plant, to the air surrounding the leaves. The main driving force is the water potential gradient between the soil and the air in the leaf air spaces.

Water uptake and movement across the root

The outermost layer of cells (the epidermis) of a root contains **root hair cells** (see Figure 1). These are cells with a long extension (root hair) that increases the surface area of the root. These cells absorb mineral ions and water from the soil. The water then moves across the root cortex down a water-potential gradient to the endodermis of the vascular bundle (see topic 3.3.1). Water may also travel through the apoplast pathway as far as the endodermis, but must then enter the symplast pathway, as the apoplast pathway is blocked by the **Casparian strip** (see Figure 2).

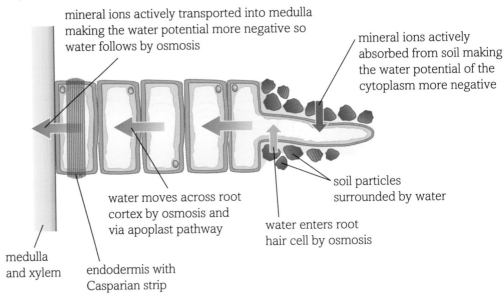

mineral ions actively transported into medulla making the water potential more negative so water follows by osmosis

mineral ions actively absorbed from soil making the water potential of the cytoplasm more negative

water moves across root cortex by osmosis and via apoplast pathway

soil particles surrounded by water

water enters root hair cell by osmosis

medulla and xylem

endodermis with Casparian strip

Figure 1 Water uptake and movement across the root.

The role of the endodermis

- The movement of water across the root is driven by an active process that occurs at the endodermis. The endodermis is a layer of cells surrounding the medulla and xylem. This layer of cells is also known as the starch sheath, as it contains granules of starch – a sign that energy is being used.

- The Casparian strip blocks the apoplast pathway between the cortex and the medulla.

- This ensures that water and dissolved mineral ions (especially nitrates) have to pass into the cell cytoplasm through the plasma membranes.

- The plasma membranes contain transporter proteins, which actively pump mineral ions from the cytoplasm of the cortex cells into the medulla and xylem. (See topic 2.5.4 for more about active transport.)

- This makes the water potential of the medulla and xylem more negative, so that water moves from the cortex cells into the medulla and xylem by osmosis.
- Once the water has entered the medulla, it cannot pass back into the cortex, as the apoplast pathway of the endodermal cells is blocked by the Casparian strip.

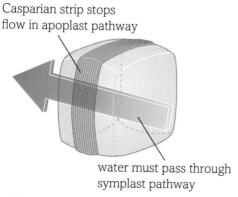

Figure 2 The Casparian strip.

Movement of water up the stem

Movement of water up through the xylem is by mass flow – a flow of water and mineral ions in the same direction. There are three processes that help to move water up the stem.

Root pressure

The action of the endodermis moving minerals into the medulla and xylem by active transport draws water into the medulla by osmosis. Pressure in the root medulla builds up and forces water into the xylem, pushing the water up the xylem. Root pressure can push water a few metres up a stem, but cannot account for water getting to the top of tall trees.

Transpiration pull

The loss of water by evaporation from the leaves must be replaced by water coming up from the xylem. Water molecules are attracted to each other by forces of **cohesion** (see Figure 3). These cohesion forces are strong enough to hold the molecules together in a long chain or column. As molecules are lost at the top of the column, the whole column is pulled up as one chain. The pull from above creates tension in the column of water. This is why the xylem vessels must be strengthened by lignin. The lignin prevents the vessel from collapsing under tension.

Because this mechanism involves cohesion between the water molecules and tension in the column of water, it is called the cohesion–tension theory. It relies on the plant maintaining an unbroken column of water all the way up the xylem. If the water column is broken in one xylem vessel, then the water column can still be maintained through another vessel via the bordered pits.

Capillary action

The same forces that hold water molecules together also attract the water molecules to the sides of the xylem vessel. This is called **adhesion**. Because the xylem vessels are very narrow, these forces of attraction can pull the water up the sides of the vessel.

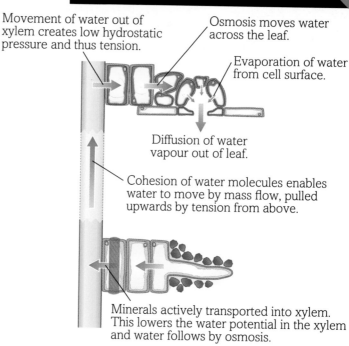

Figure 3 The transpiration stream.

How water leaves the leaf

You learned in the previous topic (3.3.4) that most of the water that leaves the leaf exits as vapour through the stomata. Only a tiny amount leaves through the waxy cuticle. Water evaporates from the cells lining the cavity immediately above the **guard cells** (the sub-stomatal air space). This lowers the water potential in these cells, causing water to enter them by osmosis from neighbouring cells. In turn, water is drawn from the xylem in the leaf by osmosis. Water may also reach these cells by the apoplast pathway from the xylem.

Questions

1. Draw a large diagram of a root-hair cell and annotate the diagram to explain how these cells are specialised.

2. Why does the presence of starch in the endodermis suggest that an active process is involved?

3. Explain the significance of the Casparian strip.

4. Explain how water can be transported up to the leaves of a tall tree.

5. What property of water causes cohesion?

6. List three features of a xylem vessel that enable the easy transport of water.

7. Explain why it is important that the column of water in the xylem vessel is not broken.

6 The adaptations of plants to the availability of water

By the end of this topic, you should be able to demonstrate and apply your knowledge and understanding of:

* adaptations of plants to the availability of water in their environment

Terrestrial plants

For most plants living on land access to water can be a problem. As you learned in topic 3.3.4, water is lost by transpiration, because plants exchange gases with the atmosphere via their stomata. During the day, plants take up a lot of carbon dioxide for use in photosynthesis. They must also remove oxygen which is a by-product of photosynthesis. So the stomata must be open during the day. While the stomata are open, there is an easy route for water to be lost. This water must be replaced.

Plants living on land must be adapted to:

* reduce this loss of water
* replace the water that is lost.

Most terrestrial plants can reduce their water losses by structural and behavioural adaptations:

* A waxy cuticle on the leaf will reduce water loss due to **evaporation** through the epidermis.
* The stomata are often found on the under-surface of leaves, not on the top surface – this reduces the evaporation due to direct heating from the sun.
* Most stomata are closed at night, when there is no light for photosynthesis.
* Deciduous plants lose their leaves in winter, when the ground may be frozen (making water less available) and when temperatures may be too low for photosynthesis.

Marram grass

Marram grass (*Ammophila*) specialises in living on sand dunes. The conditions are particularly harsh, because any water in the sand drains away quickly, the sand may be salty and the leaves are often exposed to very windy conditions. Marram grass is a **xerophyte** – a plant adapted to living in arid conditions.

The adaptations of marram grass include (see Figure 1):

* The leaf is rolled longitudinally so that air is trapped inside – this air becomes humid, which reduces water loss from the leaf. The leaf can roll more tightly in very dry conditions.
* There is a thick waxy cuticle on the outer side of the rolled leaf (upper epidermis), to reduce evaporation.

* The stomata are on the inner side of the rolled leaf (lower epidermis), so they are protected by the enclosed air space.
* The stomata are in pits in the lower epidermis, which is also folded and covered by hairs. These adaptations help to reduce air movement and therefore loss of water vapour.
* The spongy mesophyll is very dense, with few air spaces – so there is less surface area for evaporation of water.

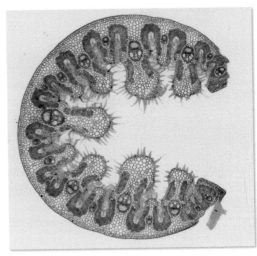

Figure 1 Transverse section of leaf from marram grass, showing the features that help it to reduce water loss.

Cacti

Cacti show other features to overcome arid conditions:

* Cacti are **succulents** – they store water in their stems which become fleshy and swollen. The stem is often ribbed or fluted so that it can expand when water is available.
* The leaves are reduced to spines. This reduces the surface area of the leaves. When the total leaf surface area is reduced, less water is lost by transpiration.
* The stem is green for photosynthesis.
* The roots are very widespread, in order to take advantage of any rain that does fall.

Other xerophytic features

* Closing the stomata when water availability is low will reduce water loss and so reduce the need to take up water.
* Some plants have a low water potential inside their leaf cells. This is achieved by maintaining a high salt concentration in the

cells. The low water potential reduces the evaporation of water from the cell surfaces as the water potential gradient between the cells and the leaf air spaces is reduced.

- A very long tap root that can reach water deep underground.

Hydrophytes

Hydrophytes are plants that live in water, e.g. water lilies (family Nymphaeales). These plants have easy access to water, but are faced with other issues such as getting oxygen to their submerged tissues and keeping afloat – they need to keep their leaves in the sunlight for photosynthesis.

The adaptations of a water lily include (see Figure 2):

- Many large air spaces in the leaf. This keeps the leaves afloat so that they are in the air and can absorb sunlight.

- The stomata are on the upper epidermis, so that they are exposed to the air to allow gaseous exchange.

- The leaf stem has many large air spaces. This helps with buoyancy, but also allows oxygen to diffuse quickly to the roots for aerobic respiration.

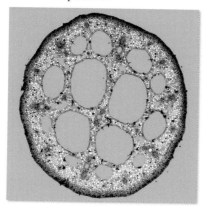

Figure 2 Transverse section of a water lily leaf stem, showing air spaces.

How do they transpire?

Transpiration is the loss of water vapour from the surfaces of the leaves – but the water will not evaporate into water or into air that has a very high humidity. If water cannot leave the plant, then the transpiration stream stops and the plant cannot transport mineral ions up to the leaves. Many plants contain specialised structures at the tips or margins of their leaves called **hydathodes.** These structures can release water droplets which may then evaporate from the leaf surface.

DID YOU KNOW?
Most dew seen on grass in the early morning is actually water released through hydathodes.

Questions

1. Many plants would not survive in arid conditions – explain why.

2. Describe two adaptations to the roots that could help plants survive in arid conditions.

3. Hairs, pits and rolling the leaf can help to reduce loss of water vapour – explain in terms of evaporation and diffusion how these features help to reduce loss of water vapour from the leaf.

4. Draw a diagram of a cactus and annotate your diagram to explain how it is adapted to living in arid conditions.

5. Many plants cannot survive if their roots are permanently submerged – explain why.

6. Explain why leaves in water cannot transpire.

By the end of this topic, you should be able to demonstrate and apply your knowledge and understanding of:

* the mechanism of translocation

KEY DEFINITIONS

assimilates: substances that have become a part of the plant.
sink: a part of the plant where those materials are removed from the transport system; for example, the roots receive sugars and store them as starch. At another time of year, the starch may be converted back to sugars and transported to a growing stem – so the roots can also be a source!
source: a part of the plant that loads materials into the transport system; for example, the leaves photosynthesise and the sugars made are moved to other parts of the plant.
translocation: the transport of assimilates throughout a plant.

Translocation occurs in the phloem, and is the movement of **assimilates** throughout the plant. Assimilates are substances made by the plant, using substances absorbed from the environment. These include sugars (mainly transported as sucrose) and amino acids. A part of the plant that loads assimilates into the phloem sieve tubes is called a **source**. A part of the plant that removes assimilates from the phloem sieve tubes is called a **sink**.

Active loading

Sucrose is loaded into the sieve tube by an active process. This involves the use of energy from ATP in the companion cells (see topic 3.3.2). The energy is used to actively transport hydrogen ions (H^+) out of the companion cells. This increases their concentration outside the cells and decreases their concentration inside the companion cells. As a result, a concentration gradient is created. The hydrogen ions diffuse back into the companion cells through special cotransporter proteins. These proteins only allow the movement of the hydrogen ions into the cell if they are accompanied by sucrose molecules. This is known as **cotransport**. It is also called secondary active transport, as it results from the active transport of the hydrogen ions out of the cell and moves the sucrose against its concentration gradient. As the concentration of sucrose in the companion cell increases, it can diffuse through the plasmodesmata into the sieve tube. Figure 1 shows this process.

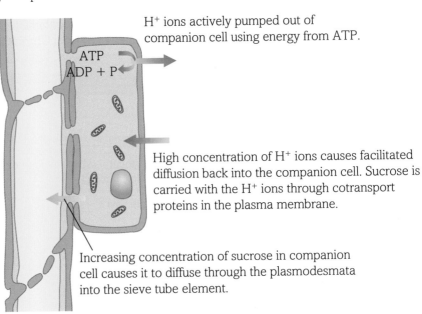

H^+ ions actively pumped out of companion cell using energy from ATP.

High concentration of H^+ ions causes facilitated diffusion back into the companion cell. Sucrose is carried with the H^+ ions through cotransport proteins in the plasma membrane.

Increasing concentration of sucrose in companion cell causes it to diffuse through the plasmodesmata into the sieve tube element.

Figure 1 Active loading of sucrose into the sieve tube.

Movement of sucrose

Movement of sucrose along the phloem is by mass flow. A solution of sucrose, amino acids and other assimilates flows along the tube. The solution is called sap, and it can be made to flow either up or down the plant as required.

The flow is caused by a difference in hydrostatic pressure between the two ends of the tube, which produces a pressure gradient. Water enters the tube at the source, increasing the pressure, and it leaves the tube at the sink, reducing the pressure. Therefore the sap flows from the source to the sink.

The source

Sucrose entering the sieve-tube element, as described in Figure 1, makes the water potential inside the sieve tube more negative (lower). As a result, water molecules move into the sieve-tube element by osmosis from the surrounding tissues. This increases the hydrostatic pressure in the sieve tube at the source.

A source is any part of the plant that loads sucrose into the sieve tube. In early spring, this could be the roots, where energy stored as starch is converted to sucrose and moved to other parts of the plant in order to enable growth in the spring. Perhaps the most obvious source is a leaf. Sugars made during photosynthesis are converted to sucrose and loaded into the phloem sieve tubes. This occurs during late spring, summer and early autumn, whilst the leaves are green. The sucrose is transported to other areas of the plant that may be growing (meristems), or to areas such as the roots – for storage.

The sink

A sink is anywhere that removes sucrose from the phloem sieve tubes. The sucrose could be used for respiration and growth in a meristem, or it could be converted to starch for storage in a root. Where sucrose is being used in the cells, it can diffuse out of the sieve tube via the plasmodesmata. It may also be removed by active transport. The removal of sucrose from the sap (see Figure 2) makes the water potential less negative (higher), so that water moves out of the sieve tube into the surrounding cells. This reduces the hydrostatic pressure in the phloem at the sink.

Along the phloem

DID YOU KNOW?

The pressure in the phloem is high enough to enable insects such as aphids to feed without the need to suck – they simply pierce the phloem and sap flows into their mouths!

Water entering the sieve tube at the source increases the hydrostatic pressure. Water leaving the sieve tube at the sink reduces the hydrostatic pressure. Therefore a pressure gradient is set up along the sieve tube, and the sap flows from higher pressure to lower pressure. This could be in either direction, depending upon where sucrose is being produced and where it is needed. It is even possible that sap could be flowing in opposite directions in different sieve tubes at the same time. Since the sap in one tube is all moving in the same direction, this is mass flow.

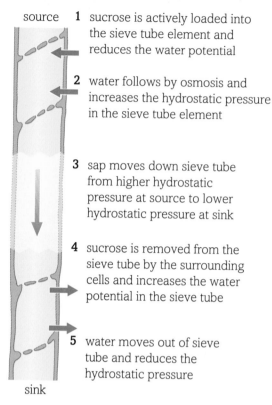

source
1 sucrose is actively loaded into the sieve tube element and reduces the water potential

2 water follows by osmosis and increases the hydrostatic pressure in the sieve tube element

3 sap moves down sieve tube from higher hydrostatic pressure at source to lower hydrostatic pressure at sink

4 sucrose is removed from the sieve tube by the surrounding cells and increases the water potential in the sieve tube

5 water moves out of sieve tube and reduces the hydrostatic pressure

sink

Figure 2 Movement of sap in the phloem.

Questions

1 Define the term 'translocation' and explain why it is important in the life of plants.

2 Explain how a sieve tube can transport sucrose upwards at one time of year and downwards at another time of year.

3 Explain how (a) a leaf and (b) a root can be a sink at one time of year and a source at another time of year.

4 At the sink, sucrose diffuses out of the sieve tube – what must happen in the cells of the sink to enable this to occur?

5 A meristem is a sink; what processes that use sucrose and amino acids occur at a meristem?

6 'Active loading' is a term used to describe how sucrose is loaded into the sieve tube. What is the difference between active loading and active transport?

7 Describe the role of hydrogen ions in active loading.

8 Why is the pH of the companion cells higher than in the surrounding cells?

GUTTATION

This topic shows an article from a gardening magazine about how plants lose water. It aims to explain to gardeners a phenomenon seen in the early morning. At night, some plants release drops of water from their leaves to maintain the flow of the transpiration stream. The questions below should help you to make links between this and other areas of the specification.

WHAT HAPPENS WHEN THE IMPOSSIBLE IS ESSENTIAL?

Early in the morning many small plants are covered by tiny water droplets. Most people believe this is dew. Dew is formed when moisture in the air condenses on a cool surface. Warm air can hold more moisture than cool air and condensation occurs where warm air meets a cooler surface. The moisture in warm air may condense during cool still nights. When this happens the moisture will condense all over a cool leaf.

However, if you only see droplets of water at the tip of a leaf, along its margins or along the veins, this is not dew. This moisture is produced by the activity of the plant itself in a phenomenon known as guttation.

Plants take water from the soil and use it to move minerals and nutrients around the plant. The main vehicles of transport are the xylem and phloem. Water pulled in from the soil is moved up the plant to the leaves by the xylem. The leaves use the sun to create energy and food for the plant which is then moved down to the roots by the phloem.

During the day the warmth of the sun evaporates water through the stomata. This is known as transpiration. Transpiration drives the movement of water up the stem. At night the stomata close and transpiration slows down. However, the xylem must still flow in order to keep the cells turgid and to supply essential minerals to the leaves. The roots allow minerals to build up so that water is brought into the roots from the soil. This creates a pressure which continues to push the xylem up to the leaves (called root pressure). If this water continues to build up in the leaves it would prevent any further flow of the xylem. Instead the pressure forces excess water out of special structures at the end of the xylem. These structures are called hydathodes and they can be found at the tips and margins of leaves, as well as along the veins in larger leaves.

A possible cause of leaf burn?
This loss of water is not usually a problem for plants unless your soil has a high mineral content. When the water does evaporate these minerals get left behind on the leaf surface and can burn the tips of the leaves.

Where transpiration is impossible
In terrestrial plants guttation occurs mainly at night, but it can occur during the day if the humidity is high.

The leaves of aquatic plants are either submerged in water or sit on the water surface – therefore evaporation and, hence, transpiration are not possible. However, the leaves of aquatic plants still need the xylem to flow in order to maintain turgidity and the supply of essential minerals to the leaves. Therefore, the role of hydathodes and guttation are essential for many aquatic plants to survive.

Figure 1 Guttation in leaves.

Where else will I encounter these themes?

1.1　　2.1　　2.2　　2.3　　2.4　　2.5

Let's start by considering the nature of the writing in the article.

1. The writer describes xylem as a 'vehicle' (line 13) and as 'flowing' (line 21) and describes root pressure 'pushing the xylem up to the leaves' (line 25). Explain why these examples might not be appropriate terminology in a more scientific context.

2. Read the article again and pick out two further examples where the terminology is not specific enough for a scientific article.

3. The terminology used in scientific articles can be very important. What might the writer mean by minerals are 'pulled in' by plant roots? Why was such a term used here rather than the correct term?

Now we will look at the biology in, or connected to, this article. Don't worry if you are not ready to give answers to these questions yet. You might like to return to the questions once you have covered other topics later in the book. Use the timeline at the bottom of the page to help you put this work in context with what you have already learned and what is ahead in your course.

4. Explain why transpiration occurs most quickly in dry, windy conditions.

5. Why are water droplets formed at night or in humid conditions?

6. Describe the full mechanism used to absorb minerals and water from the soil.

7. The article suggests that if transpiration is slow, and flow in the xylem is not maintained, then turgidity will be lost. This will cause the plant to wilt. Aquatic plants are surrounded by water and transpiration is very slow or non-existent.
Explain why aquatic plants have air spaces in the stem and why the stomata are usually on the upper surface of the leaves.

Activity 1

Leaf-tip burn depends on the concentration of minerals in the soil.

Suggest a suitable practical investigation to determine whether leaf-tip burn can act as a reliable indicator of mineral levels in the soil. Write out a clear plan and a method. Your plan should include the scientific background in order to explain clearly why salts accumulate on the leaf and why this may cause tip burn. (1000 words)

Write a detailed method and ensure your plan includes a clear scientific explanation.

Activity 2

Halophytes are plants that are adapted to living in conditions where the soil is more salty than usual. Draw a large annotated diagram to explain (in terms of water potentials) why plants find it difficult to live in salty soil. Carry out some research to find out how hydathodes may contribute to the adaptation of these plants, and add further annotations to your diagram to explain how these adaptations work.

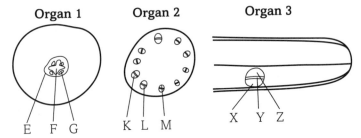

Figure 1 Sections from three plant organs.

1. Look at Figure 1. Which row correctly identifies the three plant organs? [1]
   A. 1 = root, 2 = stem; 3 = leaf
   B. 1 = stem, 2 = root, 3 = leaf
   C. 1 = leaf, 2 = stem, 3 = root
   D. 1 = root, 2 = leaf, 3 = stem

2. Which row correctly identifies the tissues labelled in the Figure 1. [1]

	Tissue			
	xylem	**phloem**	**cambium**	**endodermis**
A	F, M and X	G, K and Z	L and Y	E
B	F, M and Z	G, K and X	L and Y	E
C	F, M and Z	G, K and X	E, L and Y	
D	F, K and Y	G, M and X	L and Y	E

3. Which row correctly states the functions of the tissues labelled in Figure 1. [1]

Function	Transport of water	Transport of assimilates	Cell division	Pumping mineral ions
A	G	K	Y	E
B	K	G	E	Y
C	M	G	E	Y
D	F	K	Y	E

4. What is the correct name for the holes that connect xylem vessels? [1]
   A. plasmodesmata
   B. bordered pits
   C. sieve pores
   D. channel proteins

5. What happens to plant cells placed in a solution with a water potential that is higher than that of the cell contents? [1]
   A. become plamolysed
   B. stay the same size
   C. become turgid
   D. become flaccid

   [Total: 5]

6. (a) What is meant by the term 'transpiration'? [3]
   (b) Describe how the xylem is adapted to transporting water. [9]
   [Total: 12]

7. Translocation is the movement of assimilates around the plant.
   (a) Name the tissue that transports assimilates. [1]
   (b) Name two assimilates transported around the plant. [2]
   (c) Name one other substance transported in the same tissue. [1]
   (d) A source is an organ where substances are loaded into the sieve tubes. A sink is an organ where substances are removed from the sieve tubes.
      (i) Explain how one named plant organ can be both a source and a sink. [2]
      (ii) How might assimilates removed from the sieve tube be used in the sink? [3]
   (e) Movement of fluid in a sieve tube is caused by a hydrostatic pressure gradient between the source and the sink. Describe how the hydrostatic pressure gradient is achieved. [8]
   [Total: 17]

8. Figure 2 shows a diagram of a potometer.

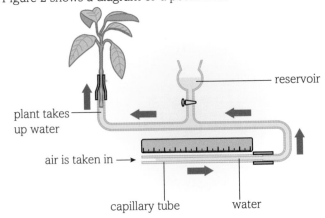

Figure 2 A diagram of a potometer.

A student used the potometer to investigate the effect of different conditions on the rate of transpiration from a leafy shoot. The tables show the results of the investigation.

Temperature (°C)	Distance moved by air/water meniscus along capillary tube in two minutes (mm)
10	4
20	9
30	14
40	18
50	24

Wind speed (m/s)	Distance moved by air/water meniscus along capillary tube in two minutes (mm)
0	2
1	4
2	7
3	11
4	4

(a) What effect does increasing the temperature have upon the rate of transpiration in this leafy shoot? [1]

(b) Describe the effect of increasing wind speed on the rate of transpiration. [3]

(c) A teacher suggested that this method provides only an estimate of transpiration rate. Explain the teacher's comment. [3]

(d) The volume of a cylinder is given by the formula $V = \pi r^2 l$ where $\pi = 3.142$ and l is the length of the cylinder. The radius of the capillary tube is 0.5 mm.

   (i) Calculate the rate of transpiration at 40 °C in mm³ per minute. Show your working and give the answer to one decimal place. [2]

   (ii) Another student suggested that this was not an accurate estimate of the true transpiration rate as a lot of water is used in photosynthesis. Do you agree with this assessment? Justify your view. [2]

[Total: 11]

9. Many plants are adapted to the availability of water in their environment.

(a) Some plants can survive where water is scarce. State the name given to a type of plant adapted to living in dry conditions. [1]

(b) The table below shows some adaptations of these plants and explains how the adaptation works. Complete the table.

Feature	How it works	
leaf rolled up		[2]
stomata in pits		[2]
thick waxy cuticle		[1]
leaf wilts when insufficient water available		[1]

(c) Aquatic plants also need special adaptations. Describe three adaptations of an aquatic plant such as a lily. [3]

[Total: 10]

10. Figure 3 shows a diagram representing water uptake in a root.

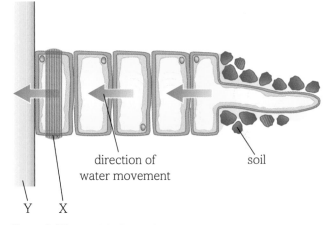

Figure 3 Water uptake in a root.

(a) Water is shown moving across the root. What pathway is shown in this diagram? [1]

(b) Name the pathway in which water does not enter the cells. [1]

(c) Tissue Y is a transport tissue. Name tissue Y. [1]

(d) X is a band of waterproof material in the cell wall. Name this band. [1]

(e) Describe the role of this band in creating root pressure. [4]

(f) Root pressure can account for water rising a few feet up the stem. Explain how the properties of water enable water to reach the top of a tall tree. [5]

[Total: 13]

Biodiversity, evolution and disease

COMMUNICABLE DISEASES

Introduction

You may have heard of an arms race in which each side in a conflict always tries to ensure that it has the best weapons. These weapons may be for attack or for defence. In this chapter we learn about an evolutionary arms race in which pathogens adapt to attack their hosts while the hosts develop ever better defences. We gain an understanding of the variety of organisms that are pathogenic and have evolved to live in or on other living things causing them harm. There are so many pathogens that all living things are surrounded by pathogenic organisms. As a result plants and animals have evolved defences to deal with disease.

We learn how plants defend themselves and about the mammalian immune system. Medical intervention can be used to support these natural defences but care must be taken as continued evolution of pathogens has an impact on the treatment of disease.

All the maths you need

To unlock the puzzles in this section you need the following maths:

* Recognise and make use of appropriate units in calculations
* Recognise and use expressions in decimal and standard form
* Use ratios, fractions and percentages
* Use an appropriate number of significant figures
* Find arithmetic means
* Construct and interpret frequency tables and diagrams, bar charts and histograms
* Understand the principles of sampling as applied to scientific data
* Use a scatter diagram to identify a correlation between two variables
* Translate information between graphical, numerical and algebraic forms
* Plot two variables from experimental or other data

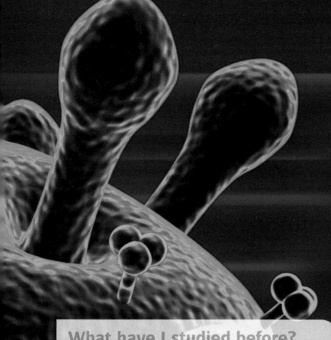

What have I studied before?

- Pathogenic organisms and the diseases they cause
- The presence of antigens on the surface of pathogens
- The role of antibodies in binding to antigens
- The structure of membranes
- How the structure of proteins can produce molecules with very specific shapes

What will I study later?

- How pathogens have managed to evolve to be so effective (AS and AL)
- About non-communicable diseases such as diabetes (AL)
- How organisms that are potentially pathogenic can be used in biotechnology and genetic manipulation (AL)

What will I study in this chapter?

- The structure of pathogenic organisms
- How pathogens can be transmitted from host to host
- How plants can defend themselves against pathogens
- How animals prevent the entry of pathogens
- How mammals can defend themselves against pathogens that have entered the body through non-specific means and through more effective specific means
- Activation of the immune system and both primary and secondary responses
- The structure and action of antibodies
- The principles of vaccination and the use of antibiotics to combat disease

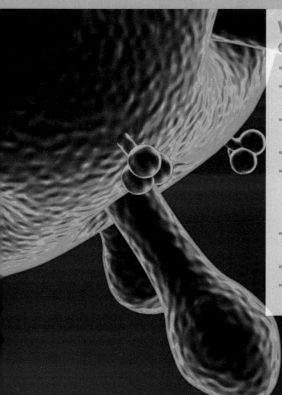

By the end of this topic, you should be able to demonstrate and apply your knowledge and understanding of:

* the different types of pathogen that can cause communicable diseases in plants and animals

What is a pathogen?

Organisms that cause disease are called pathogens. The organism in which they live is called the host. A host body creates a good habitat in which microorganisms can live. As a result, there are numerous types of microorganism that live in or on the body of another organism. Pathogens live by taking nutrition from their host, but also cause damage in the process. This damage can be considerable.

Bacteria

Bacteria (see Figure 1) belong to the kingdom Prokaryotae (see topic 4.3.2). Their cells are smaller than eukaryotic cells, but they can reproduce rapidly. In the right conditions, some types of bacteria can reproduce every 20 minutes. Once in the host body, they can multiply rapidly. Their presence can cause disease by damaging cells or by releasing waste products and/or toxins that are toxic to the host. In plants, the bacteria often live in the vascular tissues and cause blackening and death of these tissues (see Figure 2). See Table 1 for examples of diseases caused by bacteria.

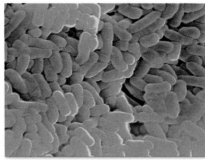

Figure 1 Tuberculosis bacteria (×5000).

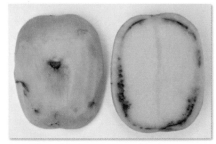

Figure 2 A potato tuber with ring rot caused by bacteria.

Fungi

Fungi can also cause a variety of diseases in both plants and animals. There are common fungal infections where the fungus lives in the skin of an animal, and where its hyphae, which form a **mycelium**, grow under the skin surface (see Figure 3). The fungus can send out specialised reproductive **hyphae**, which grow to the surface of the skin to release spores. This causes redness and irritation.

In plants, the fungus often lives in the vascular tissue, where it can gain nutrients. The hyphae release extracellular enzymes, such as cellulases, to digest the surrounding tissue, which causes decay. Leaves will often become mottled in colour, curl up and shrivel, before dying. Fruit and storage organs, such as tubers (potatoes), will turn black and decay. See Table 1 for examples of diseases caused by fungi.

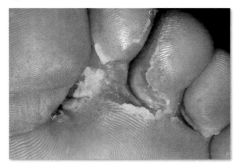

Figure 3 Athlete's foot – caused by a fungus.

Viruses

Viruses cause many well-known diseases in both plants and animals (e.g. HIV, see Figure 4). Viruses invade cells and take over the genetic machinery and other organelles of the cell. They then cause the cell to manufacture more copies of the virus. The host cell eventually bursts, releasing many new viruses which will infect healthy cells. See Table 1 for examples of diseases caused by viruses.

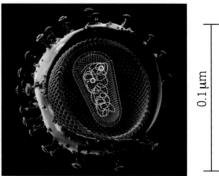

0.1 μm

Figure 4 The HIV virus.

Protoctista

There are a number of diseases caused by animal-like protoctists. These organisms usually cause harm by entering host cells and feeding on the contents as they grow. The malarial parasite *Plasmodium* (see Figure 5) has immature forms that feed on the haemoglobin inside red blood cells.

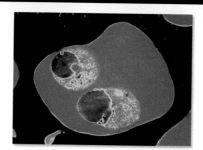

Figure 5 Two immature forms of the malarial parasite *Plasmodium* inside a red blood cell (×4700).

Name of disease	Characteristics of disease	Organism that causes disease
Tuberculosis	a disease that affects many parts of the body, killing the cells and tissues; the lungs are most often affected	bacteria: *Mycobacterium tuberculosis* and *M. bovis*
Bacterial meningitis	infection of the meninges – the membranes that surround the brain and spinal cord; the membranes become swollen and may cause damage to the brain and nerves	bacteria: *Neisseria meningitidis* or *Streptococcus pneumonia*
Ring rot (in plants)	ring of decay in the vascular tissue of a potato tuber or tomato, accompanied by leaf wilting	bacterium: *Clavibacter michiganensis* subsp. *sepedonicus*
HIV/AIDS	attacks cells in the immune system and compromises the immune response	virus: human immunodeficiency virus
Influenza	attacks respiratory system and causes muscle pains and headaches	virus: from family Orthomyxoviridae – 'flu' viruses
Tobacco mosaic virus	causes mottling and discolouration of leaves	virus: tobacco mosaic virus
Black sigatoka (bananas)	causes leaf spots on banana plants reducing yield	fungus: *Mycosphaerella fijiensis*
Blight (tomatoes and potatoes)	affects both leaves and potato tubers	protoctistan: *Phytophthora infestans*
Ringworm (cattle)	growth of fungus in skin with spore cases erupting through skin to cause a rash	fungus: *Trichophyton verrucosum*
Athlete's foot (humans)	growth under skin of feet – particularly between the toes	fungus: *Trichophyton rubrum*
Malaria	parasite in the blood that causes headache and fever and may progress to coma and death	protoctistan: *Plasmodium falciparum, P. vivax, P. ovale, P. malariae*

Table 1 Some communicable diseases and the organisms that cause them.

Questions

1. List two animal diseases caused by each of the following: viruses, bacteria, fungi and protoctista.

2. Give one plant disease caused by each of the following: viruses, bacteria and fungi.

3. Bacteria and fungi in plants most often live in the vascular tissue. Suggest why.

4. In order to enter plant cells, a pathogen must get through the cell wall. Suggest what type of molecule these pathogens must release.

5. HIV is a retrovirus – a type of virus that does not contain DNA. In place of DNA it carries RNA. Explain why it also needs to carry the enzyme reverse transcriptase.

6. Draw a diagram of a fungus and annotate the diagram to show how the fungus causes damage to the host body as it gains nutrition.

DID YOU KNOW?

Some organisms can be difficult to classify. *Phytophthora* (blight) has been classified as a fungus for many years. Recently, it was moved to the kingdom Protoctista, before being placed in a new kingdom (Stramenopila), because it has many features that do not fit with other fungi (it does not fit the mould!).

② Transmission of pathogens

By the end of this topic, you should be able to demonstrate and apply your knowledge and understanding of:

* the means of transmission of animal- and plant-communicable pathogens

Pathogens have a life cycle that involves living in or on other living things. The by-product of this life cycle is that they cause harm to their host. The life cycle involves the following stages:

* travel from one host to another (transmission)
* entering the host's tissues
* reproducing
* leaving the host's tissues.

In this topic we will concentrate on the first stage – transmission.

Transmission between animals

Direct transmission

Pathogenic organisms can be transmitted between animals in a variety of ways. The most common forms of transmission are known as **direct transmission** (see Figure 1). These are shown in Table 1.

Figure 1 Pathogens can be transmitted in droplets of water.

Means of transmission	Factors that affect transmission
Direct physical contact, such as touching a person who is infected or touching contaminated surfaces (including soil) that harbour the pathogens. For example, HIV, bacterial meningitis, ringworm, athlete's foot.	Hygiene: washing hands regularly – especially after using the toilet. Keeping surfaces clean – especially door handles. Cleaning and disinfecting cuts and abrasions. Sterilising surgical instruments. Using condoms during sexual intercourse.
Faecal–oral transmission, usually by eating food or drinking water contaminated by the pathogen. For example, cholera, food poisoning.	Using human sewage to fertilise crops is a common practice in some parts of the world. Treatment of waste-water and treatment of drinking water are important ways to reduce the risk. Thorough washing of all fresh food (using treated water). Careful preparation and thorough cooking of all food.
Droplet infection, in which the pathogen is carried in tiny water droplets in the air. For example, tuberculosis, influenza.	Catch it – bin it – kill it. Cover your mouth when coughing or sneezing. Use a tissue and ensure the tissue is disposed of correctly.
Transmission by spores, which are a resistant stage of the pathogen. These can be carried in the air or reside on surfaces or in the soil. For example, anthrax, tetanus.	Use of a mask. Washing skin after contact with soil.

Table 1 Factors that affect direct transmission.

KEY DEFINITIONS

direct transmission: passing a pathogen from host to new host, with no intermediary.
indirect transmission: passing a pathogen from host to new host, via a vector.
transmission: passing a pathogen from an infected individual to an uninfected individual.
vector: an organism that carries a pathogen from one host to another.

LEARNING TIP

Remember that it is the pathogen that is transmitted, not the disease.

Other factors that affect transmission include social factors such as:

- overcrowding – many people living and sleeping together in one house

- poor ventilation

- poor health – particularly if a person has HIV/AIDS, as they are more likely to contract other diseases

- poor diet

- homelessness

- living or working with people who have migrated from areas where a disease is more common.

Indirect transmission

Some pathogens are transmitted indirectly via a vector. A vector is another organism that may be used by the pathogen to gain entry to the primary host. For example, the *Plasmodium* parasite that causes malaria enters the human host via a bite from a female *Anopheles* mosquito (see Figure 2).

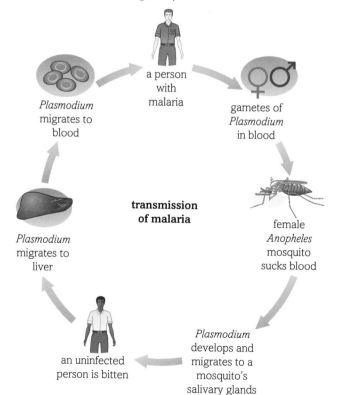

transmission of malaria

Plasmodium migrates to blood

a person with malaria

gametes of *Plasmodium* in blood

female *Anopheles* mosquito sucks blood

Plasmodium migrates to liver

an uninfected person is bitten

Plasmodium develops and migrates to a mosquito's salivary glands

Figure 2 The life cycle of *Plasmodium,* which uses a vector.

Transmission of plant pathogens

Plant pathogens can also be spread by direct and indirect means.

Many pathogens are present in the soil and will infect plants by entering the roots – especially if these have been damaged as a result of replanting, burrowing animals or movement caused by a storm.

Many fungi produce spores as a means of sexual or asexual reproduction. These spores may be carried in the wind – airborne transmission.

Once a pathogen is inside the plant, it may infect all the vascular tissue. Pathogens in the leaves are distributed when the leaves are shed and carry the pathogen back to the soil where it can grow and infect another plant. Pathogens can also enter the fruit and seeds, and will then be distributed with the seeds – so that many or all of the offspring are infected.

Indirect transmission of plant pathogens often occurs as a result of insect attack. Spores or bacteria become attached to a burrowing insect, such as a beetle, which attacks an infected plant. When that beetle attacks another plant, the pathogen is transmitted to the uninfected plant. The beetle is acting as a vector. For example, the fungus that causes Dutch elm disease is carried by the beetle *Scolytus multistriatus*.

Disease and climate

Many protoctists, bacteria and fungi can grow and reproduce more rapidly in warm and moist conditions. Therefore, they tend to be more common in warmer climates. In cooler climates, these pathogens may be damaged or even killed by cold winter weather – such weather will certainly reduce their ability to grow and reproduce. As a result, there is a greater variety of diseases to be found in warmer climates, and animals or plants living in these regions are more likely to become infected.

Questions

1. Explain why regularly cleaning door handles helps to reduce transmission of pathogens such as MRSA.

2. Suggest three reasons why many people suffer from food poisoning when they travel to tropical or sub-tropical regions.

3. In recent years, tuberculosis has increased in some major European cities. Explain why.

4. Suggest why diseases such as malaria and Dutch elm disease do not harm the vector that is involved in transmission of the pathogen.

5. Suggest why pathogens living in the soil cannot infect plant roots unless some damage has occurred to the roots.

6. Explain why large trees may die some years after a major storm.

By the end of this topic, you should be able to demonstrate and apply your knowledge and understanding of:

∗ plant defences against pathogens

callose: a large polysaccharide deposit that blocks old phloem sieve tubes.

Plants manufacture sugars in photosynthesis and convert those sugars to a wide variety of compounds such as proteins and oils. Therefore they represent a rich source of nutrients for many organisms such as bacteria, fungi, protoctists, viruses, insects and vertebrates. The bacteria, fungi, protoctists and viruses may be pathogenic, and the insects and vertebrates may act as vectors to help transmit these pathogens.

Plants do not have an immune system comparable with animals. But they have developed a wide range of structural, chemical and protein-based defences which can detect invading organisms and prevent them from causing extensive damage. This includes both passive defences to prevent entry and active defences which are induced when the pathogen is detected.

Passive defences

These are defences present before infection, and their role is to prevent entry and spread of the pathogen. Passive defences include physical barriers and chemicals.

Physical defences

- Cellulose cell wall – this not only acts as a physical barrier but most plant cell walls contain a variety of chemical defences that can be activated when a pathogen is detected.
- Lignin thickening of cell walls – lignin (a phenolic compound) is waterproof and almost completely indigestible.
- Waxy cuticles – these prevent water collecting on the cell surfaces. Since pathogens collect in water and need water to survive, the absence of water is a passive defence.
- Bark – most bark contains a variety of chemical defences that work against pathogenic organisms.
- Stomatal closure – stomata are possible points of entry for pathogens. Stomatal aperture is controlled by the guard cells. When pathogenic organisms are detected, the guard cells will close the stomata in that part of the plant.
- Callose – callose is a large polysaccharide that is deposited in the sieve tubes at the end of a growing season. It is deposited around the sieve plates and blocks the flow in the sieve tube. This can prevent a pathogen spreading around the plant. See Figure 4 in topic 3.3.2.
- **Tylose** formation – a tylose is a balloon-like swelling or projection that fills the xylem vessel. When a tylose is fully formed, it plugs the vessel and the vessel can no longer

carry water. Blocking the xylem vessels prevents spread of pathogens through the heartwood. The tylose contains a high concentration of chemicals such as terpenes that are toxic to pathogens.

In active xylem, the vessels are surrounded by living parenchyma cells. As the wood ages, the parenchyma cells become filled with terpenes. Eventually, the contents of the parenchyma cell burst into the dead vessel through a pit connecting the two. The parenchyma cell dies as its contents enter the empty space of the xylem vessel.

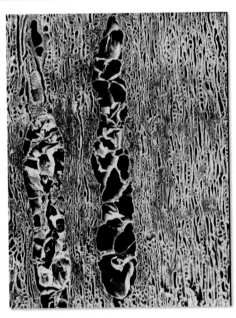

Figure 1 A xylem vessel filled with tyloses.

Chemical defences

Plant tissues contain a variety of chemicals that have anti-pathogenic properties. These include terpenoids, phenols, alkaloids and hydrolytic enzymes.

Some of these chemicals, such as the terpenes in tyloses and tannins in bark, are present before infection. However, because the production of chemicals requires a lot of energy, many chemicals are not produced until the plant detects an infection.

Active defences

When pathogens attack, specific chemicals in their cell walls can be detected by the plant cells. These chemicals include specific proteins and glycolipids. The plant responds by fortifying the defences already present. This includes increasing the physical defences and producing defensive chemicals. Active defences include the following.

- Cell walls become thickened and strengthened with additional cellulose.

- Deposition of callose between the plant cell wall and cell membrane near the invading pathogen. Callose deposits are polysaccharide polymers that impede cellular penetration at the site of infection. It strengthens the cell wall and blocks plasmodesmata.

- Oxidative bursts that produce highly reactive oxygen molecules capable of damaging the cells of invading organisms.

- An increase in production of chemicals.

Chemical used	Action
Terpenoids	A range of essential oils that have antibacterial and antifungal properties. They may also create scent, for example, the menthols and menthones produced by mint plants.
Phenols	These also have antibiotic and antifungal properties. Tannins found in bark inhibit attack by insects. These compounds bind to salivary proteins and digestive enzymes such as trypsin and chymotrypsin, deactivating the enzymes. Insects that ingest high amounts of tannins do not grow and will eventually die. This helps to prevent the transmission of pathogens.
Alkaloids	Nitrogen-containing compounds such as caffeine, nicotine, cocaine, morphine, solanine. These give a bitter taste to inhibit herbivores feeding. They also act on a variety of metabolic reactions via inhibiting or activating enzyme action. Some alkaloids inhibit protein synthesis. If the plant can reduce grazing by larger animals, then it will suffer less damage that can allow pathogens to enter the plant.
Defensive proteins (defensins)	These are small cysteine-rich proteins that have broad anti-microbial activity. They appear to act upon molecules in the plasma membrane of pathogens, possibly inhibiting the action of ion transport channels.
Hydrolytic enzymes	These are found in the spaces between cells. They include chitinases (which break down the chitin found in fungal cell walls), glucanases (which hydrolyse the glycosidic bonds in glucans) and lysozymes (which are capable of degrading bacterial cell walls).

Table 1 lists the chemicals that are commonly used.

- **Necrosis** – deliberate cell suicide. A few cells are sacrificed to save the rest of the plant. By killing cells surrounding the infection, the plant can limit the pathogen's access to water and nutrients and can therefore stop it spreading further around the

plant. Necrosis is brought about by intracellular enzymes that are activated by injury. These enzymes destroy damaged cells and produce brown spots on leaves or dieback (see Figure 2).

Figure 2 Necrotic lesions on leaves and dieback in ash.

- **Canker** – a sunken necrotic lesion in the woody tissue such as the main stem or branch. It causes death of the cambium tissue in the bark (see Figure 3).

Figure 3 Canker in bark.

Questions

1. Many plant defences concentrate on preventing the spread of the pathogen through the plant – list these defences.

2. Explain the differences between passive and active protection.

3. How do tyloses protect plants from disease?

4. Describe two ways in which callose helps to protect plants from disease.

5. Why are many protective chemicals only manufactured when infection is detected?

6. Explain how necrosis can help to save the whole plant.

Primary defences against disease

By the end of this topic, you should be able to demonstrate and apply your knowledge and understanding of:

* the primary non-specific defences against pathogens in animals

> **KEY DEFINITIONS**
>
> **inflammation:** swelling and redness of tissue caused by infection.
> **mucous membrane:** specialised epithelial tissue that is covered by mucus.
> **primary defences:** those that prevent pathogens entering the body.

Pathogenic organisms need to enter the body of their host before they can cause harm. Evolution has selected hosts adapted to defend themselves against such invasions. The mechanisms that have evolved to prevent entry of pathogenic organisms are called primary defences. They are non-specific, as they will prevent the entry of any pathogen.

The skin

The body is covered by the skin. This is the main primary defence. The outer layer of the skin is called the epidermis, and it consists of layers of cells. Most of these cells are called **keratinocytes**. These cells are produced by mitosis at the base of the epidermis. They then migrate out to the surface of the skin. As they migrate, they dry out and the cytoplasm is replaced by the protein **keratin**. This process is called keratinisation, and it takes about 30 days. By the time that the cells reach the surface, they are no longer alive. The keratinised layer of dead cells acts as an effective barrier to pathogens. Eventually, the dead cells slough off.

> **DID YOU KNOW?**
>
> Much of the dust found in houses is dead skin cells.

Blood clotting and skin repair

Of course the skin is only protective as long as it is complete. Abrasions or lacerations damage the skin and open the body to infection. When this occurs, the body must prevent excess blood loss by forming a clot, making a temporary seal to prevent infection, and repairing the skin.

Blood clotting is a complex process, as it is important to prevent clots forming in the blood vessels, where they are not needed. It involves calcium ions and at least 12 factors – known as clotting factors. Many of the clotting factors are released from platelets and from the damaged tissue. These factors activate an **enzyme cascade** (see Figure 1).

Once the clot has formed, it begins to dry out and form a scab. The scab shrinks as it dries, drawing the sides of a cut together. This makes a temporary seal, under which the skin is repaired. The first stage is the deposition of fibrous collagen under the scab. Stem cells in the epidermis then divide by mitosis to form new cells, which migrate to the edges of the cut and differentiate to form new skin. New blood vessels grow to supply oxygen and nutrients to the new tissues. The tissues contract to help draw the edges of a cut together so that the repair can be completed. As the new skin is completed, the scab will be released.

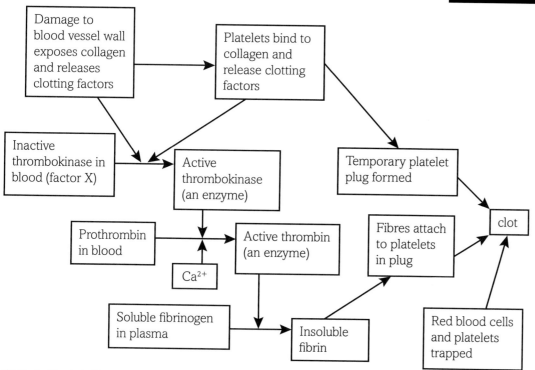

Figure 1 Blood clotting.

Mucous membranes

Certain substances, such as oxygen and the nutrients in our food, must enter our blood. The exchange surfaces where this occurs must be thinner and are less well protected from pathogens. The air and food that we take in from our environment may harbour microorganisms. Therefore, the airways, lungs and the digestive system are at risk of infection.

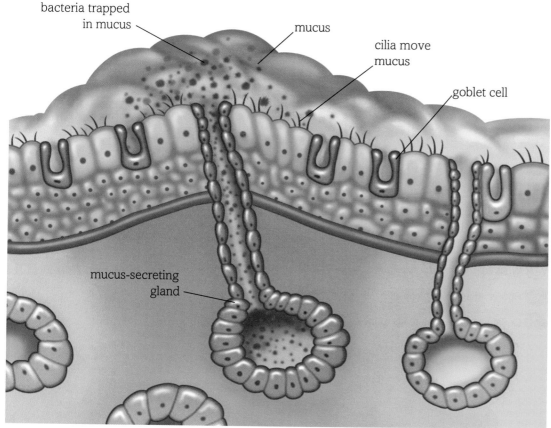

Figure 2 How a mucous membrane protects the body.

These areas are protected by **mucous membranes** (see Figure 2). The epithelial layer contains mucus-secreting cells called goblet cells. There are also extra mucus-secreting glands under the epithelium. In the airways, the mucus lines the passages and traps any pathogens that may be in the air. The epithelium also has ciliated cells. The cilia are tiny, hair-like organelles that can move. They move in a coordinated fashion to waft the layer of mucus along. They move the mucus up to the top of the trachea, where it can enter the oesophagus. It is swallowed and passes down the digestive system. Most pathogens in the digestive system are killed by the acidity of the stomach, which can be pH 1–2. This denatures the pathogen's enzymes. Mucous membranes are also found in the gut, genital areas, anus, ears and nose.

LEARNING TIP
Remember that it is the mucus that traps the pathogens, not the cilia.

Coughing and sneezing

Areas that are prone to attack are also sensitive. They respond to the irritation that may be caused by the presence of microorganisms or the toxins that they release. These reflexes include coughing, sneezing and vomiting. In a cough or sneeze the sudden expulsion of air will carry with it the microorganisms causing the irritation (see topic 4.1.2, Figure 1).

Inflammation

One of the signs that a tissue is infected is swelling and redness known as **inflammation**. The tissue may also feel hot and painful.

The presence of microorganisms in the tissue is detected by specialised cells called *mast cells*. These cells release a cell signalling substance called **histamine**. Histamine has a range of effects on the surrounding tissue, which act to help combat the infection. The main effect is to cause **vasodilation** and make the capillary walls more permeable to white blood cells and some proteins. Blood plasma and phagocytic white blood cells leave the blood and enter the tissue fluid. This leads to increased production of tissue fluid, which causes the swelling (oedema). Excess tissue fluid is drained into the lymphatic system where lymphocytes are stored. This can lead to the pathogens coming into contact with the lymphocytes and initiating specific immune responses (see topic 4.1.6).

Other primary defences

- The eyes are protected by antibodies and enzymes in the tear fluid.
- The ear canal is lined by wax, which traps pathogens.
- The female reproductive system is protected by a mucus plug in the cervix and by maintaining relatively acidic conditions in the vagina.

Questions

1. Explain why primary defences are non-specific.
2. Why is it important that blood clots are not formed in the blood vessels?
3. Prothrombin is activated by removal of some amino acids. Explain how this can activate the enzyme.
4. What is meant by the term 'expulsive reflex'?
5. What are the symptoms relieved by the drug antihistamine?
6. How is inflammation caused?
7. Explain why smoking leads to increased lung infections.

By the end of this topic, you should be able to demonstrate and apply your knowledge and understanding of:

* the structure and mode of action of phagocytes
* examination and drawing of cells observed in blood smears

KEY DEFINITIONS

antigen-presenting cell: a cell that isolates the antigen from a pathogen and places it on the plasma membrane so that it can be recognised by other cells in the immune system.
clonal selection: selection of a specific B or T cell that is specific to the antigen.
cytokines: hormone-like molecules used in cell signalling to stimulate the immune response.
neutrophil: a type of white blood cell that engulfs foreign matter and traps it in a large vacuole (phagosome), which fuses with lysosomes to digest the foreign matter.
opsonins: proteins that bind to the antigen on a pathogen and then allow phagocytes to bind.

Antigens and opsonins

Secondary defences are used to combat pathogens that have entered the body. When a pathogen invades the body, it is recognised as foreign by the chemical markers on its outer membrane. These markers are called **antigens**. They are proteins or glycoproteins intrinsic to the plasma membrane. Antigens are specific to the organism. Our own cells have antigens, but these are recognised as our own and do not produce a response.

Opsonins are protein molecules that attach to the antigens on the surface of a pathogen. They are a type of antibody (see topic 4.1.7). Some opsonins are not very specific – so that they can attach to a variety of pathogenic cells. The role of the opsonin is to enhance the ability of phagocytic cells to bind and engulf the pathogen.

Phagocytes

The first line of secondary defence is phagocytosis. Specialised cells in the blood and tissue fluid engulf and digest the pathogens.

Neutrophils

The most common phagocytes are neutrophils. You can recognise these cells by their multi-lobed nucleus. Neutrophils are manufactured in the bone marrow. They travel in the blood and often squeeze out of the blood into tissue fluid. Neutrophils are short-lived, but they will be released in large numbers as a result of an infection. Neutrophils contain a large number of lysosomes (see also topic 2.5.4). They engulf and digest pathogens, as shown in Figure 1. Neutrophils usually die soon after digesting a few pathogens. Dead neutrophils may collect in an area of infection, to form pus.

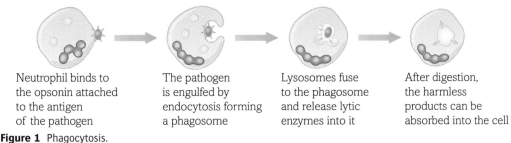

Neutrophil binds to the opsonin attached to the antigen of the pathogen

The pathogen is engulfed by endocytosis forming a phagosome

Lysosomes fuse to the phagosome and release lytic enzymes into it

After digestion, the harmless products can be absorbed into the cell

Figure 1 Phagocytosis.

Cells are specialised for their function. Phagocytes are active cells that can select specific pathogenic cells to attack. These specialisations include:

- receptors on the plasma membrane that can bind to the opsonin or a specific antigen
- a lobed nucleus that allows the cell to squeeze through narrow gaps
- a well-developed cytoskeleton that helps the cell to change shape to engulf the pathogen and to move lysosomes and vacuoles around inside the cell
- many lysosomes containing lysin
- many mitochondria, to release energy from glucose
- a lot of ribosomes to synthesise the enzymes involved.

Macrophages

Macrophages are larger cells manufactured in the bone marrow. They travel in the blood as **monocytes** (see the investigation box below 'Looking at blood cells') before settling in the body tissues. Many are found in the lymph nodes where they mature into macrophages. Dendritic cells – a type of macrophage – are found in the more peripheral tissues. Macrophages play an important role in initiating the specific responses to invading pathogens. When a macrophage engulfs a pathogen, it does not fully digest it. The antigen from the surface of the pathogen is saved and moved to a special protein complex on the surface of the cell. The cell becomes an antigen-presenting cell. It exposes the antigen on its surface, so that other cells of the immune system can recognise the antigen. The special protein complex ensures that the antigen-presenting cell is not mistaken for a foreign cell and attacked by other phagocytes.

Active immunity

Antigen presentation

The antigen-presenting cell moves around the body where it can come into contact with specific cells that can activate the full immune response. These are the T lymphocytes and B lymphocytes. There may be only one T cell and one B cell with the correct recognition site for the antigen. Therefore the role of the antigen-presenting cells is to increase the chances that the antigen will come in contact with them.

Specific immune response

Activation of the specific B and T cells is called clonal selection. This brings into play a complex series of events that leads to the production of antibodies that can combat the specific pathogen and memory cells that will provide long-term immunity. The specific immune response is explained in detail in topic 4.1.6.

The whole series of events is stimulated and coordinated by a number of hormone-like chemicals called cytokines. These stimulate the differentiation and activity of macrophages, B cells and T cells.

Looking at blood cells

Wear eye protection.

Blood cells can be viewed in a blood smear. You can use your own blood and look at your own blood cells, or blood smears may be provided ready prepared. A thin layer of blood is spread out on a slide, and is stained to make the white blood cells more visible. Figure 2 shows a blood smear. The red blood cells are pink in colour. Note that red blood cells do not always look like a perfect biconcave disc. Monocytes are the largest white blood cells and usually have a large kidney-shaped nucleus. Neutrophils have a multilobed nucleus. Lymphocytes are smaller, and the nucleus almost fills the cell.

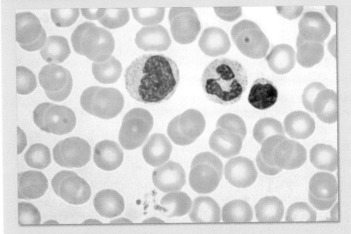

Figure 2 A blood smear (×935).

Questions

1. Opsonins are non-specific. Explain why they need to be non-specific.

2. Explain what is meant by a non-specific response.

3. Describe how a neutrophil performs phagocytosis.

4. Describe how neutrophils are specialised for their role.

5. Why must antigen-presenting cells present the antigens of pathogenic organisms on a special protein complex?

6. Explain why it is important that the antigens on our own body cells are not recognised by phagocytes.

(6) The specific immune response

By the end of this topic, you should be able to demonstrate and apply your knowledge and understanding of:

* the structure, different roles and modes of action of B and T lymphocytes in the specific immune response
* autoimmune diseases

KEY DEFINITIONS

antibodies: specific proteins released by plasma cells that can attach to pathogenic antigens.

B memory cells: cells that remain in the blood for a long time, providing long-term immunity.

clonal expansion: an increase in the number of cells by mitotic cell division.

interleukins: signalling molecules that are used to communicate between different white blood cells.

plasma cells: derived from the B lymphocytes, these are cells that manufacture antibodies.

T helper cells: cells that release signalling molecules to stimulate the immune response.

T killer cells: cells that attack and destroy our own body cells that are infected by a pathogen.

T memory cells: cells that remain in the blood for a long time, providing long-term immunity.

T regulator cells: cells that are involved with inhibiting or ending the immune response.

The specific immune response involves B lymphocytes (B cells) and T lymphocytes (T cells). These are white blood cells with a large nucleus and specialised receptors on their plasma membranes (cell surface membranes). The immune response produces antibodies. It is the antibodies that actually neutralise the foreign antigens. The immune response also provides long-term protection from the disease. It produces immunological memory through the release of memory cells, which circulate in the body for a number of years. (See topic 4.1.7 for more about antibodies and long-term protection.)

Cells produced in the immune response

T lymphocytes develop or differentiate into four types of cell:

* T helper cells (T_h), which release cytokines (chemical messengers) that stimulate the B cells to develop and stimulate phagocytosis by the phagocytes.

* T killer cells (T_k), which attack and kill host-body cells that display the foreign antigen.

* T memory cells (T_m), which provide long-term immunity.

* T regulator cells (T_r), which shut down the immune response after the pathogen has been successfully removed. They are also involved in preventing autoimmunity (see below).

The B lymphocytes (**B cells**) develop into two types of cell:

* Plasma cells, which circulate in the blood, manufacturing and releasing the antibodies.

* B memory cells (B_m), which remain in the body for a number of years and act as the immunological memory.

Cell signalling

The specific immune response involves the coordinated action of a range of cells. In order to work together effectively, these cells need to communicate. This is known as cell signalling. Figure 1 shows the sequence of events in the immune response. You can see how many cell types are involved, and why it is so important that they can communicate effectively.

This communication is achieved through the release of hormone-like chemicals called cytokines. There is a huge range of signalling molecules, each performing a different role. And, in order to detect a signal, the target cell must have a cell surface receptor complementary in shape to the shape of the signalling molecule.

The following are examples of communication using cytokines.

* Macrophages release monokines. Some monokines attract neutrophils (by chemotaxis – the movement of cells towards a particular chemical) and others stimulate B cells to differentiate and release antibodies.

* T cells and macrophages release interleukins, which can stimulate the clonal expansion (proliferation) and differentiation of B and T cells.

* Many cells can release interferon, which inhibits virus replication and stimulates the activity of T killer cells.

Autoimmune diseases

An autoimmune disease occurs when the immune system attacks a part of the body. Normally, any B or T cells that are specific to our own antigens are destroyed during early development of the immune system. An autoimmune disease arises when antibodies start to attack our own antigens – possibly because antigens that are not normally exposed become exposed to attack.

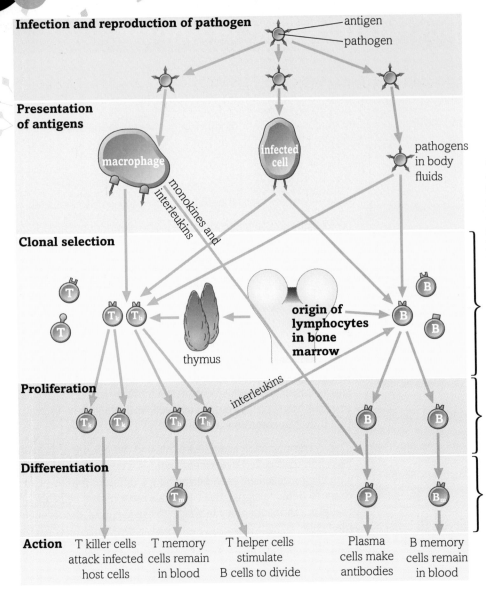

Infection and reproduction of pathogen

antigen
pathogen

Presentation of antigens

macrophage

infected cell

pathogens in body fluids

monokines and interleukins

Clonal selection

thymus

origin of lymphocytes in bone marrow

Activation – clonal selection
An invading pathogen has specific antigens. In order to trigger the immune response, these must be detected by T and B lymphocytes that carry the specific receptor molecules on their plasma membranes. The receptor molecules are proteins that have a shape that is complementary to the shape of the antigen.
Contact between the antigen and lymphocytes can be achieved directly when pathogenic cells enter the lymph nodes or by the action of antigen-presenting cells.

Proliferation

interleukins

Clonal expansion (proliferation)
Once the correct lymphocytes have been activated they must increase in numbers to become effective. This is achieved by mitotic cell division.

Differentiation

Differentiation
The B and T lymphocytes do not manufacture the antibodies directly. Once selected, clones of the lymphocytes develop into a range of useful cells.

Action T killer cells attack infected host cells   T memory cells remain in blood   T helper cells stimulate B cells to divide   Plasma cells make antibodies   B memory cells remain in blood

Figure 1 The specific immune response.

The causes of autoimmune disease are unknown, but seem to include both genetic and environmental factors. Examples include the following:

- Arthritis is a painful inflammation of a joint (see Figure 2). The cause is uncertain, but it starts with antibodies attacking the membranes around the joint.
- Lupus can affect any part of the body, causing swelling and pain. It may be associated with antibodies that attack certain proteins in the nucleus in cells and affected tissues.

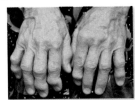

Figure 2 Arthritic hands.

Questions

1. Suggest how T killer cells might recognise infected host cells.

2. What is the role of the memory cells?

3. Why are T regulator cells needed?

4. Explain why cell-signalling molecules such as interleukins have a very specific shape.

5. Suggest why the immune system does not normally attack our own cells.

6. What might cause the onset of an autoimmune disease?

By the end of this topic, you should be able to demonstrate and apply your knowledge and understanding of:

* the structure and general functions of antibodies

* an outline of the action of opsonins, agglutinins and anti-toxins

* the primary and secondary immune response

KEY DEFINITIONS

agglutinins: antibodies that cause pathogens to stick together.
anti-toxins: antibodies that render toxins harmless.
opsonins: antibodies that make it easier for phagocytes to engulf the pathogen.
primary immune response: the initial response caused by a first infection.
secondary immune response: a more rapid and vigorous response caused by a second or subsequent infection by the same pathogen.

Antigens and antibodies

As introduced in topic 4.1.5, antigens are molecules that can stimulate an immune response. Almost any molecule could act as an antigen, but they are usually proteins or glycoproteins in the plasma membrane of the pathogen. A foreign antigen will be detected by the immune system and will stimulate the production of antibodies. Antibodies are specific to the antigen. As the antigen is specific to the organism, we can think of the antibody as being specific to the pathogen. Our own antigens are recognised by our immune system and do not usually stimulate any response (topic 4.1.6, on autoimmune diseases).

Antibodies are immunoglobulins – complex proteins produced by the plasma cells in the immune system. They are released in response to an infection. They have a region with a specific shape that is complementary to that of a particular antigen. Our immune system must manufacture one type of antibody for every antigen that is detected. Antibodies attach to antigens and render them harmless.

DID YOU KNOW?

Each B cell carries a slightly different DNA code, so that the antibodies it can produce are slightly different from those of other B cells. The plasma cells cloned from one B cell will all make identical antibodies.

The structure of an antibody

Antibody molecules are Y-shaped and have two distinct regions. They consist of four polypeptide chains. The structure of a generalised antibody is shown in Figure 1.

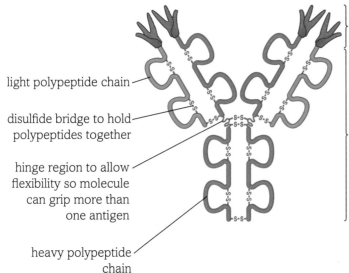

Figure 1 The structure of a generalised antibody.

How antibodies work

Antibodies work in a variety of ways, but most act by attaching to the antigens on a pathogen. There are three main groups (see Figure 2):

- **Opsonins** (see also topic 4.1.5) are a group of antibodies that bind to the antigens on a pathogen. They then act as binding sites for phagocytic cells, so that these can more easily bind and destroy the pathogen.

  Some opsonins are not very specific and stick to types of molecules that are not found in the host cell, e.g. the peptidoglycans found in the cell walls of bacteria.

  Other opsonins are produced as part of the specific immune response and bind to very specific antigens. The pathogen may have another use for this antigen molecule. For example, it may be a binding site used for attachment to the host cell. In this case, the opsonin bound to the antigen renders the antigen useless – a process known as neutralisation. The opsonin assists in phagocytosis, but also prevents the pathogen entering a host cell before it can be attacked by phagocytes.

- **Agglutinins** – because each antibody molecule has two identical binding sites it is able to 'crosslink' pathogens by binding an antigen on one pathogen with one binding site and then an antigen on another pathogen with its other binding site. When many antibodies perform this crosslinking they clump together (agglutinate) pathogens. This has two advantages: the agglutinated pathogens are physically impeded from carrying out some functions, such as entering host cells, and the agglutinated pathogens are readily engulfed by phagocytes. This is particularly effective against viruses.

- **Anti-toxins** – some antibodies bind to molecules that are released by pathogenic cells. These molecules may be toxic and the action of anti-toxins renders them harmless.

(a) opsonisation

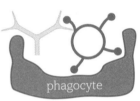

| antibody molecule has two binding sites | it also has an end that can stimulate phagocytosis | a pathogen with several antigens on its surface | opsonisation: the end of the antibody stimulates phagocytosis |

(b) agglutination

by using each binding site to bind a different pathogen the antibodies crosslink pathogens into a big clump which is non-infective and easily phagocytosed

(c) action of anti-toxin

toxin

anti-toxin

Figure 2 How antibodies work.

Primary and secondary responses

Antibodies are produced in response to infection (see Figure 3). When an infecting agent is first detected, the immune system starts to produce antibodies (see topic 4.1.6, on the specific immune response). But it takes a few days before the number of antibodies in the blood rises to a level that can combat the infection successfully. This is known as the **primary immune response**. Once the pathogens have been dealt with, the number of antibodies in the blood drops rapidly.

Antibodies do not stay in the blood. If the body is infected a second time by the same pathogen, the antibodies must be made again. However, as described in topic 4.1.6, as a result of the specific immune response there will be B memory cells and T memory cells circulating in the blood. These cells can recognise the specific antigens and the immune system can swing into action more quickly. This time the production of antibodies starts sooner and is much more rapid. So the concentration of antibodies rises sooner and reaches a higher concentration. This is known as the **secondary immune response** – it is usually quick enough to prevent any symptoms being detected by the host.

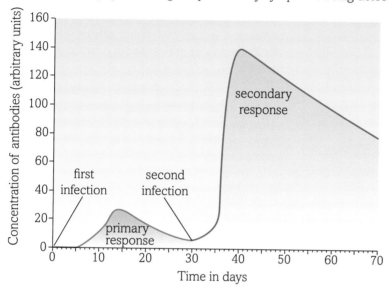

Figure 3 Primary and secondary response to infection.

Questions

1 Distinguish clearly between an antigen and an antibody.

2 Why is it important that some opsonins are not very specific?

3 What is meant by the term immunoglobulin?

4 Antibodies are made by plasma cells. Explain how plasma cells are specialised for their role.

5 Explain what is meant by having a shape that is complementary to the shape of the antigen.

6 Describe how the structure of an antibody enables it to perform its function.

7 Explain why it may take several days for the primary immune response to become effective.

Vaccination

By the end of this topic, you should be able to demonstrate and apply your knowledge and understanding of:

* the principles of vaccination, and the role of vaccination programmes in the prevention of epidemics

* the differences between active and passive immunity, and between natural and artificial immunity

---

KEY DEFINITIONS

active immunity: where the immune system is activated and manufactures its own antibodies.
artificial immunity: immunity that is achieved as a result of medical intervention.
epidemic: a rapid spread of disease through a high proportion of the population.
natural immunity: immunity achieved through normal life processes.
passive immunity: immunity achieved when antibodies are passed to the individual through breast feeding or injection.
vaccination: a way of stimulating an immune response so that immunity is achieved.

What is a vaccination?

Vaccination provides immunity to specific diseases. This is created by deliberate exposure to antigenic material that has been rendered harmless. The antigenic material is usually injected, but in some cases it can be taken orally. The immune system treats the antigenic material as a real disease (see topics 4.1.5−4.1.7). As a result, the immune system is activated and manufactures antibodies and memory cells. The memory cells provide the long-term immunity.

The antigenic material used in vaccines can take a variety of forms:

* Whole, live microorganisms – usually ones that are not as harmful as those that cause the real disease. But they must have very similar antigens, so that the antibodies produced will be effective against the real pathogen (e.g. the smallpox vaccine, which uses a similar virus that causes cowpox).
* A harmless or attenuated (weakened) version of the pathogenic organism (e.g. measles and TB vaccines).
* A dead pathogen (e.g. typhoid and cholera vaccines).
* A preparation of the antigens from a pathogen (e.g. the hepatitis B vaccine).
* A toxoid, which is a harmless version of a toxin (e.g. the tetanus vaccine).

Application of vaccines

Herd vaccination

Herd vaccination is using a vaccine to provide immunity to all or almost all of the population at risk. Once enough people are immune, the disease can no longer be spread through the population and you achieve 'herd immunity'. In order to be effective, it is essential to vaccinate almost all the population. To eradicate smallpox, it was necessary to vaccinate 80–85% of the population. It is estimated that at least 95% of the population would need to be immunised in order to prevent the spread of measles.

In the UK, there is a vaccination programme to immunise young children against the following diseases: diphtheria, tetanus, whooping cough, polio, meningitis, measles, mumps and rubella. These vaccines are given to the majority of children at the appropriate age.

Ring vaccination

Ring vaccination is used when a new case of a disease is reported. Ring vaccination involves vaccinating all the people in the immediate vicinity of the new case(s). This may mean vaccinating the people in the surrounding houses, or even in the whole village or town. Ring vaccination is also used in many parts of the world to control the spread of livestock disease.

Control of epidemics

Once a disease has been eradicated, or reduced to such a low incidence that it is unlikely to spread, the routine vaccination programme can be relaxed. This has occurred with smallpox. TB vaccinations have also been stopped for most children in the UK. However, some pathogens can undergo genetic mutations which change their antigens. The memory cells produced by vaccination may not recognise the new antigens. When this occurs, the pathogen may be transmitted, and the incidence of the disease increases.

Certain pathogens, such as the influenza virus, are relatively unstable and regularly undergo changes in their antigens. When this occurs, an epidemic may arise.

Threats from epidemics must be monitored so that new strains of pathogens can be identified. This enables the health authorities to prepare for an impending epidemic by stockpiling suitable vaccines and vaccinating people who are at particular risk from the disease.

Influenza

Influenza (also known as 'flu) is a killer disease caused by a virus. People over 65 years of age and those with respiratory tract conditions are particularly at risk. Occasionally, a new strain of flu virus arises that is particularly virulent. This may cause an epidemic. For example, in 1918 a flu epidemic killed at least 40 million people worldwide (some estimates are as high as 100 million). More recently, in 1968/69, about 1 million people were killed by Hong Kong flu, also known as the H3N2 strain. In 2009–2010 about 540 000 cases of swine flu (H1N1) were reported in the UK. This strain of flu affected many people worldwide, but the exact numbers are uncertain, as the strain was less virulent than expected and many cases may have gone unreported.

In attempts to avoid another worldwide epidemic (a **pandemic**), people at risk are immunised. In the UK, there is a vaccination programme to immunise all those aged over 65 and those who are at risk for any other reason. In 2013–2014, about 73% of people over 65 were vaccinated, along with about 52% of younger people in 'at-risk' groups. New versions of the influenza vaccination have been developed which can be administered via a nasal spray. It is proposed that these will be offered to all children in the coming years. The strains of flu used in this immunisation programme change each year. Worldwide research is undertaken to determine which of the strains of flu are most likely to spread in any given year.

> **DID YOU KNOW?**
>
> Swine flu took many health authorities by surprise, as they were expecting a version of bird flu to spread. This meant that vaccinations for seasonal flu were ineffective against the swine flu, and new vaccines had to be prepared.

Different types of immunity

Immunity can be achieved naturally or artificially (see Table 1). Natural immunity is achieved through normal life processes. Artificial immunity is achieved through medical intervention.

Immunity can be achieved actively or passively. Active immunity is achieved when the immune system is activated and manufactures its own antibodies. Passive immunity is achieved when the antibodies are supplied from another source.

	Natural	Artificial
Active	Immunity provided by antibodies made in the immune system as a result of infection. A person suffers from the disease once and is then immune (e.g. immunity to chickenpox).	Immunity provided by antibodies made in the immune system as a result of vaccination. A person is injected with a weakened, dead or similar pathogen, or with antigens, and this activates his/her immune system (e.g. immunity to TB and influenza).
Passive	Antibodies provided via the placenta or via breast milk. This makes the baby immune to diseases to which the mother is immune. It is very useful in the first year of the baby's life, when its immune system is developing.	Immunity provided by injection of antibodies made by another individual (e.g. hepatitis A and B). Tetanus can also be treated this way when vaccination using a toxoid has not worked well.

Table 1 Types of immunity.

Questions

1. In many countries, it is important to have a reporting procedure in place in order to report new cases of a disease to the health authorities. Explain why such a reporting procedure is essential in order to use vaccines effectively.

2. Vaccines using dead pathogens or preparations of the antigen are not as effective as those using live organisms. Such vaccinations often require a second or 'booster' jab. Suggest why using live organisms provides better immunity.

3. Explain how herd vaccination can protect us from a disease such as measles.

4. Explain why passive immunity only provides short-term immunity.

5. What diseases or conditions may make a person 'at risk' from influenza?

6. Use an annotated diagram to explain the difference between herd vaccination and ring vaccination.

(9) Development and use of drugs

By the end of this topic, you should be able to demonstrate and apply your knowledge and understanding of:

* possible sources of medicines
* the benefits and risks of using antibiotics to manage bacterial infection

Sources of new medicines

There are currently over 6000 different kinds of medicine available in the UK. But new drugs are needed because:

* new diseases are emerging
* there are still many diseases for which there are no effective treatments
* some antibiotic treatments are becoming less effective.

New medicines can be discovered in a number of ways.

Accidental discovery

The accidental discovery of the antibiotic **penicillin** by Alexander Fleming is well documented (see Figure 1). The fungus *Penicillium* releases compounds that kill bacteria. This is a classic example of how science works – a scientist makes an observation and sets out to explain what he or she has seen. In this case, it was the work of Florey and Chain, who purified penicillin, that really demonstrated the potential value of antibiotics. This shows how important it is for scientists to work together.

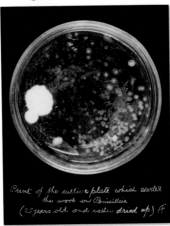

Figure 1 The original agar plate from which Fleming observed the action of penicillin.

Traditional remedies

Many drugs have been used for centuries. They are used because people have noted that certain plants or extracts have a beneficial effect. The World Health Organization calculates that 80% of the world's population relies on traditional medicines. In India, some 7000 different plants are used for their medicinal properties, and in China they use about 5000 different plants. In Europe, many of our modern drugs have their origins in traditional medicine.

* Morphine has its origins in the use of sap from unripe poppy seed-heads as long ago as Neolithic times. In the 12th century, opium from poppies was used as an anaesthetic and, by the 19th century, morphine and opium were being used. These opiate drugs reduce nervous action in the central nervous system. If the nerves cannot carry impulses, then no pain is felt.

* Medicinal use of willow-bark extract to relieve pain and fever has a long history (see topic 2.4.9). After discovery of its active ingredient, a way was later found (in 1897) to reduce the side effect of stomach bleeding, by adding an acetyl group. This led to the development of the drugs aspirin and ibuprofen.

Observation of wildlife

Many animals make use of plants with medicinal properties. For example:

* monkeys, bears and other animals rub citrus oils on their coats as insecticides and antiseptics in order to prevent insect bites and infection

* birds line their nests with medicinal leaves in order to protect chicks from blood-sucking mites.

DID YOU KNOW?

Chimpanzees swallow leaves folded in a particular way, in order to remove parasites from their digestive tract.

Further plant research

Scientists have used traditional plant medicines and animal behaviour as a starting point in their search for new drugs. As you will have seen in the case of aspirin, research into the plants used for traditional remedies enables scientists to isolate the active ingredient. This molecule can be analysed, and similar molecules can be manufactured.

In recent decades, discovery of natural drugs has concentrated on tropical plants. Owing to their great diversity, there are hopes that many may contain molecules that could form new medicinal drugs. But it is important to remember that there may be many potential uses of wild and cultivated plants in the UK. New chemical fingerprinting technology is enabling scientists to screen natural chemicals more effectively for their activity as potential medicines.

> **LEARNING TIP**
>
> The need to develop new drugs is a strong argument in favour of conserving biodiversity. See topic 4.2.6 on conservation of biodiversity.

Research into disease-causing mechanisms

Pharmaceutical companies have been conducting research into the way that microorganisms cause disease. Many make use of receptors on plasma membranes (cell surface membranes). For example, the HIV virus binds to the CD4 and CCR5 receptors on the surface of T helper cells. If the binding between the pathogen and the receptor site can be blocked, then the disease-causing pathogen cannot gain access to the cell. The glycoprotein receptor molecules can be isolated and sequenced. Once the amino acid sequence is known, molecular modelling can be used to determine the shape of the receptor. The next step is to find a drug that mimics the shape of the receptor and could be used to bind to the virus itself, which would block the virus from entering the T helper cell. In a similar way, drugs that inhibit the action of certain enzymes can also be developed (see topic 2.4.9).

Personalised medicine

Sequencing technology and molecular modelling have huge potential for future medicines. It is possible to screen the genomes of plants or microorganisms to identify potential medicinal compounds from the DNA sequences. It is hoped that this technology can even be taken a step further. Once the technology is fully developed, it may be possible to sequence the genes from individuals with a particular condition and develop specific drugs for the condition. This is known as personalised medicine.

Synthetic biology

The development of new molecules – in particular, enzymes – that mimic biological systems is one form of synthetic biology. Another way that synthetic biology is used is to design and construct new devices and systems that may be useful in research, healthcare or in manufacturing. For example, there is the development of tomatoes which contain the pigment anthocyanin. This pigment is found in fruit such as blueberries, and has specific health benefits. Anthocyanins are antioxidants and help protect against coronary heart disease.

Antibiotic use and abuse

Antibiotics are compounds that prevent the growth of fungi or bacteria. As mentioned above, the first antibiotic was discovered by Alexander Fleming in 1928. Since then, many different compounds with antibiotic properties have been discovered. Most antibiotics currently in use are derivatives of a compound made by bacteria from the genus *Streptomyces*. Antibiotics have been used widely to treat bacterial infections.

Before the use of antibiotics, many people died as a result of wounds or surgery when they became infected. Antibiotic use became very widespread during World War II, in order to prevent the infection of wounds. However, over-use and misuse of antibiotics have enabled microorganisms to develop resistance, and many of the current antibiotics have limited effectiveness as a result. Some bacteria have become infamous for their multiple resistance to a range of antibiotics. These include *Clostridium difficile* (*C. diff*) and methicillin-resistant *Staphylococcus aureus* (MRSA) (see also topic 4.3.9).

> **DID YOU KNOW?**
>
> Leaf-cutter ants farm fungi, which they feed to their growing larvae. In order to prevent infection of their fungal gardens, they carry symbiotic bacteria, including *Streptomyces,* which produce antibiotics. Scientists are studying the bacteria carried by the leaf-cutter ants to see if they have potential to produce new drugs to treat human bacterial infections.

Questions

1. Suggest how our ancestors may have discovered that sap from unripe poppy seed-heads could act as an anaesthetic.

2. Describe how a drug that binds to antigens on the HIV virus can prevent transmission of the virus.

3. Explain the link between the base sequence in a gene and the shape of a protein molecule such as a receptor or an enzyme.

4. Explain how a drug can inhibit the action of an enzyme.

5. Explain why it is important to use antibiotics correctly – especially the need to complete a full course as prescribed by the doctor.

6. Describe how a microorganism can become resistant to an antibiotic.

HIV

Reducing the spread of HIV has been at the forefront of medical research ever since AIDS spread during the 1980s. This article looks at an unexpected way in which the spread of HIV/AIDS can be reduced.

HIV AND CIRCUMCISION

The scourge of human immunodeficiency virus (see Figure 1) and acquired immune deficiency syndrome (HIV/AIDS) continues to wreak havoc in the world more than 25 years after the pandemic took hold. A number of preventative programmes aimed at behavioural changes are in place worldwide, and there have been some reports of successful outcomes, despite the challenges of programmes promoting behavioural change. The promotion of abstinence, single-partner relationships and consistent use of condoms have been reported to successfully curb the spread of HIV infection in Thailand, Senegal and Uganda. Recently, some researchers have called for mass male circumcision for prevention of HIV infection, from three randomised controlled trials (RCTs), which have shown that the risk of female-to-male transmission is reduced by 50–60% in heterosexual circumcised men in countries with high HIV prevalence rates. Although there have been published reviews and commentaries on circumcision as a prevention strategy against female-to-male HIV transmission, some authors suggest a cautious approach to widespread circumcision, because of unresolved ethical, economic and social issues that require debate.

Male circumcision and HIV: biological plausibility

The actual mechanism by which circumcision may prevent HIV is unknown. However, the results of a number of studies suggest that the foreskin acts as a reservoir for HIV-containing secretions, thus increasing the contact time between the virus and target cells lining the foreskin's inner mucosa; the inner mucosa of the foreskin is poorly keratinised and contains a high density of Langerhans cells (target cells for HIV); and the foreskin is susceptible to trauma, microabrasions and ulceration, which facilitate acquisition of HIV. It is also proposed that male circumcision decreases the risk of infection by removing the Langerhans cells and reducing the synergy that normally exists between HIV and other sexually transmitted infections. There is a view, however, that the Langerhans cells are protective, in that they neutralise the virus at the onset of infection by initial sampling of foreign antigens and presenting the processed antigens to naïve T cells. Therefore, Langerhans cells are very effective in preventing infection, unless overwhelmed by high viral loads.

Timing of male circumcision

There are three distinct time points at which circumcision can be instituted, namely neonatal, pre-adolescent/prepubertal and post-adolescent/adulthood.

Cost implications

Estimates of the cost of circumcision vary, but Kahn *et al.* (2006) estimated that it could cost anything between US$117 and US$306 to perform a single procedure in South Africa. At this cost, circumcision may be prohibitive in most resource-constrained countries, especially as estimates of circumcision costs should take into account any costs incurred by the treatment of complications. In contrast, a male condom costs two US cents, and is almost 100% effective, while circumcision, even at quoted conservative costs, is only 50% to 60% effective. It is plausible that circumcision in clinical settings, with effective anaesthesia, post-operative care and counselling, will initially be affordable for the relatively privileged groups in sub-Saharan Africa. However, it is doubtful whether governments in poor countries will be able to afford the roll-out of mass circumcision without assistance from donor countries. It is also feasible that the less-privileged groups may only be able to access cheaper and inappropriate circumcision services, and consequently may face risks of higher complications and higher potential risk of HIV infection.

Source

- Kahn, J.G., Marseille, E. and Auvert, B. (2006) Cost-Effectiveness of Male Circumcision for HIV Prevention in a South African Setting. *PLoS Med*, 3(12): e517. doi:10.1371/journal.pmed.0030517
- http://reference.sabinet.co.za/webx/access/electronic_journals/m_samj/m_samj_v98_n10_a17.pdf.

Where else will I encounter these themes?

1.1 2.1 2.2 2.3 2.4 2.5

Let's start by considering some ethical issues arising from this article.

1. What ethical and social issues may affect the decision to conduct mass male circumcision?

Now we will look at the biology in, or connected to, this article. Don't worry if you are not ready to give answers to these questions yet. You might like to return to the questions once you have covered other topics later in the book. Use the timeline at the bottom of the page to help you put this work in context with what you have already learned and what is ahead in your course.

> You might like to consider what people would feel if told that they or their children must be circumcised. Are there any risks associated with male circumcision?

2. Of the three suggested timings for circumcision, the last (i.e. in adulthood) is least likely to be effective as a way to reduce HIV transmission. Explain why this is the case.

3. Langerhans cells are antigen-presenting cells. What is the role of antigen-presenting cells in the immune response?

4. An alternative treatment for HIV/AIDS is an aggressive cocktail of antiretroviral drugs. The highest prevalence of HIV/AIDS is in sub-Saharan Africa. Why might this affect the development of drugs to combat HIV?

5. Explain why the life cycle of viruses makes it difficult for the immune system to combat the infection.

6. HIV enters the T cells by using cell-membrane receptors known as CD4 and CCR5 receptors. Some modern drugs act to block specific cell-surface-membrane receptors. How are such drugs able to be specific to the receptor? Explain how such a drug would work to reduce transmission.

7. How do programmes to promote behavioural change help to reduce the transmission of HIV?

8. Some scientists believe that mass circumcision may lead to an increase in HIV transmission, due to disinhibition – the relaxation of attitudes taught in behavioural-change programmes. Evaluate the effectiveness of male circumcision using the information provided.

Figure 1 A digital illustration of the HIV virus.

Activity

Design and produce an A3-sized wall poster for public health education. Your poster should explain how HIV is transmitted. It should also indicate where preventative measures could be effective and how effective each measure can be.

Include: abstinence, use of condoms, male circumcision, antiretroviral drugs and drugs that block membrane receptors. In your annotations explain how each preventative measure works.

1. What is the role of callose? [1]

    A. Storage molecule which stores sugars in the root

    B. Additional thickening of the cell wall

    C. Large polysaccharide that blocks sieve tubes

    D. A toxic molecule that inhibits bacterial reproduction

2. What type of pathogen causes malaria? [1]

    A. Bacterium

    B. Virus

    C. Fungus

    D. Protoctist

3. Which row correctly describes the sequence of events during phagocytosis? [1]

	Step 1	Step 2	Step 3	Step 4	Step 5
A	pathogen recognised	pathogen engulfed	pathogen trapped in phagosome	lysosomes fuse with phagosome	digestion
B	pathogen recognised	lysosomes fuse with pathogen	digestion	components engulfed by neutrophil	
C	pathogen engulfed by neutrophil	pathogen recognised as foreign	pathogen trapped in phagosome	lysosomes fuse with phagosome	digestion
D	lysosomes fuse with phagosome	pathogen recognised as foreign	pathogen trapped in phagosome	digestion	

4. Which row correctly identifies the functions of the cells listed in the table. [1]

	neutrophils	B lymphocytes	T lymphocytes	macrophages
A	engulf and digest pathogens	release cytokines	manufacture antibodies	digest pathogens and present antigens
B	digest pathogens and present antigens	manufacture antibodies	release cytokines	engulf and digest pathogens
C	engulf and digest pathogens	manufacture antibodies	release cytokines	digest pathogens and present antigens
D	digest pathogens and present antigens	release cytokines	manufacture antibodies	engulf and digest pathogens

5. What term is used to describe immunity achieved as a result of being injected with dead pathogens? [1]

    A. natural passive immunity

    B. artificial passive immunity

    C. natural active immunity

    D. artificial active immunity

    [Total: 5]

6. The following questions are about vaccination.

    (a) What is meant by the term 'vaccination'? [2]

    (b) Describe how a vaccination programme can be used to prevent an epidemic. [4]

    (c) Scientists carried out an investigation into the effectiveness of vaccination against influenza. The scientists enrolled 850 volunteers. 425 were given a vaccine and 425 were given a placebo. Over the following six months the scientists recorded the incidence of illness. The table below shows the results of the investigation.

	Given vaccination	Given placebo
Episodes of upper respiratory illness	117	149
Number of days off work due to respiratory illness	58	132
Number of visits to a doctor for respiratory problems	30	65
Number of days of illness	679	1045

    (i) State whether the vaccine is effective or not. [1]

    (ii) Justify your answer above. [3]

    (iii) Write an evaluation of the investigation. [3]

    [Total: 13]

7. Plants may suffer from infection.

    (a) How do plants reduce the chances of infection? [5]

    (b) Describe how plants prevent the spread of infection around the whole organism. [8]

    (c) Distinguish between primary defences and secondary defences to disease in mammals. [2]

    [Total: 15]

8. Figure 1 shows the concentration of immunoglobulins in the blood in response to exposure to antigen material on day 0.

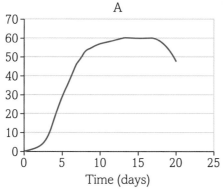

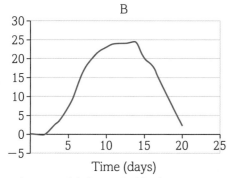

Figure 1 Concentration of immunoglobulins in the blood in response to exposure to antigen material.

(a) Which graph shows a secondary immune response? Explain the reasons for your choice. [3]

(b) Explain why there is a delay in graph B before the concentration of molecules starts to rise. [7]

[Total: 10]

9. Figure 2 is a diagram of an immunoglobulin.

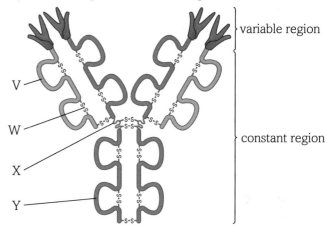

Figure 2 A diagram of an immunoglobulin.

(a) Identify the parts labelled V, W, X and Y. [4]

(b) Explain why the variable region of different immunoglobulins is a different shape. [4]

(c) Describe how the different types of immunoglobulin in the blood combat disease. [5]

[Total: 13]

10. An autoimmune disease is one in which the immune system starts to attack your own tissues.

Rheumatoid arthritis is an autoimmune disease in which the immune system attacks the joints causing pain and swelling. The first part of the joint affected is the lining.

(a) Name the tissue lining the joint. [1]

(b) What is the normal function of this tissue? [1]

(c) Suggest why the joints may be more susceptible to autoimmune disease than other parts of the body. [2]

(d) Suggest two factors that might affect the chances of developing rheumatoid arthritis. [2]

(e) Name another autoimmune disease. [1]

[Total: 7]

CHAPTER 4.2

BIODIVERSITY

Introduction

When Charles Darwin visited the Galapagos Islands he wrote about his amazement over the number of different organisms he found. The biodiversity of these islands arose from the varying conditions they presented to living species which needed to adapt to those conditions. Biodiversity refers to the variety and complexity of life. Biodiversity can be considered at a number of levels: (a) diversity within a species which is known as genetic diversity, (b) diversity of species in a habitat – which is an important factor in the study of habitats and (c) diversity of habitats – the range of different habitats found.

Maintaining biodiversity is important for many reasons. Many of our actions cause loss of biodiversity. Notably, the increase in the human population gives rise to the need for more intense agricultural practices and global warming. Conservation includes attempts to maintain biodiversity. These efforts can be made both *in situ* and *ex situ*. But to be successful these efforts must be made at local, national and global levels.

All the maths you need

To unlock the puzzles in this section you need the following maths:

- Recognise and use expressions in decimal and standard form
- Use an appropriate number of significant figures
- Construct and interpret frequency tables and diagrams, bar charts and histograms
- Understand simple probability
- Understand the principles of sampling as applied to scientific data
- Understand the terms mean, median and mode
- Use a scatter diagram to identify a correlation between two variables
- Select and use a statistical test
- Understand measures of dispersion, including standard deviation and range
- Substitute numerical values into algebraic equations using appropriate units for physical quantities
- Solve algebraic equations
- Translate information between graphical, numerical and algebraic forms
- Plot two variables from experimental or other data

What have I studied before?

- Sampling organisms in a habitat
- Human population growth and its effect on the environment
- Endangered species and the efforts that can be made to conserve them

What will I study later?

- How biodiversity is created by natural selection (AS and AL)
- How competition can affect biodiversity (AL)
- How biodiversity affects the stability of food webs and habitats (AL)
- How biodiversity can affect food production (AL)
- How we can produce food and other requirements sustainably (AL)

What will I study in this chapter?

- The different levels of biodiversity
- How to make quantitative studies of biodiversity at both genetic and species level
- Things that affect biodiversity
- *In situ* and *ex situ* efforts to maintain biodiversity
- The need for local, national and international agreements to maintain biodiversity

(1) Biodiversity

By the end of this topic, you should be able to demonstrate and apply your knowledge and understanding of:

* how biodiversity may be considered at different levels
* how sampling is used in measuring the biodiversity of a habitat and the importance of sampling

Figure 1 A wildflower meadow (left) and a grass pasture (right).

Biodiversity

Biodiversity is a measure of all the different plant, animal, fungus and other microorganism species worldwide, the genes they contain, and the ecosystems of which they form a part.

Biodiversity is about the structural and functional variety in the living world. We can consider it at a number of levels.

Habitat biodiversity

A habitat is the place where individuals in a species live. The range of habitats in which different species live is known as the habitat biodiversity. Common habitats found in the UK include sand dunes, woodland, meadows and streams. Even in your school grounds or in a local park, there may be a variety of habitats. Carefully manicured lawns, ponds, dark corners between buildings or a small patch of trees are all different habitats. Each habitat will be occupied by a range of organisms.

Species biodiversity

A species consists of individual organisms that are very similar in appearance, anatomy, physiology, biochemistry and genetics. As a result, individuals in a species can interbreed freely to produce fertile offspring (offspring that can breed to give rise to more offspring). The range of organisms found in a habitat contributes to the species biodiversity. However, it is not a simple matter of counting the number of different species. Two habitats may have an equal number of different species, but they may not be considered as equally diverse. For example, a wild meadow might have 25 species of grasses and herbs (see Figure 1(a)). In any small sample you might find most of these species. Compare that with a garden lawn, or a managed cow pasture (see Figure 1(b)). Here there may be 25 plant species, but one or two grass species dominate, with just a few individuals of the other species dotted about. The wild meadow is much more diverse, because the 25 species are more evenly represented.

We call the number of plant species the *species richness*. The degree to which the species are represented is known as the *species evenness*. These terms are further used in topic 4.2.4.

Genetic biodiversity

Genetic biodiversity is the variation between individuals belonging to the same species. This is the variation found within any species that ensures we do not all look identical. Genetic variation can create breeds within a species, as shown by the difference between breeds of cattle or dogs (see Figure 2).

Figure 2 Genetic diversity in dogs.

Using samples to measure biodiversity of a habitat

In order to measure the biodiversity of a habitat, you need to observe all the species present, identify them, and count how many individuals of each species there are. Ideally, you should do this for all the plants, animals, fungi, bacteria and other single-celled organisms living in the habitat. Obviously, this is not practical, as it would be impossible to count all the fungi, bacteria and single-celled organisms. One estimate suggests that there may be billions of single-celled organisms per square metre of soil, and possibly hundreds of thousands of mites per square metre! Some microorganisms can be cultured on a nutrient medium in the laboratory to gain an estimate of numbers, but not all will grow like this.

Instead, you can sample a habitat. This means you select a small portion and study that carefully. Then you can multiply up the numbers of individuals of each species found, in order to estimate the number in the whole habitat.

It is important that the samples taken are representative of the habitat. There are a number of sampling strategies that can be adopted, and each has advantages and disadvantages (see Table 1).

Type of sampling		How sampling is carried out	Advantages	Disadvantages
Random		Sample sites inside the habitat are randomly selected. You can do this by deciding where to take samples before you study any area in detail. This can be achieved by using randomly generated numbers as coordinates for your samples, or possibly selecting coordinates from a map and using a portable global-positioning satellite system to find the exact position inside the habitat.	Ensures that the data are not biased by selective sampling.	May not cover all areas of a habitat equally. Species with a low presence may be missed, leading to an underestimate of biodiversity.
Non-random	Opportunistic	This is when the researcher makes sampling decisions based on prior knowledge or during the process of collecting data. The researcher may deliberately sample an area that he or she knows (or can see) contains a particular species.	Easier and quicker than random sampling.	The data may be biased. The presence of large or colourful species may entice the researcher to include that species. This may lead to an overestimate of its importance and therefore an overestimate of biodiversity.
	Stratified	Dividing a habitat into areas which appear different, and sampling each area separately. For example, patches of bracken in heathland might be sampled separately from the heather or gorse patches.	Ensures that all different areas of a habitat are sampled and species are not under-represented due to the possibility that random sampling misses certain areas.	There is a possibility that this may lead to over-representation of some areas in the sample – i.e. a disproportionate number of samples are taken in small areas that look different.
	Systematic	This is when samples are taken at fixed intervals across the habitat. Line transects and belt transects are systematic techniques.	Particularly useful when the habitat shows a clear gradient in some environmental factor such as getting drier further from a pond.	Only the species on the line or within the belt can be recorded. Other species may be missed, leading to an underestimate of biodiversity.

Table 1 A summary of sampling strategies.

Questions

1 Make a list of the habitats found in your school/college or in a local park.

2 Describe what is meant by the terms 'species richness' and 'species evenness'.

3 Explain why it is important to sample a habitat in order to estimate the biodiversity.

4 Why should sampling be as random as possible?

5 Explain why opportunistic sampling may lead to an overestimate of biodiversity.

6 Describe what type of sampling would be best to sample a field such as the wild meadow in Figure 1.

(2) Sampling plants

By the end of this topic, you should be able to demonstrate and apply your knowledge and understanding of:

* practical investigations collecting random and non-random samples in the field

Sampling a habitat

Preparation

It is important to be properly prepared for any fieldwork. Your planning should include:

* suitable clothing – this will depend upon the type of habitat and the expected weather conditions
* suitable footwear
* apparatus needed to carry out the sampling
* clipboard, pen and paper to record your observations
* appropriate keys to identify plants
* camera or smartphone to record specimens and grid location.

Ideally, you will have considered the number of samples that you will collect, and you will have prepared a results table ready to record your observations.

> **LEARNING TIP**
>
> It is important that you can recognise all species correctly, so that you do not record several different species as one. Plants can be identified using a dichotomous key. Before heading out to the sample site, you should be familiar with using this type of key. If you are unable to identify the plant on site, then you can take photographs and give the plant a label for your record sheet. The plant can be identified later. At some times of year, some plants may be visible only as a leaf or two (or not even visible at all). Features such as shape and size of leaf, hairs and colour will all help to identify it. You may find that it helps to look at other plants nearby to see if they have more leaves or flowers that may help identification. Some plants look different at different times of year – in spring only a few small leaves may be visible. It may be necessary to visit a site several times to get a full estimate of biodiversity.

At the site

When visiting a site to measure its biodiversity, it may be best to use a range of techniques. Random sampling is important, but it may be helpful to modify the sampling technique if the habitat is not homogeneous (even). Moving some of the sample sites into areas that look different would be classed as opportunistic sampling (as you are making decisions during the sampling process) and also stratified sampling (as you are treating parts of the habitat differently). It may also be helpful to combine random sampling with systematic sampling, as described below for the point frame.

One important aspect of your work should be to consider the effect your presence will have on the habitat. Any sampling should cause as little disturbance as possible. Trampling, picking flowers, placing quadrats, etc. will all cause some disturbance.

Sampling plants

Large plants such as trees in a wood or in a field can be identified and counted individually. However, many plants may be too small or too numerous. In this case it may be best to calculate a value of percentage ground cover occupied by each species.

Using random quadrats

A **quadrat** is a square frame used to define the size of the sample area. A quadrat may be any size, but it often measures 50 cm or 1 m on each side. For random sampling, you can generate random numbers and then use these numbers as coordinates to place the quadrats within the habitat. A tape measure will help with placing the quadrat accurately. Alternatively, counting even paces will help to locate the correct coordinates.

Inside the quadrat you will need to identify the plants found and then calculate the percentage cover as a measure of their abundance.

- It may be possible to estimate the percentage cover of each species, although most students tend to underestimate. Some quadrats have a grid of string that divides the quadrat into a number of smaller squares (usually 100). This grid can help to make your estimates more accurate.

- You can measure percentage cover using a point frame (see Figure 1). This is a frame holding a number of long needles or pointers. You lower the frame into the quadrat and record any plant touching the needles. If the frame has 10 needles, and is used 10 times in each quadrat, you will have 100 readings. So each plant recorded as touching the needle will have 1% cover. As one needle may touch several plants, it is possible to find you have 300–400% cover in some habitats. Don't forget to record bare ground. It is easy to bias your readings by using the point frame non-randomly within the quadrat. Therefore, it may be best to use the point frame at regular intervals across the quadrat (systematic sampling within the randomly placed quadrat).

Figure 1 A point frame.

Using a transect

A **transect** is a line taken across the habitat. You stretch a long string or tape measure across the habitat and take samples along the line.

In a large habitat, you might use a line transect (see Figure 2). In this case you would record the plants touching the line at set intervals along it.

You may also decide to use a quadrat at set intervals along the line. This is called an interrupted belt transect. This will provide quantitative data at intervals across the habitat.

Alternatively, you may use a continuous belt transect. In this case you place a quadrat beside the line, and move it along the line so you can study a band or belt in detail. You should study each quadrat as described above. This will provide quantitative data in a band or belt across the habitat.

Figure 2 A line transect.

Questions

1. Describe how you would sample a meadow to find out what plants live there.

2. Explain why it is necessary to take samples.

3. Describe the ways that sampling a habitat may disturb that habitat.

4. When would you use a belt transect?

5. Explain why sampling in spring may give a very different measure of plant biodiversity when compared with sampling in summer.

6. Why should you always use a key to identify plants?

(3) Sampling animals

By the end of this topic, you should be able to demonstrate and apply your knowledge and understanding of:

* practical investigations collecting random and non-random samples in the field

Sampling animals by observation

Many animals are not easy to spot and are more difficult to count. Larger animals can detect the presence of humans before we see them, and will hide away. Small animals will also hide, and often move too quickly to count accurately. Therefore, obtaining quantitative data on animals is difficult. You can note the presence of many larger animals by careful observation – looking for signs that they have left behind. For example, many animals leave footprints or easily identified droppings; owls deposit pellets of undigested food; rabbits have obvious burrows; and deer damage the bark of trees in a particular way.

Ecologists often rely on these signs to estimate population sizes. Recent advances allow scientists to use DNA sequencing to distinguish droppings from different individuals. This provides a more accurate way to calculate the population size.

Collecting samples of live animals

Preparation for a site visit is described in the previous topic. Planning should include bringing along the appropriate apparatus for the chosen sampling techniques.

Catching invertebrates

* The technique of sweep netting involves walking through the habitat with a stout net (see Figure 1). You sweep the net through the vegetation in wide arcs. Any small animals, such as insects, will be caught in the net. Then you can empty the contents on to a white sheet to identify them. You need to be careful, as many of the animals may crawl or fly away as soon as you release them from the net. You can use a device called a pooter (see Figure 2) to collect the animals before they fly away. This type of sampling is suitable for low vegetation that is not too woody. You can use a similar technique to take samples in water.

Figure 1 Using sweep nets.

Figure 2 A pooter.

* Collecting from trees using a sweep net is unlikely to work well – here it is better to spread a white sheet out under a branch, and knock the branch with a stout stick. The vibrations dislodge any small animals, which then drop on to the sheet. Again, you will need to be quick to identify and count the animals before they crawl or fly away.

* A pitfall trap (see Figure 3) is a trap set in the soil to catch small animals. It consists of a small container buried in the soil so that its rim is just below the surface. Any animals moving through the plants or leaf litter on the soil surface will fall into the container. The trap should contain a little water or scrunched paper to stop the animals crawling out again. In rainy weather, the trap should be sheltered from the rain so that it does not fill up.

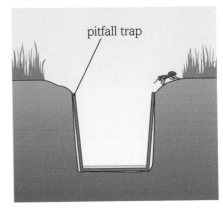

pitfall trap

Figure 3 A pitfall trap.

* A Tullgren funnel (see Figure 4) is a device for collecting small animals from leaf litter. You place the leaf litter in a funnel. A light above the litter drives the animals downwards as the litter dries out and warms up. They fall through the mesh screen to be collected in a jar underneath the funnel.

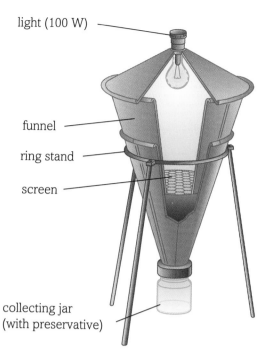

light (100 W)

funnel

ring stand

screen

collecting jar
(with preservative)

Figure 4 A Tullgren funnel.

- A light trap can be used to collect flying insects at night. It consists of an ultraviolet light that attracts the insects. Under the light is a collecting vessel containing alcohol. Moths and other insects attracted to the light eventually fall into the alcohol.

Trapping small animals

Small animals can be trapped and population estimates calculated. The technique that you use will depend on the habitat and the type of animals you are hoping to catch. Beware though, you may need a licence to trap some animals, and care should also be taken not to harm the animals in any way.

Small mammals can be trapped using a Longworth trap. This is a humane trap that does not harm the animal. These traps must be monitored regularly to release any trapped animals. Trapping with a Longworth trap enables the population size to be calculated using the mark-and-recapture technique.

- First you need to capture a sample of animals.
- Mark each individual in some way that causes it no harm. The number captured will be C_1.
- Release the marked animals and leave the traps for another period of time.
- The number captured on this second occasion will be C_2. The number of already marked animals captured on the second occasion is C_3.
- You can then calculate the total population using the formula:

$$\text{Total population} = \frac{(C_1 \times C_2)}{C_3}$$

However, the estimate calculated can be affected by animals that learn that the trap is harmless and contains food, or by animals that do not like the experience and therefore keep away from traps after the first capture.

It is also possible to calculate the size of some insect populations (such as grasshoppers) by the mark and recapture technique. Populations of birds can be estimated by using a ringing technique to identify individuals, and some larger mammals can be tagged. These techniques require skill and experience, and should only be carried out with a suitable permit issued by the relevant authority (the British Trust for Ornithology issues permits to ring birds for research purposes).

Questions

1. Describe how you might sample animals from (a) a meadow, and (b) under a hedge.

2. Plan an investigation to determine whether sweet chestnut trees support a higher biodiversity than hazel. Describe how you would carry out the investigation.

3. How could owl pellets be useful in determining what animals live in a habitat?

4. How could you mark a small mammal, causing no harm to it?

5. How would animals that *like* to be trapped affect the estimated size of the population when using the mark-and-recapture technique?

Calculating biodiversity

By the end of this topic, you should be able to demonstrate and apply your knowledge and understanding of:

* how to measure species richness and species evenness in a habitat
* the use and interpretation of Simpson's index of diversity (*D*) to calculate the biodiversity of a habitat
* how genetic diversity may be assessed, including calculations

> **KEY DEFINITIONS**
>
> **allele** or **gene variant**: a version of a gene.
>
> **locus**: the position of that gene on a chromosome.
>
> **polymorphic gene locus**: a locus that has more than two alleles.
>
> **Simpson's index of biodiversity**: a measure of the diversity of a habitat.
>
> **species evenness**: a measure of how evenly represented the species are.
>
> **species richness**: a measure of how many different species are present.

Estimating biodiversity

When measuring biodiversity, we have to consider species richness – the number of species found in a habitat. The more species present, the richer the habitat. However, richness is not sufficiently quantitative to be a measure of biodiversity on its own. It does not take into account the number of individuals in each species. For this, we need to estimate species evenness.

Species evenness is a measure of the relative numbers or abundance of individuals in each species. A habitat in which there are even numbers of individuals in each species is likely to be more diverse than one in which individuals of one species outnumber all the others. As discussed in topic 4.2.2, we can measure abundance in plants as percentage cover, rather than as numbers of individuals.

Table 1 shows an example of two simple surveys. Fields A and B have equal richness, as they both contain six species. However, field B has greater evenness. Therefore field B would be considered more diverse.

Species observed	Percentage cover	
	Field A	Field B
Cocksfoot grass	57	38
Timothy grass	32	16
Meadow buttercup	3	14
White clover	3	22
Creeping thistle	1	5
Dandelion	4	5
Total	100	100

Table 1 Results of a simple survey of two fields.

Measuring biodiversity within a habitat

Species richness can be measured by counting all the species present in the habitat. Measuring species evenness is more difficult. For this you need to carry out a quantitative survey. Once a full quantitative survey has been carried out as described in topic 4.2.1, the data can be used to calculate the biodiversity.

Surveying the frequency of plants

First use the sampling techniques described in topic 4.2.2 to take your samples. Record the percentage cover of each plant species. With large plants it is better to count the number of individuals per unit area. You can use similar techniques in both terrestrial and aquatic habitats. You may need to take extra precautions when sampling aquatic habitats!

Measuring the density of animals in a habitat

This means calculating how many animals of each species there are per unit area of the habitat. Larger animals can be counted by observation, and smaller ones can be counted by using the sampling techniques described in topic 4.2.3. The population size of smaller animals can be calculated using the mark-and-recapture technique, but this will not work for the numerous tiny animals living in soil. Here, the only way to estimate population size is to take a sample of soil and sift through it to find all the individuals and count them.

Sampling in water is a similar process. You can use a net to sample in the body of the water and to sift through the mud at the bottom. Then you can estimate population size and density.

Simpson's index of diversity

Simpson's index of diversity (see Table 2) is a measure of the diversity of a habitat. It takes into account both species richness and species evenness. It is calculated by the formula:

$$D = 1 - \left[\Sigma \left(\frac{n}{N} \right)^2 \right]$$

where *n* is the number of individuals of a particular species (or the percentage cover for plants), and *N* is the total number of all individuals of all species (or the total percentage cover for plants).

Species	Field A		
	n	n/N	(n/N)²
Cocksfoot grass	57	0.57	0.3249
Timothy grass	32	0.32	0.1024
Meadow buttercup	3	0.03	0.0009
White clover	3	0.03	0.0009
Creeping thistle	1	0.01	0.0001
Dandelion	4	0.04	0.0016
Sum (Σ)	–	–	0.4308
1 – Σ	–	–	0.5692

Table 2 Applying Simpson's diversity index to the results for field A.

Interpreting the data

A high value of Simpson's index indicates a diverse habitat. Such a habitat provides a place for many different species and many organisms to live. A small change to the environment may affect one species. If this species is only a small part of the habitat, the total number of individuals affected is a small proportion of the total number present. Therefore the effect on the whole habitat is small. The habitat tends to be stable and able to withstand change.

A low value for diversity suggests a habitat dominated by a few species. In this case, a small change to the environment that affects one of those species could damage or destroy the whole habitat. Such a small change could be a disease or predator, or even something that humans have done nearby.

Measuring genetic diversity

Isolated populations, such as captive animals in a zoo, rare breeds or pedigree animals, may be small (e.g. Przewalski's horses, see Figure 1 (a)). Therefore their genetic diversity may be limited. Assessing their genetic diversity can help to assess the value of that population as a resource for conservation. A simple assessment of genetic diversity can be made by looking at the observable features of the individuals. If a particular feature shows variation between individuals (see, for example, the horses in Figure 1 (b)), then it suggests there may be genetic diversity.

Figure 1 A herd of Przewalski's horses (*Equus ferus przewalskii*) (left) all look identical due to low genetic diversity, while greater diversity is shown between members of the subspecies *Equus ferus caballus* (right).

Genetic diversity is found where there is more than one **allele** for a particular **locus**. This will lead to variations between individuals that may be easily observable. More importantly, perhaps, it means that there will be more genetic differences between the gametes (sperm and eggs) produced by members of the population.

Calculating genetic diversity

Genetic diversity can be estimated by calculating the number of loci in one individual that are heterozygous (i.e. that contain two different alleles). However, this does not give a good measure of the value of the population as a genetic resource.

Another measurement of genetic diversity involves calculating the percentage of loci in the population that have more than one allele. For example, in a species that has 10 000 loci, if 2000 loci have more than one allele, then the genetic diversity would be given by the equation:

$$\text{Genetic diversity} = (2000 \div 10\,000) \times 100\% = 20\%$$

Some loci have more than two alleles, these are known as **polymorphic gene loci**. Having more than two alleles at a locus increases genetic diversity.

> **DID YOU KNOW?**
>
> An example of a polymorphic gene locus is the ABO blood grouping system found in humans, apes and monkeys. An individual has one locus for this gene on each of a pair of chromosomes. Therefore an individual can have two copies of the gene. These two copies may the same (as in AA or BB or OO), or they may be two different versions of the gene (AB, AO or BO). This significantly increases potential genetic diversity. Another estimate of genetic diversity is given by the equation:
>
> $$\text{proportion of polymorphic gene loci} = \frac{\text{number of polymorphic gene loci}}{\text{total number of loci}}$$

> **DID YOU KNOW?**
>
> In 2013 a giraffe called Marius was deliberately killed at Copenhagen Zoo, because he did not contribute any extra genetic diversity to the population of giraffes at the zoo. This caused some controversy, as many people believed that Marius could have been sent to another zoo.

Questions

1. Calculate the value of Simpson's index for field B (see Table 1).

2. Explain why field B is likely to be a better field to keep as a nature reserve.

3. Explain why a habitat with higher diversity is more stable than one with low diversity.

4. Why is it important to be able to identify plants and animals accurately?

5. Why would wild animal parks need to monitor the genetic diversity of their animal populations?

6. Explain why genetic variations may be easy to identify.

7. Explain why genetic variations may not be easy to identify.

(5) What affects biodiversity?

By the end of this topic, you should be able to demonstrate and apply your knowledge and understanding of:

* the factors affecting biodiversity

> **KEY DEFINITIONS**
>
> **climate change:** significant, long-lasting changes in weather patterns.
> **monoculture:** a crop consisting of one strain of one species.

Current estimates of biodiversity vary, as it is impossible to know how many species currently in existence have not yet been discovered. What is certain is that the number of species is declining, and the genetic diversity of many species is also declining, as a result of human activities.

Human population growth

Several thousand years ago, humans lived as hunter–gatherers in small numbers and had little effect on natural processes. However, as the human population grows (see Figure 1) and we demand more food and consumer goods, we have a greater and greater effect upon other species.

* We have learned to use the environment to our advantage.

* We alter **ecosystems** to provide ourselves with food.

* We destroy and fragment habitats.

* We are using more and more of the Earth's resources.

* We pollute the atmosphere.

As a result of our activities we often harm other species either directly or indirectly, and this can lead to **extinction** (see below).

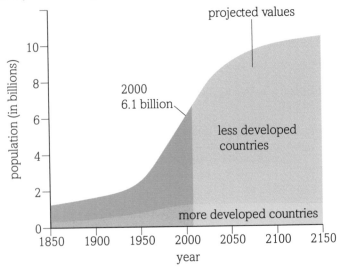

Figure 1 The growth of the human population.

Agriculture

Agriculture has a huge effect on the biodiversity of natural habitats. As we clear natural vegetation, we reduce the size of habitats and the population size of any wild species living in those habitats. This reduces the genetic diversity of the species as their population is reduced. This means that the species has less capacity to adapt to changing conditions through evolution. It may also leave isolated and fragmented populations that are too small to survive.

Modern agriculture relies upon monoculture and **selective breeding** to increase efficiency. A monoculture is a crop consisting of one strain of the species – it has very limited genetic diversity. This makes the product easier to harvest. The oil palm, which is grown for palm oil (see Figure 2), is a good example: 85% of palm oil is produced in Indonesia and Malaysia. Rainforests with huge natural biodiversity are cut down at the rate of 150 hectares an hour. These are replaced by huge stands of a single strain of one species. Indonesia's oil palm plantations already cover 9 million hectares and this is set to rise to 26 million hectares by 2025.

Figure 2 A monoculture of palm oil trees.

Selective breeding also reduces genetic diversity, because farmers select particular traits such as rapid growth or high protein content. Concentrating on these characteristics means that other characteristics may be ignored altogether. Again the genetic diversity of the species declines.

Selecting for specific breeds of domesticated plants and animals means that other breeds become rare and may die out. Again, loss of these varieties reduces the genetic diversity of the species – a process known as **genetic erosion**.

Climate change

Human activities appear to be altering the climate. Species that have lost their genetic diversity show less variation between individuals. As the climate changes, they are less able to adapt to the changes in temperature and rainfall in the area where they live.

The only alternative will be for them to move and follow the climate patterns to which they are most suited. This will mean a slow migration of populations, communities and whole ecosystems towards the poles – plants currently growing in southern Europe many soon grow in northern Europe.

However, there will be obstructions to this migration. Possible obstructions include:

- major human developments
- agricultural land
- large bodies of water
- mountain ranges.

DID YOU KNOW?

Consider the plight of the golden toad of the Costa Rican cloud forest. This amphibian is already facing extinction due to climate change. As the climate warms, the toad moves uphill to stay in the most suitable habitat. What happens when it reaches the top of the hill? The toad will be faced with having to migrate through unsuitable habitats to reach another area with a suitable habitat. This is unlikely to happen. Consider the effect on protected areas such as national parks, where hunting is not allowed. As the climate changes, the selected site no longer maintains the conditions and plants that the protected animals need. The animals will migrate to live outside their protected area.

Domesticated plants and animals are particularly at risk. We have selectively bred our crop plants and animals to provide the best yield in specific conditions, which means that they have little variation. As a result, our agricultural species are unlikely to be able to adapt to changing conditions and are vulnerable to disease. The efficiency of agriculture will decline and less food will be available. Farmers will need to change the crops that they grow and the varieties of animals that they keep. Crops from southern Europe may be grown in Britain, while parts of southern Europe may become desert.

Extinction

Extinction occurs when the last living member of a species dies and the species ceases to exist. Since humans started to spread widely over the Earth about 100 000 years ago, the rate of extinction has risen dramatically. Some scientists believe that increasing human activity caused the extinction of animals such as the giant sloth (*Megatherium*) and the mammoths (*Mammuthus*) 10 000–14 000 years ago. These animals were hunted for food.

What is more certain is that:

- There have been over 800 recorded extinctions since the year 1500.
- Up to 20% of the species alive today could be extinct by 2030.
- One-third of the world's primate species now face extinction – even our closest relatives, the great apes, could be extinct in 20 years.
- Some scientists believe that up to half the species alive today could be extinct by the year 2100.
- The current rate of extinction is 100–1000 times the normal 'background' rate.
- The current rate of extinction is at least as fast as in any previous extinction event.

Many scientists believe that we are at the start of a great mass-extinction event. There have been other mass-extinction events in the past – for example, when the dinosaurs and many other species became extinct. But this extinction event is being caused by human activity, rather than by natural climate change or natural disaster.

Questions

1. Why has the human population risen so quickly?
2. Describe how using a monoculture contributes to loss of biodiversity.
3. Describe how selective breeding contributes to loss of biodiversity.
4. What is meant by genetic erosion?
5. Explain how genetic erosion occurs.
6. Explain why domesticated species of animals are more at risk of extinction than wild populations.

By the end of this topic, you should be able to demonstrate and apply your knowledge and understanding of:

* the ecological, economic and aesthetic reasons for maintaining biodiversity

KEY DEFINITIONS

keystone species: one that has a disproportionate effect upon its environment relative to its abundance.
soil depletion: the loss of soil fertility caused by removal of minerals by continuous cropping.

Ecological reasons to maintain biodiversity
Interdependence of organisms

Natural ecosystems are complex. They have developed over millions of years as species have evolved to live with each other and depend upon one another. All the organisms in a habitat are linked together in a food chain or food web. The range of relationships between organisms includes predator–prey, intra- and inter-species competition, and parasitic and mutualistic relationships.

When one species is affected by human activity and its numbers decline, then this will affect other species. In some cases the effect may not be great. For example, if birds feed on a variety of insects and one insect species falls in number, the birds may simply be able to eat more of the other insects. This is why habitats with higher species diversity tend to be more stable. They can withstand a certain amount of change. However, in a simple habitat with lower species diversity, the loss in numbers of one prey-insect species may mean that the bird has less food and will, itself, decline in numbers.

Some species have a disproportionate effect upon their environment relative to their abundance. The decline of such a keystone species will have a dramatic or even catastrophic effect on the habitat. Keystone species may be predators that limit the populations of herbivores so that the vegetation is not overgrazed. For example, during the early part of the 20th century the mountain lions of the Kaibab Plateau in Arizona were hunted in order to protect the deer population (see Figure 1). As a result, the deer population increased dramatically, ate all the vegetation, and then the population plummeted as the deer starved.

Many plants are also keystone species. A plant that is dominant in the habitat has many effects upon the other species in the habitat, for example, the sugar maple tree in the deciduous forests of northern USA. The sugar maple is a large tree with deep roots. Its roots can access water deep in the soil and transfer the water to shallower regions that are dry, making it available to other plants. The foliage produces a covering canopy, which keeps the soil cool and moist. This allows many soil organisms to thrive, including earthworms and many soil-dwelling insects. The shade also prevents excess undergrowth, which allows other tree species to take root and grow. The sap is very sweet and provides food for a variety of insects (and the manufacture of maple syrup!). Many insects, birds and small mammals can live in the tree canopy.

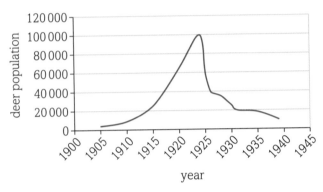

Figure 1 Changes in the deer population of the Kaibab Plateau.

Another example of a keystone species is the beaver (see Figure 2). There may be a relatively small population of beavers, but the dams that they build have a huge effect on the habitat. Large areas can be flooded, and this creates still water in which many other species can live.

Figure 2 A beaver dam can create a whole new habitat.

Genetic resource

Allowing biodiversity to decline means that genetic diversity also declines. This means that we could lose the natural solutions to some of our problems.

Wild animals and plants may hold the answers to problems caused by climate change. Populations of wild plants growing in an area have had thousands of years to evolve. They have adapted to overcome the problems presented by the environment. They may also have adapted to the pests and diseases found in that area. By careful selection and breeding from wild strains and wild species, we may be able to breed new crop varieties that can cope with the new conditions created by climate change. Genetic engineering to produce transgenic species could also be used to good effect.

The number of potential new medicines and vaccines to be found in native plants, animals, fungi and bacteria is unknown. Plants have evolved a wide range of molecules that combat diseases (see topic 4.1.3 on plant defences), and fungi have evolved molecules that help them compete with bacteria in the soil. Any of these molecules could be of value in developing a new and effective pharmaceutical product. It is important to maintain the genetic diversity of wild species because of the potential that exists in the wide range of species currently **extant**.

Economic reasons to maintain biodiversity

In 1997, an international team of economists and environmental scientists attempted to quantify the economic value of natural ecosystems. They came up with a figure of $\$33 \times 10^{12}$. They looked at all the ways in which natural ecosystems perform processes that are of value to humans. These include:

- regulation of the atmosphere and climate
- purification and retention of fresh water
- formation and fertilisation of soil
- recycling of nutrients
- detoxification and recycling of wastes
- crop pollination
- growth of timber, food and fuel
- discovery of molecules with potential as medicines.

All of these factors impact upon food production. One of the most immediate effects that loss of biodiversity has on food production is the depletion of soil. Soils that are subjected to continuous monoculture become less and less fertile. As a result, agricultural yields decline. This is because the crop takes minerals out of the soil, and when the crop is harvested these minerals are removed from the ecosystem. When a monoculture is grown, the plants always take the same minerals, and the effect is observed more quickly. The effects of soil depletion were seen in the huge dust storms experienced in the dust bowl of America during the 1930s (see Figure 3). Years of cropping without replacing the organic matter, followed by drought, led to the soil turning to dust. Similar soil depletion is occurring in the tropics as a result of tropical forest being converted to monoculture.

Figure 3 Overcropping led to soil depletion and dust storms during the 20th century.

Aesthetic reasons to maintain biodiversity

We experience a feeling of joy and wellbeing when observing the infinite variations of nature. It is said that no human art or design can compete. Studies have shown that patients recover more rapidly from stress and injury when they are exposed to pleasing natural environmental conditions. It is clear that natural systems are very important for our wellbeing and for our physical, intellectual and emotional health.

> **DID YOU KNOW?**
>
> 'Chlorophyllia' is a term that has been used to describe people's love of all things green.

However, there are more basic reasons to maintain the aesthetic value created by biodiversity. Landscapes are formed by the action of climatic factors on the land. But the living biosphere also has its effect on the landscape. Diverse tropical forests protect the soil from climatic factors such as rainfall that could erode and wash away exposed soils. Woodland or forest acts as a reservoir when it rains. The trees take up water, and the organic matter in the soil holds water. This means that run-off and drainage are reduced. The water that collects in forested hills will slowly drain away and supply water downstream for some time after the rainfall has stopped.

Reducing biodiversity exposes the soil and changes the landscape. Deforestation has been linked to severe flooding in many areas of the world. The protection of the soil is reduced when the trees are removed and replaced by buildings or agriculture. Rainfall drains more quickly and there is more run-off. This means that flooding is more likely. In an extreme case, the rainfall will erode the soil and wash it downstream. The soil is deposited where the water flow is reduced and further reduces the drainage – this increases the chances of flooding. In regions where rainfall is very seasonal, this means that water is no longer stored in the hills and, when the rain stops, the flooding is quickly followed by drought.

Questions

1. Why did the deer on the Kaibab Plateau starve after the mountain lions were removed?

2. There has been concern over recent reductions in the number of bees. Explain why bees could be considered to be a keystone species.

3. Why might new varieties of crops be needed to overcome the problems caused by climate change?

4. What sorts of features might be selectively bred into a crop to overcome the effects of climate change?

5. What caused the huge dust storms in the American dust bowl of the 1930s?

6. How does a forested landscape help to manage water flow?

(7) Conservation *in situ*

By the end of this topic, you should be able to demonstrate and apply your knowledge and understanding of:

* *in situ* methods of maintaining biodiversity

Conservation is not a passive process. It involves active management to maintain habitats and the species that live in those habitats. The aim is to enable endangered species to survive and maintain biodiversity.

Conservation *in situ* means conserving species in their natural habitat. It involves attempting to minimise the human impact on the natural environment and protecting the natural environment. This can be carried out in a number of ways.

Legislation

It is possible to pass legislation to stop such activities as hunting, logging and clearing land for development or agriculture. The legislation is specific to a particular country. It can be difficult to persuade some countries that legislation is necessary. It can also be difficult to enforce such legislation – especially if the government is not in favour of it. Topic 4.2.9 describes some international agreements that aim to conserve biodiversity. International law governs what people are allowed to import and export.

Wildlife reserves

Wildlife reserves (see Figure 1) are designated areas established for the conservation of habitats and species. Large reserves are an important part of the conservation efforts in many parts of the world, including Africa. We should remember that wildlife reserves are not the only sites of *in situ* conservation. Land management agreements on private land and farm sites can also be used for conservation.

The principles for choosing a wildlife reserve must include the following.

1. Comprehensiveness – how many species are represented in the area and what are the prevailing environmental conditions?

2. Adequacy – is the area large enough to provide for the long-term survival of all the species, populations and communities represented?

3. Representativeness – is there a full range of diversity within each species and each set of environmental conditions?

Conservation should not mean excluding all human activity. A reserve should meet the needs of the indigenous people. They might use the land for traditional hunting, or for spiritual and religious activities. In the past, reserves have been set up without consideration of the local people, and this has led to conflict. The reasons why conflict arises could be due to:

* protected animals coming out of the reserve to raid crops – primates often raid farms for maize, mangoes and sugar cane

* people continuing to hunt protected animals for food (poaching)

* illegal harvesting of timber and other plant products

* tourists feeding protected animals or leaving litter.

Figure 1 A nature reserve can successfully conserve biodiversity.

Wildlife reserves in the UK

Various bodies in the UK work to conserve and enhance the natural environment – including landscape, biodiversity, natural resources, geology and soils. Many parts of the UK are protected by designated status.

* National Parks – there are 15 in the UK, covering many of the most beautiful and valued landscapes. They are areas of protected countryside that everyone can visit, and where people live, work and shape the landscape.

* National Nature Reserves (NNR) – in 2014 there were nearly 400 NNRs in the UK, covering nearly every type of vegetation found here, and occupying over 94 000 hectares of land. These areas are set up to protect sensitive features of the environment and to enable research and education.

- Sites of Special Scientific Interest (SSSIs) – there are over 6000 SSSIs in the UK. These are the country's very best wildlife and geological sites. They include some of our most spectacular and beautiful habitats – large wetlands teeming with waders and waterfowl; winding chalk rivers; gorse- and heather-clad heathlands; flower-rich meadows, windswept shingle beaches, and remote upland moors and peat bogs.

- Local Nature Reserves – often run by County Wildlife Trusts.

- Marine conservation zones – 27 sites around the UK coast were designated as marine conservation zones in 2013. These are areas that are important to conserve the biodiversity of nationally rare, threatened and representative habitats and species in our seas.

Advantages and disadvantages of *in situ* conservation

Designating an area as a wildlife reserve or marine conservation zone has a number of advantages:

- Plants and animals are conserved in their natural environment.
- It permanently protects biodiversity and representative examples of ecosystems.
- It permanently protects significant elements of natural and cultural heritage.
- It allows management of these areas to ensure that ecological integrity is maintained.
- It may provide opportunities for ecologically sustainable land uses, including traditional outdoor heritage activities and the associated economic benefits.
- It facilitates scientific research.
- It may be possible to improve and restore the ecological integrity of the area.

Unfortunately, there are also some disadvantages of *in situ* conservation:

- Endangered habitats may be fragmented, and each small area may not be large enough to ensure survival.
- The population may already have lost much of its genetic diversity.
- The conditions that caused the habitat or species to become endangered may still be present.
- The area can act as a 'honeypot' to poachers and ecotourists, who inadvertently cause disturbance.

Repopulation

Where biodiversity has been lost, it is possible to rebuild it. There are many examples of sites where recreated wildlife habitats have been made to work. In the UK, the numbers of bitterns and otters are increasing in new reed beds. Conifer crops are being cleared for wildlife habitat recovery, and large areas of grazing land are being helped to revert to traditional meadow grassland.

In the Phinda Reserve of South Africa, work began in 1990 to clear away livestock and reintroduce natural fauna. More than 1000 wildebeest, zebras, giraffes and other ungulates were released between 1990 and 1992. Nearly 30 white rhinos and 56 elephants followed. Later, in 1992, 13 lions and 17 cheetahs were released. This was a start towards recreating the rich mammal community that existed in the region before European colonisation.

Questions

1 Explain the meaning of the term '*in situ* conservation'.

2 Why does conservation require active management?

3 Suggest why the sites of many SSSIs are not openly signposted.

4 Suggest what difficulties may need to be overcome when attempting to repopulate an area with its original flora and fauna.

5 Suggest why the lions and cheetahs were introduced to the Phinda Valley some time after the herbivore species.

Conservation *ex situ*

By the end of this topic, you should be able to demonstrate and apply your knowledge and understanding of:

* *ex situ* methods of maintaining biodiversity

> **KEY DEFINITION**
>
> **conservation *ex situ*:** conservation outside the normal habitat of the species.

Conservation *ex situ* means conserving an endangered species by activities that take place outside its normal habitat.

Zoos

Traditional zoological collections held any animals that were collected by their owner or, perhaps, were unusual to the public. More recently, the role of zoos has changed, and many prefer to be known as 'wildlife parks'. These wildlife parks now play an important role in conservation. Many concentrate on captive breeding; breeding endangered species (such as the giant panda – see Figure 1); and conducting research that should benefit endangered species.

Modern reproductive technologies such as freezing sperm, eggs or embryos can preserve large amounts of genetic material. **Artificial insemination**, *in vitro* **fertilisation** and **embryo-transfer** techniques are also being used with wild animals. Reproductive physiology is quite species-specific, and further research into each endangered species is needed to ensure that the techniques are used effectively. Some zoos may carry out research on domestic species or common wild species that are very similar to the target species. This means that individuals from endangered species can be spared from the experimental research, but will benefit in the long term.

Figure 1 The giant panda – an endangered species.

Advantages of *ex situ* conservation

* The organisms are protected from predation and poaching.
* The health of individuals can be monitored and medical assistance given as required.
* Populations can be divided, so that if a disaster strikes one population, then the other still survives.
* The genetic diversity of the population can be measured.

* Selective breeding can be carried out to increase genetic diversity.
* Modern reproductive technology such as *in vitro* fertilisation can be used to increase the chances of reproductive success.
* Animals (and plants) can be bred to increase the numbers of an endangered species.
* Research into the reproductive physiology, lifestyle and ecology of endangered species is made easier.
* Conservation sites can be used as attractions to raise funds for further conservation efforts, including fundraising for iconic animal species (such as the panda).
* Conservation sites can be used for education.

Disadvantages of *ex situ* conservation

* A captive population is always likely to have a limited genetic diversity.
* The animals can be exposed to a wide range of different diseases.
* The organisms are living outside their natural habitat.
* Nutritional issues can be difficult to manage.
* Animals may not behave as normal, and reproduction may be difficult.
* The correct environmental conditions for survival may be difficult to achieve.
* It may be expensive to maintain suitable environmental conditions.
* Even if reproduction is successful, the animals have to survive reintroduction to the wild, where they need to find food and survive predation.
* When reintroduced, there can also be difficulties with acceptance by the existing wild members of their species.

Botanic gardens

As with traditional zoos, many botanic gardens started life because of an affluent collector. Today, most botanic gardens are involved with the conservation of endangered species.

The *ex situ* conservation of plants is, perhaps, a little easier than that of animals.

* As part of their life cycle, most plants naturally have a dormant stage – the seed.
* As seeds are produced in large numbers, they can be collected from the wild without causing too much disturbance to the ecosystem or damaging the wild population.

- These seeds can be stored and germinated in protected surroundings.

- Seeds can be stored in huge numbers without occupying too much space.

- Plants can often be bred asexually.

- The botanic garden can increase the numbers of individual plants very quickly, through techniques such as tissue culture, which provides an ample supply for research or for reintroduction to the wild.

- The captive-bred individuals can be replanted in the wild.

However, there are problems.

- Funding a botanic garden can be difficult. Public perception of plants is not the same as with animals, and fewer people are willing to sponsor a plant or give money to save a particular 'iconic' species.

- Collecting wild seeds will always cause some disturbance.

- The collected samples may not have a representative level of genetic diversity.

- Seeds collected from one area may be genetically different from those collected elsewhere, and may not succeed in a different area.

- Seeds stored for any length of time may not be viable.

- Plants bred asexually will be genetically identical – reducing genetic diversity further.

- Conclusions from research based on a small sample may not be valid for the whole species.

Seed banks

A seed bank is a collection of seed samples. The Kew Millennium Seed Bank Project at Wakehurst in Sussex (see Figure 2) is the largest *ex situ* conservation project yet conceived. Its aim is to store a representative sample of seeds from every known species of plant. These will include examples of the rarest, most useful and most threatened species.

Seed banks contain seeds that can remain **viable** for decades and possibly hundreds of years. However, the seeds are not simply being stored. Some of them are being used to provide a wide

Figure 2 The Millennium Seed Bank.

range of benefits to humanity. These benefits include providing seeds for food crops and building materials for rural communities, and disease-resistant crops for agriculture. The seeds can also be used for habitat reclamation and repopulation of endangered habitats. The collections held in the Millennium Seed Bank, and the knowledge derived from them, provide almost infinite options for their conservation and use.

Storage of seeds

In order to prolong their viability, seeds are stored in very dry or freezing conditions. Seeds are resistant to desiccation, and the level of moisture in each seed has a direct effect on storage. For every 1% decrease in seed moisture level, the life span doubles. For every 5 °C reduction in temperature, the life span also doubles. But seeds stored for decades may deteriorate. There is little use in storing seeds that die and will not be able to germinate. So it is essential to test the seeds at regular intervals to check their viability.

Scientists at the Millennium Seed Bank carry out some 10 000 germination tests each year. They periodically remove samples and germinate them in petri dishes of nutrient agar (see Figure 3), keeping them in controlled conditions. Germination rates are monitored, and research into the physiology of seed dormancy and germination is carried out. This should lead to discovery of the most effective methods of storage.

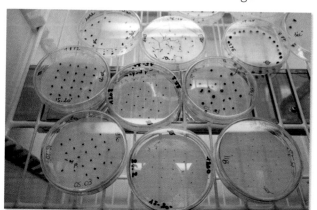

Figure 3 Testing seeds for germination.

Questions

1. Explain why some animals may not reproduce as normal in a wildlife park.

2. Explain why it is important to monitor the genetic diversity of plant and animal collections and attempt to increase the genetic diversity if possible.

3. What are the advantages of carrying out research on species closely related to an endangered species?

4. Why is it better to increase plant populations by sexual reproduction than by asexual reproduction?

5. Explain why reducing the moisture content and the temperature allow seeds to be stored for longer.

By the end of this topic, you should be able to demonstrate and apply your knowledge and understanding of:

* international and local conservation agreements made to protect species and habitats

An international problem

The loss of habitats and the increasing number of endangered species is a worldwide problem. It can be tackled at a local level but these efforts may be insignificant if other countries do not contribute. It therefore needs a worldwide solution. International cooperation in conservation can be achieved in a number of ways.

CITES

The Convention on International Trade in Endangered Species of Wild Fauna and Flora (CITES) is an international agreement between the majority of governments in the world. It was first agreed in 1973. The overall aim is to ensure that international trade in specimens of wildlife does not threaten their survival. Over 25 000 species of plants and animals have been identified as being at risk from international trade.

CITES aims to:

* regulate and monitor international trade in selected species of plants and animals
* ensure that international trade does not endanger the survival of populations in the wild
* ensure that trade in wild plants for commercial purposes is prohibited
* ensure that trade in artificially propagated plants is allowed, subject to permit
* ensure that some slightly less endangered wild species may be traded subject to a permit, as agreed between the exporting and importing countries.

International trade policies can be very hard to enforce. Where there is demand for a product, there will be attempts to supply it. Smuggling of live plants and animals and their products, such as ivory, is a constant problem (see Figure 1).

KEY DEFINITIONS

CITES: the Convention on International Trade in Endangered Species.

Countryside Stewardship Scheme: a scheme to encourage farmers and other landowners to manage parts of their land in a way that promotes conservation.

Figure 1 Checking for illegal trade in ivory at airports.

Rio Convention on Biological Diversity

The Convention on Biological Diversity was signed by 150 government leaders at the 1992 Rio Earth Summit. The Convention is dedicated to promoting sustainable development. It recognises that biological diversity is about more than plants, animals, microorganisms and their ecosystems. It is also about people and our need for secure sources of food, medicines, fresh air and water, shelter and a clean and healthy environment in which to live.

The aims of the convention are:

- conservation of biological diversity
- sustainable use of its components
- appropriate shared access to genetic resources
- appropriate sharing and transfer of scientific knowledge and technologies
- fair and equitable sharing of the benefits arising out of the use of genetic resources.

The Convention encourages cooperation between countries and states. It encourages each partner to develop a national strategy for conservation and the sustainable use of biological diversity. More specifically, it states that partner states must adopt *ex situ* conservation facilities, mainly to complement *in situ* measures.

The role of zoos, botanic gardens and seed banks

International cooperation involves the sharing of research, genetic information and technology. *Ex situ* conservation facilities in different member states provide support for each other and share their technologies and genetic material.

Breeding programmes in zoos are strengthened by importing animals from parks or zoos in other countries. Different wildlife parks and zoos specialise in the breeding of different animals. They will hold the 'studbooks' for that species for the world zoo community. Time, expense and distress to rare animals can be reduced by importing genetic material. This means transporting the sperm, eggs or embryos and using artificial insemination or *in vitro* fertilisation techniques. Research and technology is shared between member states to help improve the chances of success. Reintroduction of animals bred in zoos will fail unless there is cooperation from the countries where the animals are reintroduced. Wildlife reserves with suitable protection for the animals and plants are essential for successful reintroduction.

Similarly, plant-breeding programmes can be enhanced by sharing stored specimens. The Kew Millennium Seed Bank has partner projects in about 50 countries around the world, including Australia, Botswana, Chile, China, Egypt, Jordan, Kenya, Lebanon, Madagascar, Malawi, Mali, Mexico, Namibia, Saudi Arabia, Tanzania, South Africa and the USA. These partners also duplicate the collections in case of unforeseen disaster.

The level of sharing between the partners is demonstrated by statistics from the Millennium Seed Bank. The seed information database contains hundreds of thousands of records that are available to other projects. As of 21 October 2014, the Millennium Seed Bank had collected seeds from 34 088 wild plant species, and had 1 980 405 036 seeds in storage. Many of these seeds are duplicated in at least one of the 14 000 other seeds banks around the world.

Local conservation schemes

As stated in topic 4.2.7, conservation schemes are not limited to international agreements. Many governments make agreements with local land owners and tenants. These agreements are designed to enhance the biodiversity and conservation value of land at a local level.

The Countryside Stewardship Scheme

The Countryside Stewardship Scheme was introduced in England in 1991. It was applied to land that was not considered to be in an environmentally sensitive area. Payments were made to farmers and other land managers in order to enhance and conserve English landscapes. Grants were also available for capital works such as hedge laying, planting and repairing dry-stone walls. The aims of the scheme were to:

- improve the natural beauty and diversity of the countryside
- enhance, restore and re-create targeted landscapes, their wildlife habitats and historical features
- improve opportunities for public access.

This scheme was replaced by the Environmental Stewardship Scheme in 2005.

The Environmental Stewardship Scheme provides funding to farmers and other land managers in England to deliver effective environmental management on their land. The aim is to provide funding and advice to help land managers to conserve, enhance and promote the countryside by:

- looking after wildlife, species and their many habitats
- ensuring land is well managed and retains its traditional character
- protecting historic features and natural resources
- ensuring traditional livestock and crops are conserved
- providing opportunities for people to visit and learn about the countryside.

Questions

1. Suggest why international agreements on conservation may be hard to enforce.

2. Explain why products from protected species must be destroyed when they are discovered being shipped from one country to another.

3. Explain why *ex situ* conservation efforts have limited value if the country of origin of an endangered species does not support the conservation of that species.

4. State three reasons why the species conserved in the Millennium Seed Bank should also be conserved at another location.

5. Explain why it is important to replicate the research carried out in one country by repeating it elsewhere.

6. What are the advantages to the farmer of the Environmental Stewardship Scheme?

CONSERVATION

Western lowland gorillas (see Figure 1) are one of many species on the endangered list. A species becomes endangered when its numbers drop too low and genetic diversity is lost. This article explores the efforts to count gorillas in the wild.

GORILLA 'PARADISE' FOUND; MAY DOUBLE WORLD NUMBERS

Deep in the hinterlands of the Republic of the Congo lies a secret ape paradise that is home to 125 000 western lowland gorillas, researchers announced today. The findings, if confirmed, would more than double the world's estimated population of gorillas.

Western lowland gorillas are a subspecies classified as critically endangered by the International Union for Conservation of Nature (IUCN). Their numbers have been devastated in recent years by illegal hunting for bush meat and the spread of the Ebola virus. [In 2007] scientists projected the animals' population could fall as low as 50 000 by 2011. Now those predictions may have to be dramatically reworked to incorporate findings released today by the Wildlife Conservation Society (WCS).

A first ever ape census in northern Congo found 73 000 of the gorillas in that country's Ntokou-Pikounda region and 52 500 more in the Ndoki-Likouala area.

Wary of humans, gorillas are notoriously hard to tally in the wild. To assess their populations, WCS researchers instead used data on the numbers and ages of so-called sleep nests, temporary bedding made of leaves and branches.

Each group of lowland gorillas has a range of about 7.7 square miles (20 square kilometres), and the animals build the nests to sleep in each night before moving on in the morning. The census work involved crossing hundreds of miles to count nests, then loading data into a mathematical model that estimated the number of gorillas living within a defined area.

In the 28 000-square-kilometre Ndoki-Likouala region, for example, the nest census found an estimated population density of 1.65 gorillas per square kilometre. This means that about 46 200 western lowland gorillas likely live in the area.

'That's the highest density I've seen', [Emma] Stokes [one of the WCS's lead researchers] said, adding that the data suggest Ndoki-Likouala is the subspecies' 'largest remaining stronghold.' The

discovery 'shows that conservation in the Republic of Congo is working,' said WCS president Steven Sanderson.

Perils of Counting Apes

Several experts greeted the survey findings with a mix of excitement and caution.

'If these new gorilla census figures are confirmed by further surveys, it would be the most exciting ape conservation news in years,' said Craig Stanford of the Jane Goodall Research Center at the University of Southern California. 'Nest census data are notorious for varying from one method to the next, however, and I think we should be cautious before assuming the world's known gorilla population has just doubled.'

Nesting data were among the factors used in a 2007 IUCN population assessment that placed the western lowland gorilla on the organisation's Red List of Threatened species. That report [raised concerns] about the reliability of nest counts.

Peter Walsh of the Max Planck Institute for Evolutionary Anthropology in Leipzig, Germany, led [that assessment]. 'It is not that I think that the numbers are necessarily too high,' Walsh said. 'It is just that I do not trust the assumptions made by the estimation models that are being used.'

In general, the WCS findings demonstrate that our intensely observed planet still has its biological secrets, added Richard Bergl, curator of research at the North Carolina Zoo. 'It is extraordinary that in this day and age,' he said, 'there could be a population of a hundred thousand or more gorillas that were essentially unknown to science.'

Source

● Morrison, D. (2008) *Gorilla 'Paradise' Found; May Double World Numbers*, National Geographic News. (http://news.nationalgeographic.com/news/2008/08/080805-gorillas-congo.html, last accessed February 2015)

Where else will I encounter these themes?

1.1   2.1   2.2   2.3   2.4   2.5

Let's start by considering the nature of the writing in the article.

1. In paragraph 6 the writer refers to 'an estimated population density'. Explain why it is necessary to make an estimate and suggest how the estimate for the whole area was made.

2. The writer states that the discovery 'shows that conservation in the Republic of Congo is working' and that 'several experts greeted the survey findings with a mix of excitement and caution'. Explain why experts may feel cautious about the findings.

Now we will look at the biology in, or connected to, this article.

3. Describe the factors that cause species to become endangered.

4. Explain the importance of international agreements such as CITES and the Rio Convention on Biodiversity to the conservation of species such as western lowland gorillas.

5. Conserving species in their natural habitat is called *in situ* conservation. What are the advantages and disadvantages of *in situ* conservation?

6. The Aspinall Foundation is a charity that runs two wild animal parks in Kent. These parks are dedicated to the conservation of endangered species. What contributions can such organisations make to international conservation efforts?

7. Between 1996 and 2006 the Aspinall Foundation reintroduced 51 gorillas to the wild. In 2007 43 of these gorillas were still surviving. What is the percentage success rate achieved by the Foundation?

8. Ebola is caused by a virus. Describe the structure of a typical virus. How do viruses reproduce?

Activity

Successful conservation efforts often require international cooperation. Agreements such as CITES and the Rio Convention on Biodiversity are examples of such agreements. However, international efforts are unlikely to succeed unless the local people are also on-side. Design an A3-sized poster to explain the importance of conserving species to people living locally to gorilla populations in Africa.

Try to include as many scientific points as you can including economic, ecological and aesthetic reasons for conservation, perhaps with examples that explain why they are important.

Don't forget that you should consider all parties who may be involved, including the local people and their government.

Figure 1 A family of lowland gorillas.

1. A student calculated Simpson's index of diversity for a piece of woodland. The value was 0.84. Which statement provides the best interpretation of this value? [1]

   A. High species richness but low species evenness.

   B. Low species richness but high species evenness.

   C. High species richness and high species evenness.

   D. Low species richness and low species evenness.

2. The genetic diversity of a population can be calculated using the equation:

   $$\text{proportion of polymorphic gene loci} = \frac{\text{number of polymorphic gene loci}}{\text{total number of loci}}$$

   An organism was found to have 20 034 loci. In one population researchers found 234 features that had three or more alleles. What is the value of the genetic diversity of this population? [1]

   A. 0.012

   B. 85.6

   C. 19 800

   D. 20 268

3. A student set out to investigate the diversity of plants in a field. Which of the following correctly describes the apparatus the student would need in the field? [1]

   A. Pooter, quadrat, plastic sample bags, identification key.

   B. Quadrat, identification key, paper and clip board, point frame.

   C. Quadrat, pit fall trap, paper and clip board, point frame.

   D. Point frame, plastic sample bags, identification key, sweep net.

4. Which of the following species is a keystone species? [1]

   A. An antelope that eats grass in the African plains.

   B. A squirrel that eats acorns in deciduous woodland.

   C. A fox that eats fruit and small mammals.

   D. A sea otter than eats sea urchins which damage seaweed.

5. Which of following are international conservation agreements? [1]

   (i) CITES

   (ii) Countryside Stewardship Scheme

   (iii) The EEC

   (iv) The Rio Convention

   (v) NATO

   [Total: 5]

6. Many species are now endangered.

   (a) State three reasons why species become endangered. [3]

   (b) Distinguish between *in situ* conservation and *ex situ* conservation. [2]

   (c) State three advantages of *in situ* conservation. [3]

   (d) Explain why successful conservation of an endangered species usually relies upon a combination of *in situ* and *ex situ* techniques. [6]

   [Total: 14]

7. The table below shows the results of an investigation into the diversity of two habitats.

Species	Per cent cover	
	Habitat A	**Habitat B**
Festuca rubra	68	15
Poa pratensis	60	25
Agropyron repens	0	40
Agrostis tenuis	10	35
Holcus lanatus	0	55
Bellis perennis	10	0
Trifolium repens	15	0
N	163	170

   (a) The equation used to calculate Simpson's index of diversity is:

   $$D = 1 - \left[\Sigma \left(\frac{n}{N} \right)^2 \right]$$

   Use the equation to calculate the diversity of Habitat B. [3]

   (b) Species diversity depends upon species richness and species evenness.

   (i) Which habitat has the higher species richness? [1]

   (ii) Which habitat has the higher species evenness? [1]

   (iii) Identify which habitat is more likely to be highly managed and explain your choice. [3]

   [Total: 8]

8. In 2014 a giraffe called Marius was deliberately killed at Copenhagen zoo. The zoo is part of a European captive breeding programme. The decision was reported in the newspapers and led to a public outcry.

The zoo stated that Marius was too genetically similar to other animals and was therefore not needed for the breeding programme.

MARIUS KILLED AT COPENHAGEN ZOO DESPITE WORLDWIDE PROTESTS

Figure 1

(a) Members of the public suggested that Marius could have been returned to the wild. State three problems that may have arisen to prevent successful reintroduction to the wild. [3]

(b) Suggest one other way in which Marius could have been used. [1]

(c) The Aspinall Foundation is a charity set up for the conservation of species. Between 1996 and 2005 the Foundation released 51 western lowland gorillas back into the wild. The majority of gorillas in the group were orphans reared *in situ*, but the group also included 7 gorillas born in captivity in the UK. In 2014, 43 of the gorillas were still alive.
  (i) Calculate the percentage of the introduced gorillas that have survived. [2]
  (ii) Which of the gorillas was more likely to survive – those reared *in situ* or those born in the UK? Give reasons for your answer. [2]

(d) Use the examples of conservation efforts described above to explain the importance of international agreements such as the Rio Convention on Biodiversity. [8]

[Total: 16]

9. Changes in the human population affect biodiversity.
  (a) Describe how changes in agriculture affect biodiversity. [3]
  (b) Describe how climate change affects biodiversity. [4]
  (c) Describe the reasons for conserving biodiversity. [6]

[Total: 13]

10. Biodiversity can be considered at three different levels.
  (a) One of these levels is species diversity. What are the other levels at which biodiversity can be considered? [2]
  (b) Species diversity involves both species richness and species evenness. Explain what is meant by these terms. [2]
  (c) One strategy for measuring species diversity is to take samples at random.
    (i) State one advantage and one disadvantage of random sampling. [2]
    (ii) One alternative to random sampling is opportunistic sampling. What is meant by opportunistic sampling? [1]
    (iii) What are the disadvantages of opportunistic sampling? [2]
    (iv) Name two other sampling strategies and give one advantage of each. [4]

[Total: 13]

Biodiversity, evolution and disease

CLASSIFICATION AND EVOLUTION

Introduction

The range of diverse forms that can be seen in different habitats has caught the attention of biologists for many years. In particular, it can be very striking how well an organism is adapted to its particular habitat. Adaptation to the habitat is achieved by evolution. The mechanism involves selection of the most well adapted from a variety of different forms. The wide diversity of habitats available has generated a very wide variety of organisms that can exploit those habitats.

All organisms share a common ancestry and this allows them to be classified. Classification is an attempt to place this wide variety of living things into groups. It imposes a hierarchy on the huge complexity and variety of life. Early classification systems relied solely on features that could be observed. However, these systems are not always accurate. Classification systems have changed and will continue to change as our research reveals more and more information about the biology of living organisms.

All the maths you need

To unlock the puzzles in this section you need the following maths:

- Use ratios, fractions and percentages
- Find arithmetic means
- Construct and interpret frequency tables and diagrams, bar charts and histograms
- Understand the terms mean, median and mode
- Use a scatter diagram to identify a correlation between two variables
- Understand measures of dispersion, including standard deviation and range

What have I studied before?

- Variation between organisms within a species and between species
- Classification into the five kingdoms
- How genetic material codes for the observable characteristics
- The structure of biological molecules including nucleic acids and proteins

What will I study later?

- DNA sequencing technology (AL)
- Mutation and introduction of genetic variation through inheritance (AL)
- How genetic modification can blur the boundaries between species (AL)
- How selective breeding can lead to loss of genetic diversity (AL)

What will I study in this chapter?

- The biological classification of species
- The binomial system of naming species
- The evidence used to classify organisms into a new three domain system of classification
- The relationship between classification and phylogeny
- Different types of variation and how organisms are adapted to their environment
- The mechanism of natural selection and the evidence for this mechanism
- How the evolution of some species has an impact on human population

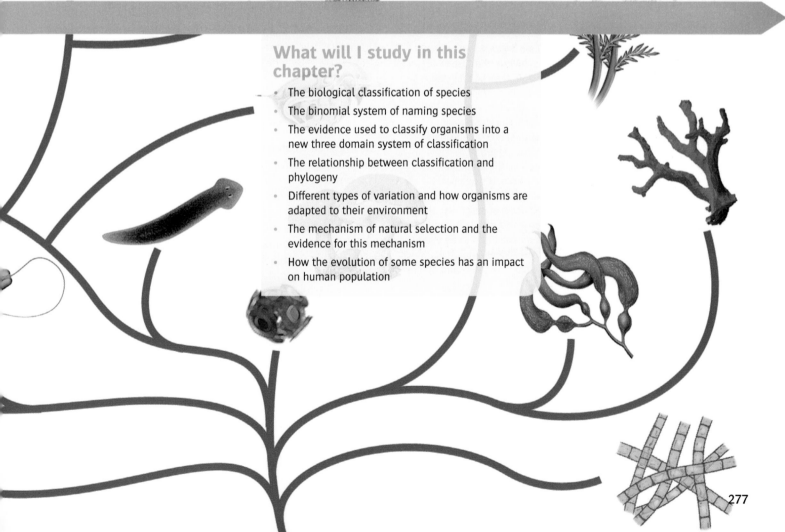

By the end of this topic, you should be able to demonstrate and apply your knowledge and understanding of:

* the biological classification of species
* the binomial system of naming species, and the advantage of such a system

KEY DEFINITIONS
binomial system: a system that uses the genus name and the species name to avoid confusion when naming organisms.
classification: the process of placing living things into groups.

Classifying living things

Humans have always wanted to put things in order or to sort them out. To place all living things into categories is a huge task, as current estimates suggest that there are nearly 2 million different species of organisms alive today. Each species must be studied in detail before it can be correctly placed in a group of similar organisms. The reasons why we do this include the following:

* it is for our convenience
* to make the study of living things more manageable
* to make it easier to identify organisms
* to help us see the relationships between species.

DID YOU KNOW?

The very earliest classification system classified organisms into those that lived in air, water and soil. Carl Linnaeus devised the system of classification that is still used today. Around 250 years ago, Linnaeus classified about 70 000 organisms. He was more concerned with how these organisms could be classified methodically than with simply collecting and describing them. He studied each one closely and organised them according to their visible features. Two organisms with many similar visible features were grouped closer together.
Linnaeus created a hierarchy of ranked categories. These categories are called taxonomic groups or taxa (singular taxon). His original classification contained five levels: kingdom, class, order, genus and species. As more organisms have been discovered and described, this original system of classification has had to be modified and expanded. We now have more kingdoms, and these are grouped into three even larger categories called domains. Kingdoms are divided into phyla, which are then divided into classes. Classes are divided into orders and orders are divided into families, etc.

The modern classification hierarchy

The current system of **classification** uses eight taxonomic levels:

* **Domain** – the domain is the highest taxonomic rank (see also topic 4.3.3). There are three domains: Archaea, Eubacteria and Eukaryotae.
* **Kingdom** – traditionally there are five main kingdoms (see also topic 4.3.2). Plantae, Animalia, Fungi and Protoctista are all eukaryotes, which possess a nucleus. All those single-celled organisms that do not possess a nucleus are grouped into the kingdom Prokaryotae.

* **Phylum** – a major subdivision of the kingdom. A phylum contains all the groups of organisms that have the same body plan, e.g. possession of a backbone.
* **Class** – a group of organisms that all possess the same general traits, e.g. the same number of legs.
* **Order** – a subdivision of the class using additional information about the organisms, e.g. the class mammal is divided into meat-eating animals (order Carnivora) and vegetation-eating animals (order Herbivora).
* **Family** – a group of closely related genera, e.g. within the order Carnivora we might recognise the 'dog' family and the 'cat' family.
* **Genus** – a group of closely related species.
* **Species** – the basic unit of classification. All members of a species show some variations, but all are essentially the same.

Some examples of classification are shown in Table 1.

LEARNING TIP
Devise your own mnemonic to help you recall this hierarchy correctly.

Classifying species

At the higher levels of this ranked system, the differences between the organisms can be very great. It is therefore quite easy to place a species into its domain, kingdom or phylum. For example, two phyla in the animal kingdom are the Chordata and the Arthropoda. The Chordata have a nervous system with a central bundle of nerves running along their back, usually protected by a series of bones called the vertebral column – these are the vertebrates. In contrast, the Arthropoda have a hard exoskeleton (skeleton on the outside of the body rather than the inside) and jointed limbs.

Within a phylum, the species must be placed in a class. This becomes a little more difficult, as the differences between the classes in one phylum may not be very great. A longer description of the species may be needed. For example, two classes in the phylum Arthropoda are the insects and the arachnids (spiders). All members of the class Insecta (see Figure 1) can easily be recognised as having three body parts (a head, a thorax and an abdomen), six legs and usually two pairs of wings (flies have one pair of wings though). All members of the class Arachnida can be recognised as having two body parts (a cephalothorax and an abdomen) and eight legs (see Figure 2).

Figure 1 A member of the class Insecta.

Figure 2 A member of the class Arachnida.

As you descend to the lower taxonomic groups, it becomes increasingly difficult to separate closely related species and to place a species accurately. A more and more detailed description of the species is needed.

Taxonomic rank	Common name		
	Human	Gorilla	Fruit fly
Kingdom	Animalia	Animalia	Animalia
Phylum	Chordata	Chordata	Arthropoda
Class	Mammalia	Mammalia	Insecta
Order	Primate	Primate	Diptera
Family	Hominidae	Hominidae	Drosophilidae
Genus	*Homo*	*Gorilla*	*Drosophila*
Species	*sapiens*	*gorilla*	*melanogaster*

Table 1 Some examples of classification.

The binomial naming system

Binomial means 'two names'. In the binomial system of naming organisms, the genus name and the species name are used. Thus, humans become *Homo sapiens*. *Homo* refers to the genus to which humans belong. The genus name is always given an upper-case first letter. The species name is *sapiens*. This can be abbreviated to *H. sapiens*. The binomial Latin name is always written in a style that makes it stand out. In printed text this is in italics, in handwritten text it is underlined.

Carl Linnaeus devised this binomial system. Before Linnaeus, species were identified by a common name, or a long and detailed description. Some scientists even used long descriptive names in Latin. Using a common name does not work well because:

- the same organism may have a completely different common name in different parts of one country

- different common names are used in different countries

- translation of languages or dialects may give different names

- the same common name may be used for different species in other parts of the world.

Linnaeus used Latin as a universal language. This means that whenever a species is named, it is given a universal name. Every scientist in every country will use the same name. This avoids the potential confusion caused by using common names.

Questions

1. When classifying an organism, why is it important to study each specimen in detail?

2. Write out a full classification for a common chimpanzee (*Pan troglodytes*) and a wolf (*Canis lupus*).

3. Explain why a spider is not an insect.

4. What is meant by the binomial system of naming organisms?

5. Explain why a standard way of naming organisms is useful.

By the end of this topic, you should be able to demonstrate and apply your knowledge and understanding of:

* the features used to classify organisms into the five kingdoms: Prokaryotae, Protoctista, Fungi, Plantae, Animalia

Using observable features

The biological definition of a species is:

'a group of organisms that can freely interbreed to produce fertile offspring'.

However, this definition does not work for organisms that reproduce asexually, and it is very hard to apply to organisms that are known only as fossils.

The phylogenetic definition of a species is:

'a group of individual organisms that are very similar in appearance, anatomy, physiology, biochemistry and genetics'.

Being so similar, the members of a species occupy the same **niche** in an ecosystem. This fact was used in early classification systems which were based only on appearance and **anatomy**. For many species, this provided enough information to allow accurate classification. But it is easy to make mistakes. In the earliest attempts at classification, Aristotle classified all living things as either plant or animal. He further subdivided the animals into three groups – those that:

* live and move in water
* live and move on land
* move through the air.

This was based on the similarities that he observed – some animals have fins, some have legs and some have wings. Unfortunately, this grouped fish with turtles, birds with insects, and mammals with frogs.

Such early classifications have been adapted and made more accurate as more research is carried out and more information becomes available.

The early classification systems of Linnaeus and other scientists were based on observable features. This means they were limited to those features of organisms that you can see. By the 17th century, scientists had microscopes to help.

Using more detailed evidence

Traditionally, all living things have been grouped into a number of kingdoms. For many years, the generally accepted number of kingdoms was two. All living things were grouped into either plants or animals. As more living things were discovered and studied closely, it became clear that not all could fit easily into one of these categories.

In the early two-kingdom classification systems, the animal kingdom included single-celled organisms that had some animal-like features, and the plant kingdom included single-celled organisms that had plant-like features. Later, electron microscopes revealed further details inside cells. These better microscopes made it clear that many single-celled organisms share some of the features of both plants and animals. *Euglena* (see Figure 1) is a single-celled organism that has chloroplasts to photosynthesise, but it also has the ability to move around using a flagellum. It does not clearly fit into either the plant or the animal kingdom.

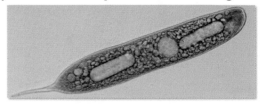

Figure 1 *Euglena* – a single-celled organism that can move about and photosynthesise.

Figure 2 Ergot fungus – has root-like structures but no chlorophyll.

Also, where do fungi (see Figure 2) fit in? Like plants, they do not move about, and their **hyphae** grow into the surrounding substrate in the same way as roots do. However, they do not photosynthesise. They digest organic matter and absorb the nutrients – like animals. Fungi were originally classified as plants.

The resulting upheaval in the world of taxonomy and classification led to the adoption of the five-kingdom classification. These five kingdoms are still based on the observable features of their anatomy, but at a microscopic level. Table 1 describes the diagnostic features of the five kingdoms.

LEARNING TIP

Remember to describe the diagnostic features of a kingdom – those features that make it different from the other kingdoms.

Kingdom	Description	Typical example
Prokaryotae	• have no nucleus • have a loop of DNA that is not arranged in linear chromosomes • have naked DNA (DNA that is *not* associated with histone proteins) • have no membrane-bound organelles • have smaller ribosomes than in other groups • have cells smaller than those of eukaryotes • may be free-living or parasitic (some cause diseases)	Figure 3 Bacteria.
Protoctista	• are eukaryotic • are mostly single-celled (but many algae are multicellular) • show a wide variety of forms (the only thing that all protoctists have in common is that they do not qualify to belong to any of the other four kingdoms!) • show various plant-like or animal-like features • are mostly free-living • have autotrophic or heterotrophic nutrition – some photosynthesise, some ingest prey, some feed using extracellular enzymes (like fungi do), and some are parasites	Figure 4 Paramecium.
Fungi	• are eukaryotic • can exist as single cells (called yeasts) or they have a **mycelium** that consists of hyphae • have walls made of chitin • have cytoplasm that is multinucleate • are mostly free-living and saprophytic – this means that they cause decay of organic matter	Figure 5 Fungus on an orange.
Plantae	• are eukaryotic • are multicellular • have cells surrounded by a cellulose cell wall • are autotrophic (absorb simple molecules and build them into larger organic molecules) • contain chlorophyll	Figure 6 Bean seedling.
Animalia	• are eukaryotic • are multicellular • heterotrophic (digest large organic molecules to form smaller molecules for absorption) • are usually able to move around	Figure 7 A frog.

Table 1 The five kingdoms of classification.

DID YOU KNOW?

Recent developments in taxonomy have moved the old kingdom Prokaryotae to become two domains. That leaves four of the original five kingdoms in the third domain – the Eukaryotae. However, continued research has blurred the distinction between the kingdoms, and modern taxonomists have created more kingdoms in an effort to accurately reflect the evolutionary relationships between organisms in a modern system of classification.

Questions

1. Explain why it was necessary to expand the classification system from two to five kingdoms.

2. Why is it difficult to classify organisms such as fungi?

3. Aristotle originally classified birds and insects in the same group. Explain why this is not an accurate classification.

4. Explain why better microscopes led to improved classification.

5. Describe the differences between fungi and plants.

6. Draw a table to compare the prokaryotes with the protoctists.

By the end of this topic, you should be able to demonstrate and apply your knowledge and understanding of:

* the evidence that has led to new classification systems, such as the three domains of life, which clarifies relationships

Using biological molecules in classification

Using observable features has created a largely successful classification of living things. However, since organisms adapt to their environment, it is possible that two unrelated species could adapt in similar ways and therefore look very similar (this is called **convergent evolution**, see topic 4.3.8). These two species might be classified in the same taxonomic group according to their observable features. The most recent research uses a wider range of techniques, and has produced even more detailed evidence for classification.

Evidence from biological molecules can help to determine how closely related one species is to another. Certain large biological molecules are found in all living things, although they may not be identical in every species. These are molecules that are involved in the most fundamental characteristics of life, such as respiration and protein synthesis. If we assume that the earliest living things all had identical versions of these molecules, then the differences seen today are a result of **evolution**. Two organisms with similar molecules will be closely related, as they have not evolved separately for long. Two organisms with very different versions of the molecule are less closely related, as they have evolved separately for longer. The differences between these molecules in different species reflect the evolutionary relationships. Such evidence has largely backed up the evolutionary relationships that have already been worked out. However, we can also use it to clarify or correct relationships that we are unsure about.

> **DID YOU KNOW?**
> Processes such as respiration and protein synthesis are fundamental to life – all living things must carry out these processes. The proteins and enzymes involved in these processes, such as cytochrome *c* and RNA polymerase, are used as evidence for evolutionary relationships.

Cytochrome *c*

A protein called **cytochrome *c*** is used in the process of respiration. All living organisms that respire must have cytochrome *c*. But cytochrome *c* is not identical in all species. Proteins are large molecules made from a chain of smaller units called amino acids. The amino acids in cytochrome *c* can be identified. If we compare the sequence of amino acids in samples of cytochrome *c* from two different species, then we can draw certain conclusions:

* if the sequences are the same, the two species must be closely related

* if the sequences are different, the two species are not so closely related

* the more differences found between the sequences, the less closely related the two species.

The amino acid sequences found in the cytochrome *c* of humans and chimpanzees are identical – we are very closely related to chimpanzees. There is one difference between our cytochrome *c* and that of the rhesus monkey – we are fairly closely related to the rhesus monkey. However, the cytochrome *c* found in the dogfish contains 11 differences from our own cytochrome *c* (see Figure 1). We are much less closely related to the dogfish.

> **LEARNING TIP**
> This topic could easily be linked to a question on protein structure – so expect questions that link different areas of the specification.

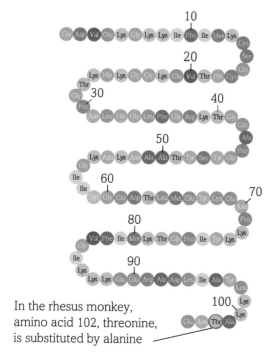

In the rhesus monkey, amino acid 102, threonine, is substituted by alanine

In dogfish, the following amino acids are changed:

44	→ Gln	92	→ Gln
46	→ Phe	101	→ Thr
50	→ Asp	102	→ Ala
58	→ Thr	103	→ Ala
65	→ Arg	104	→ Ser
89	→ Ser		

Key:

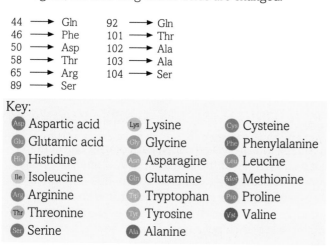

Asp Aspartic acid	Lys Lysine	Cys Cysteine
Glu Glutamic acid	Gly Glycine	Phe Phenylalanine
His Histidine	Asn Asparagine	Leu Leucine
Ile Isoleucine	Gln Glutamine	Met Methionine
Arg Arginine	Trp Tryptophan	Pro Proline
Thr Threonine	Tyr Tyrosine	Val Valine
Ser Serine	Ala Alanine	

Figure 1 Comparing the amino acid sequence of cytochrome *c* in humans and chimpanzees with that of rhesus monkeys and dogfish.

DNA

Another biological molecule that is found in all living organisms is DNA. DNA provides the genetic code – the instructions for producing proteins. The code is the same for all organisms – it is universal. This means that a particular sequence of DNA codes for the same sequence of amino acids in a bacterium as in any other organism.

Changes to the sequence of bases in DNA are called **mutations**. Mutations occur at random. Comparison of DNA sequences provides another way to classify species. The more similar the sequence in a part of the DNA, the more closely related the two species. If there are many differences, the species have evolved separately for a long time, and they can be considered as less closely related. This is probably the most accurate way to demonstrate how closely related one species is to another.

The three-domain classification

In 1990, Carl Woese suggested a new classification system (see Figure 2). He based his ideas on detailed study of the ribosomal RNA gene. He divided the kingdom Prokaryotae into two groups: the Bacteria (originally called the Eubacteria) and the Archaeae (Archaebacteria). This division is based on the fact that the Bacteria are fundamentally different from the Archaea and the Eukaryotae. Some structural differences include the fact that bacteria have:

- a different cell membrane structure
- flagella with a different internal structure
- different enzymes (RNA polymerase) for synthesising RNA
- no proteins bound to their genetic material
- different mechanisms for DNA replication and for synthesising RNA.

Archaea share certain features with eukaryotes:

- similar enzymes (RNA polymerase) for synthesising RNA
- similar mechanisms for DNA replication and synthesising RNA
- production of some proteins that bind to their DNA.

RNA and DNA are part of the basic mechanism that translates genes into visible characteristics. Woese argued that these differences between the Bacteria (Eubacteria) and the Archaea (Archaebacteria) are fundamental. He suggested that these two groups are more different from each other than the Archaea are from the Eukaryotae. Therefore an accurate classification system must reflect this difference. Woese's three-domain system of classification is now widely accepted by most biologists.

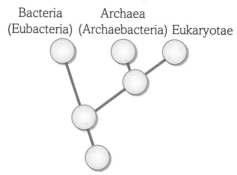

Bacteria Archaea
(Eubacteria) (Archaebacteria) Eukaryotae

Figure 2 The three-domain classification.

Questions

1. What is meant by convergent evolution?

2. How does convergent evolution make classification by observable features more difficult?

3. Why is it essential to use molecules such as DNA or cytochrome *c* when looking for evidence to be used in classification?

4. Explain why molecules such as starch or the enzyme amylase (digests starch) are not used for classification.

5. Explain why Woese felt that the differences that he observed in the RNA polymerase of different groups were so important.

Classification and phylogeny

By the end of this topic, you should be able to demonstrate and apply your knowledge and understanding of:

* the relationship between classification and phylogeny

> **KEY DEFINITION**
>
> **phylogeny:** the study of the evolutionary relationships between organisms.

Classification

As discussed in topic 4.3.1, classification is the process of sorting things into groups. We may sort things simply for our convenience, or the classification may be based on many important similarities and differences between the groups.

Artificial classification

Some classifications are done for convenience. We group things in a way that is easy to remember or in a way that makes it easy to find a particular item. For example, a do-it-yourself enthusiast might keep woodworking tools together in one place and metalworking tools together in another place. This makes it easy to find the correct saw for a particular job. In biology, a wildflower guide often has all the plants with yellow flowers on one page and those with blue flowers on another page. This enables the user to turn quickly to the section of the guide containing all the flowers of a particular colour.

This type of sorting is known as **artificial classification**. An artificial classification:

* is based on only a few characteristics

* does not reflect any evolutionary relationships

* provides limited information

* is stable.

Natural classification

Biological classification involves detailed study of the individuals in a species. Individual members of a species will show variation. For example, as discussed in topic 4.2.1 on genetic biodiversity, all the varieties of dog, from a chihuahua to a German shepherd, are members of the same species. Underneath the obvious visible differences, all dogs are very similar. As all members of the species are very similar, we can consider them to be closely related.

In the same way, different species that are very similar can be considered to be closely related. As already explained in topic 4.3.1, two closely related species will be placed in a group together – a genus. Then closely related genera will be placed together in a larger group – a family. In this way, the whole of the living world can be organised into a series of ranked groups – a hierarchy.

This is known as a **natural classification**. A natural classification:

* uses many characteristics

* reflects evolutionary relationships

* provides a lot of useful information

* may change with advancing knowledge.

A natural classification that reflects real relationships between the groups could be very useful. For example, if we want to find out more about a rare or endangered species, we may not want to risk harming any of the few surviving members of that species. However, if we know of another very similar species that is not endangered, we can carry out research on this second species to provide information that is also applicable to the endangered species. This may help us to make conservation more successful.

Phylogeny

Increasingly, modern classification has come to reflect the evolutionary history of the living world. We can think of all organisms as belonging to an evolutionary tree. Any two species living today have had a **common ancestor** at some time in the past. The time at which the two species started to evolve separately is a branch point on the tree. The common ancestor appears on the tree at that branch point. The more recent the common ancestor, the more closely related the two species are.

Phylogeny is the study of the evolutionary relationships between species. It involves studying how closely different species are related. Using the evolutionary or phylogenetic tree in Figure 1, we can see certain evolutionary relationships that indicate how closely related the species are.

1. Humans and gorillas (both mammals) share many features and are closely related.

2. We have a common ancestor in the recent past (species 1).

3. We can call humans and gorillas monophyletic – because they belong to the same phylogenetic group.

4. Humans and gorillas can be placed (classified) in the same taxonomic group.

5. The thrush is more closely related to the snake than to the mammals.

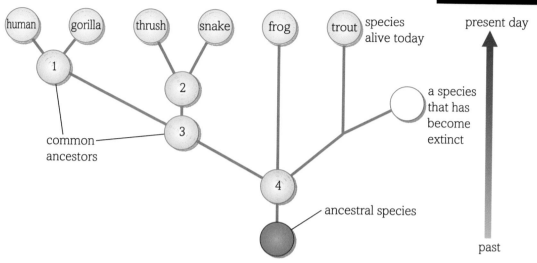

Figure 1 A phylogenetic tree.

6. We can see this because the common ancestor (species 2) shared by the thrush and the snake is more recent than the common ancestor shared by the thrush and the mammals (species 3).

7. Therefore the thrush must be placed in a different group from the mammals.

8. Similarly, the snake is more closely related to the thrush than to the frog or the trout.

9. All six species in this phylogenetic tree can be considered to be monophyletic as they all evolved from the same species (species 4).

It should be noted that the common ancestors do not survive today. We cannot say that we evolved from the apes, or from the gorillas, or even from modern-day fish. We evolved from an ancestor that lived at some time in the past. It happens that the gorillas also evolved from that same ancestor.

Questions

1. What is a common ancestor?

2. Explain the difference between phylogeny and classification.

3. Suggest why we study how closely related we are to other organisms.

4. Referring to Figure 1, state the number that denotes the common ancestor shared by:
   (a) the trout and the snake
   (b) the thrush and the gorilla.

5. What is meant by the term 'monophyletic'?

6. What arguments could be made to justify the use of live animals in testing new drugs?

By the end of this topic, you should be able to demonstrate and apply your knowledge and understanding of:

* the evidence for the theory of evolution by natural selection

---

KEY DEFINITION

natural selection: the term used to explain how features of the environment apply a selective force on the reproduction of individuals in a population.

Darwin and Wallace

Charles Darwin was a naturalist who spent much of his life observing and studying living organisms. The theory of evolution was not his idea. The idea that one species might evolve from another over time was not new, but Darwin proposed a mechanism for this process. This made it easier to believe in the theory of evolution. It also caused a certain amount of upheaval in Victorian Britain, because it countered the religious beliefs of the time. His proposed mechanism was natural selection.

Darwin's ideas began to develop during a five-year trip around the world in a ship called the HMS *Beagle*. During this trip he visited the Galapagos Islands, where he discovered a large number of unusual species. Many of these species were similar to those found on the South American mainland. What interested Darwin was that there was clear variation between members of the same species found on different islands. He also noted that what appeared to be a wide variety of bird species were actually all closely related finches. Darwin concluded that one species had arrived on the islands from the mainland and had then evolved to form many different species.

Alfred Russel Wallace was another naturalist who independently came to the same conclusions as Darwin. Wallace had made collections in both the Amazon and in South East Asia. Their first publications were joint papers on the subject of evolution by natural selection. This was soon followed by Darwin's book, best known as '*The Origin of Species*'.

DID YOU KNOW?

Herbert Spencer coined the phrase 'survival of the fittest' after reading Darwin's book '*The Origin of Species*'. He was a polymath – someone who has deep interest and knowledge in a wide range of fields. Spencer was a philosopher and a great exponent of evolution.

From observation to theory

Darwin made four particular observations:

1. Offspring generally appear similar to their parents.
2. No two individuals are identical.
3. Organisms have the ability to produce large numbers of offspring.
4. Populations in nature tend to remain fairly stable in size.

Darwin realised that variation was the key to understanding how species change. He saw that when too many young are produced, there is competition for food and resources. As all the offspring are different, some may be better adapted than others. The better-adapted individuals obtain enough food and survive long enough to reproduce. These individuals can pass on their characteristics to the next generation. Therefore the population can change or evolve to become better suited to its environment.

Darwin's conclusions can be summarised as follows:

* There is a struggle to survive.
* Better-adapted individuals survive and pass on their characteristics.
* Over time, a number of changes may give rise to a new species.

Evidence for evolution

Fossil evidence

Even in Darwin's time, known fossils clearly showed a number of interesting facts:

* In the past, the world was inhabited by species that were different from those present today.
* Old species have died out and new species have arisen.
* The new species that have appeared are often similar to the older ones found in the same place.

Darwin was fascinated by the similarities that he found between species living today and fossil species. He began to understand that fossil species gave rise to more modern species, and he felt that this must be because the more modern species had variations that meant they were better adapted to the environment. Darwin was also struck by the differences between the fossil species and the modern species. Many of the fossil species were much larger than modern species, but otherwise appeared very similar. For example, some modern species of armadillo grow to only 15 cm long, whilst the glyptodont (Figure 1) was many times this size (see Figure 2 for comparison).

Figure 1 A glyptodont, which was over a metre long.

Figure 2 An armadillo.

One of the most complete fossil records of evolution is that of the horse (see Figure 3). The evolution of the modern horse can be followed through a sequence of species that are all very similar to each other. Their similarity and their sequence in time provide evidence that one species arose from a previous one.

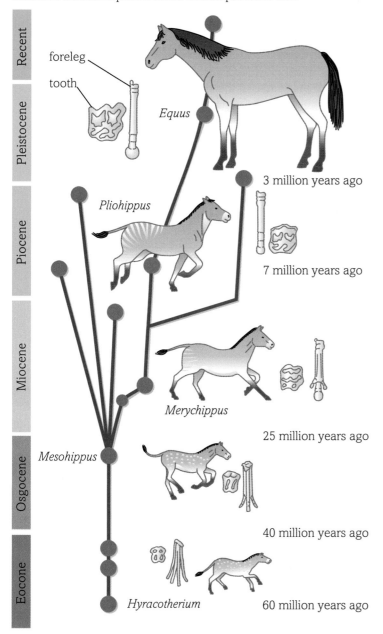

Figure 3 The evolution of the horse.

Biological molecules

Recent study of biological molecules provides very strong evidence for evolution.

- The fact that certain molecules are found throughout the living world is evidence in itself. If one species gives rise to another, both are likely to have the same biological molecules. This suggests that all species arose from one original ancestor.

- Two closely related species will have separated only relatively recently – their biological molecules are likely to be identical or very similar.

- In species that took separate evolutionary paths a long time ago, the biological molecules are likely to differ more.

- Evidence from molecules such as cytochrome *c* (see topic 4.3.3) and other proteins shows this pattern of changes.

The structure of DNA can be used in a similar way to that of cytochrome *c*. Genes can be compared by sequencing the bases in the DNA. The greater the number of similarities between the gene sequences, the more closely related the species and the more recent their evolution. It also shows that more distantly related species have more differences in their DNA. Comparison of human DNA with that of other organisms shows the following evolutionary relationships:

- 1.2% of our coding sequence is different from that of chimpanzees

- 1.6% is different from that of gorillas

- 6.6% is different from that of baboons.

A LEVEL STUDENTS NEED TO KNOW

Mitochondria contain their own DNA called mitochondrial DNA or mDNA.

During sexual reproduction mitochondria contained in the egg are passed to the offspring, therefore mDNA is passed on from the mother. This makes it uniquely useful in tracing human history. The history is not confused by DNA from the paternal line. Also, mDNA mutates more frequently than nuclear DNA as it does not have the same checking systems in place to proofread new copies. Therefore, there is plenty of variation in the sequence of mDNA between people from different parts of the world. This variation can be used to solve outstanding uncertainties about the origins of different races. For example, it has shown that the Polynesians migrated from southeast Asia and that the native Americans crossed from Siberia about 13 000 years ago rather than being descended from people in the Middle East.

The path of evolution seen in mitochondrial DNA is so clear that we can trace it back to a single female known as mitochondrial Eve who lived in Africa about 150 000 years ago. Whoever we are, and wherever we live, we are all her descendants.

Questions

1. Use Darwin's observations to explain why he believed that there was a 'struggle to survive' and why only some individuals would pass on their characteristics.

2. Spencer coined the term 'survival of the fittest'. Explain how this relates to Darwin's views.

3. Suggest why half our genes are the same as those in a very different species such as a banana.

4. What type of biological molecule (apart from DNA) is most suited to use as evidence for evolution by natural selection? Justify your answer.

5. Suggest why the fossil record of evolution is not complete.

By the end of this topic, you should be able to demonstrate and apply your knowledge and understanding of:

* the different types of variation

KEY DEFINITIONS

continuous variation: variation where there are two extremes and a full range of values in between.
discontinuous variation: where there are distinct categories and nothing in between.
environmental variation: variation caused by response to environmental factors such as light intensity.
genetic variation: variation caused by possessing a different combination of alleles.
interspecific variation: the differences between species.
intraspecific variation: the variation between members of the same species.
variation: the presence of variety – the differences between individuals.

Differences between individuals

The presence of differences between individuals is called variation. No two individuals are exactly alike, however similar they may look. Identical twins start as one cell that divides and then separates into two cells. Each of these two cells then develops into a separate person. While the two original cells had the same genetic information, the subsequent replication of DNA and cell divisions may have introduced changes to the DNA. Also, slight environmental differences in the womb or after birth can mean that the two individuals show physical differences.

Variation within species

Like any other species, humans show variation. If you think of almost any characteristic, you will be able to find differences between members of the population. For example, eye colour, hair colour, skin colour and nose shape are all characteristics that show variation between different people. These differences are known as intraspecific variation. The greater the genetic diversity of a species, the greater the intraspecific variation.

Variation between species

The variation that occurs between species is usually obvious. In fact, it is that very variation that is used to separate members of one species from another. Variation between species is called interspecific variation.

Continuous and discontinuous variation

There are two forms of variation.

Continuous variation

Continuous variation (see Figure 1) is where there are two extremes and a full range of intermediate values between those

extremes. Most individuals are close to the mean value. The number of individuals at the extremes is low. Continuous variation is often regulated by more than one gene and can be influenced by the environment in which an organism lives.

Examples of continuous variation include:

* height in humans
* length of leaves on an oak tree
* length of stalk (reproductive hypha) of a toadstool (see Figure 2)
* number of flagella on bacterium.

This type of variation is usually quantifiable. When plotting it on a graph, it is best to use a histogram.

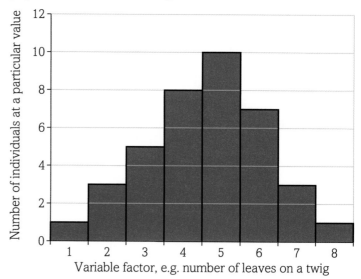

Figure 1 A histogram to show continuous data.

Figure 2 An example of continuous variation.

Discontinuous variation

Discontinuous variation is where there are two or more distinct categories with no intermediate values. The members of a species may be evenly distributed between the different forms, or there may be more of one type than the other. Discontinuous variation is usually regulated by a single gene and is not influenced by the environment in which an organism lives.

Examples of discontinuous variation include:

- gender – mammals are either male or female; plants can be male, female or hermaphrodite

- some bacteria have flagella, but others do not

- human blood groups – you are blood group A, B, AB or O.

Discontinuous variation can be shown on a bar chart (Figure 3).

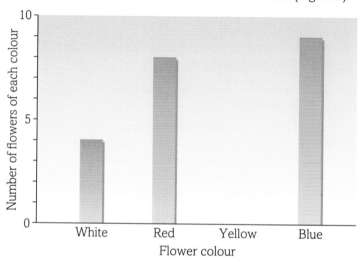

Figure 3 A bar chart to show discontinuous data.

Causes of variation

There are two general causes of variation – genetic and environmental.

Inherited or genetic variation

The genes we inherit from our parents provide information that is used to define our characteristics. The combination of alleles (versions of genes) that we inherit is not the same as that in any other living thing (unless we have an identical twin). We may share many alleles with other members of our species, and we share genes with members of other species. However, there is never a complete match. Human cells contain approximately 20 000 genes. Many of these may have more than one allele. The chances of any two individuals having exactly the same combination of alleles is remote. So the combination of characteristics that each of us possesses is unique.

Environmental variation

Many characteristics can be affected by the environment. For example, an overfed pet will become obese. A person's skin will tan and become darker with careful exposure to sunlight.

A hawthorn tree usually grows upright to a height of about 6 m, but if it is nibbled by animals or cut by a farmer, then it will become more bushy and can be used to form a hedge. In this case, the environment has affected the way it grows. If a hawthorn grows in a rock crevice, where there is very little soil or water, it may grow to a height of only 150 cm – the environment has affected the *amount* of growth.

Combined effects

Environmental variation and genetic variation are not isolated. Many characteristics are affected by both causes of variation.

- In the past century, humans have become taller as the result of a better diet. But however good your diet may be, you are unlikely to grow very tall if all the rest of your family are short. This is because the height you can reach is limited by your genes.

- Not all our genes are active at any one time. For example, when you reach puberty, many changes occur in your body, because different genes are becoming active.

- Changes in the environment can also directly affect which genes are active.

DID YOU KNOW?

Temperature can affect the expression of genes. In Himalayan rabbits, the gene responsible for development of dark pigment (called the C gene) is only active at body temperatures below 35 °C. Rabbits reared at lower temperatures (20 °C) have a specific coloration not seen in those reared at warmer temperatures (30 °C). See Figure 4.

Figure 4 The gene controlling production of pigment is only active at lower temperatures. The rabbit on the left was reared at 20 °C, the one on the right at 30 °C.

Questions

1 Explain why discontinuous variation is often caused by the action of just one gene.

2 Explain why it is less easy to spot interspecific variation between species that are closely related.

3 Explain why skin colour is a continuously variable feature.

4 Look at Figure 4. Suggest why the ears, nose, feet and tail of the rabbit reared at 20 °C are black, but this is not seen in the rabbit reared at 30 °C.

⑦ Applying statistical techniques

By the end of this topic, you should be able to demonstrate and apply your knowledge and understanding of:

* the different types of variation

Why use statistical tests?

Data about variation may provide a series of numbers. But what do they mean? A small difference between two figures may or may not be significant. Using the correct statistical test can help to determine whether the difference observed is a significant difference or simply natural variation.

Standard deviation

The standard deviation is a measure of variation. It measures the amount of variation or spread from the mean.

* A low standard deviation indicates that the data have a narrow range and the points are closely grouped to the mean. This could indicate greater reliability.

* A high standard deviation indicates that the data points have a larger range and are less well grouped. This might indicate lower reliability.

Calculating the standard deviation

Use the formula: $s = \sqrt{\dfrac{\Sigma(x - \bar{x})^2}{n - 1}}$

Where s = standard deviation, x is an individual value, $\bar{x}$ = the mean value, n = the number of data points.

For example: a student collected data about the length of leaves on an oak tree. Calculate the standard deviation of the data collected.

Length of oak tree leaf (mm) 89, 87, 65, 97, 86, 92, 88, 75, 84, 83.

n = sample size = 10, therefore, $n - 1 = 9$

$\bar{x}$ = is the mean value (sum of data points divided by n). $\bar{x} = 84.6$

$\Sigma(x - \bar{x})^2 = (89 - 84.6)^2 + (87 - 84.6)^2 + (65 - 84.6)^2 + (97 - 84.6)^2$
$+ (86 - 84.6)^2 + (92 - 84.6)^2 + (88 - 84.6)^2 + (75 - 84.6)^2$
$+ (84 - 84.6)^2 + (83 - 84.6)^2$

$= 19.36 + 5.76 + 384.16 + 153.76 + 1.96 + 54.76 + 11.56$
$+ 92.16 + 0.36 + 2.56$

$= 726.4$

$s = \sqrt{\dfrac{726.4}{9}}$

$s = 8.98$

We can therefore say that the spread of our data set is the mean ± the standard deviation.

The spread of leaf length in our oak is 84.6 ± 8.98 or 75.62 – 93.58 mm.

If the leaves show a **normal distribution**, then:

* 68% of leaves should lie within the range 75.63–93.57 mm.
* 95% of leaves should lie within two standard deviations from the mean (66.64–102.56 mm).
* Anything outside this range might be viewed as anomalous (i.e. unusually long or unusually short).

Student's t-test

The Student's t-test is used to compare two means. It will test whether the difference between the two means is a significant difference. We first state a null hypothesis that there is no significant difference between the means of these two sets of data. The t-test will then test whether we can reject this hypothesis or must accept it.

Use the formula: $t = \dfrac{(\bar{x}_1 - \bar{x}_2)}{\sqrt{\dfrac{s_1^2}{n_1} + \dfrac{s_2^2}{n_2}}}$

Where $\bar{x}_1$ is the mean of the first data set, s_1^2 is the standard deviation of the first data set squared, and n_1 is the number of data points in the first data set. $\bar{x}_2$, s_2^2 and n_2 refer to the second data set.

For example: a student measured another set of leaves from a different oak tree, and calculated the standard deviation. He now has two sets of data.

From the first set: $n_1 = 10$, $\bar{x}_1 = 84.6$ and $s_1 = 8.98$

From data set 2: $n_2 = 10$, $\bar{x}_2 = 86.3$ and $s_2 = 7.45$

Substitute these figures into the formula:

$t = \dfrac{(84.6 - 86.3)}{\sqrt{\dfrac{8.98^2}{10} + \dfrac{7.45^2}{10}}}$

$t = \dfrac{1.7}{\sqrt{\dfrac{80.64}{10} + \dfrac{55.50}{10}}}$ we can ignore the minus sign on top, as this is an absolute (number)

$t = \dfrac{1.7}{\sqrt{8.06 + 5.55}}$

$t = \dfrac{1.7}{\sqrt{13.61}}$

$t = \dfrac{1.7}{3.69}$

$t = 0.46$

Is this significant?

The degrees of freedom are defined as the number of values in a statistical calculation that are free to vary. It is usually calculated as the (sample size) – (the number of data sets). In the t-test there are two data sets, so the number of degrees of freedom for the t-test is $(n_1 + n_2) - 2 = 18$ in this case. The number of degrees of freedom must be taken into account when considering if our calculated t-value is significant.

We can now use a table of t-values to assess whether our calculated value of t indicates a significant difference. In biology we always consider the 5% significance level. If the calculated value of t is greater than the value at 5%, then we can consider the difference between the two sets of data to be significant. If the calculated value of t is lower than the 5% value, then we can consider the difference to be insignificant. Using the t table below, the value of t for 18 degrees of freedom is 2.1. Our calculated value of 0.46 is below this value. Our two sets of data are not significantly different. We must accept the null hypothesis, which stated that 'there is no significant difference between the means'.

Degrees of freedom	Value of t at 5% significance level
10	2.23
12	2.18
18	2.10
20	2.09

Figure 1 A table of t-values.

Correlation coefficient

A correlation coefficient is used to consider the relationships between two sets of the data. The Spearman rank correlation tells us whether two sets of data are correlated or not.

Use the formula: $r_s = 1 - \dfrac{6\Sigma D^2}{n(n^2 - 1)}$

where r_s is the rank coefficient, D = the difference between the ranks and n is the number of pairs of values.

For example, a student measured the light intensity and per cent plant cover at six different sites. The student wanted to assess whether higher light intensity was correlated to higher plant per cent cover. Again we need to state a null hypothesis – in this case it would be 'There is no relationship between the light intensity and the per cent plant cover'.

Table 1 shows the results recorded and how these are ranked.

Site	Light intensity	Rank 1 (R_1)	% plant cover	Rank 2 (R_2)	$D = R_1 - R_2$	D^2
1	600	3	86	4	−1	1
2	700	1	95	2	−1	1
3	400	6	100	1	5	25
4	500	4	90	3	1	1
5	650	2	80	6	−4	16
6	450	5	85	5	0	0

Table 1 A table of results, showing how the results should be ranked.

Sum of $D^2 = 44$

Substitute these figures into the formula: $r_s = 1 - \dfrac{6 \times 44}{6(6^2 - 1)}$

$r_s = -0.26$

We can now use a table of critical values to assess whether this value of r_s indicates a correlation.

Number of pairs of measurements	Critical value
6	0.89

Our value of 0.26 is lower than the critical value, indicating that there is no significant correlation. The negative value suggests a negative correlation – but it is still not a significant correlation.

Questions

1. A student collected data about the length of oak leaves in a different location. Calculate the standard deviation of the following data set: length of leaves (mm): 102, 99, 112, 114, 98, 104, 107, 103, 115, 106.

2. Use the t-test to compare the mean for this set of data with the mean for the data collected by the student above: length of leaves (mm) 89, 87, 65, 97, 86, 92, 88, 75, 84, 83.

3. The student thought that the mean length of leaf might be correlated with the height of the tree. Use the following data to test the correlation.

Height of tree (m)	Mean length of leaf (mm)
27	88
13	96
22	93
18	92
16	90
19	87
9	83

(8) Adaptation

By the end of this topic, you should be able to demonstrate and apply your knowledge and understanding of:

* the different types of adaptations of organisms to their environment

What is adaptation?

All members of a species are slightly different from one another. They show variation. Any variation that helps the organism survive is an adaptation. The organism is adapted to its environment.

Adaptations help the organism cope with environmental stresses and obtain the things that they need to survive. A well-adapted organism will be able to:

* find enough food or photosynthesise well
* find enough water
* gather enough nutrients
* defend itself from predators and diseases
* survive the physical conditions of its environment, such as changes in temperature, light and water availability
* respond to changes in its environment
* have sufficient energy to allow successful reproduction.

Adaptations can work in different ways. They may be anatomical, behavioural or physiological (or biochemical). We shall consider each of these in the context of marram grass.

Marram grass

Marram grass (*Ammophila*) (see Table 1) is a very specialised plant. It is adapted to living on sand dunes where there is little water available. Marram grass must, therefore, be adapted to take up as much water as possible and to avoid losing that water. Marram grass is a xerophyte (see topic 3.3.6).

Anatomical adaptations

Anatomical means structural. Any structure that enhances the survival of the organism is an adaptation.

Adaptation	How it works
Long roots	This enables the plant to reach water that is deep underground.
Roots also spread out over a wide area	This enables marram to absorb a lot of water when it is available. It also helps to stabilise the sand dune in which the plant lives.
Leaves are curled	This reduces the surface area exposed to the wind. It also traps air inside, against the lower epidermis, so that moisture can build up in the enclosed space.
Lower epidermis is covered in hairs	This reduces air movement so that water vapour is retained close to the lower epidermis.
Lower epidermis is folded to create pits in which the stomata are positioned	Water vapour builds up in the pits, further reducing the loss of water vapour from inside the leaf.
Low density of stomata	Fewer stomata mean that less water vapour is lost.
Leaf covered in a thick waxy cuticle	This reduces evaporation of water from the cells of the leaf.

Table 1 Structural features that enable marram grass to survive in very dry conditions.

Behavioural adaptations

A behavioural adaptation is an aspect of the behaviour of an organism that helps it to survive the conditions it lives in. For example, when you touch an earthworm it quickly contracts and withdraws into its burrow. The earthworm has no eyes, so it cannot tell that you are not a bird about to eat it. Its rapid withdrawal is a behavioural adaptation to avoid being eaten.

You may not think that plants can show behaviour, but many certainly respond to their environment. Marram grass responds to shortage of water by rolling the leaf more tightly and closing the stomata. Both changes help to reduce transpiration. When covered by sand, marram will grow more quickly to reach the sunlight.

Physiological/biochemical adaptations of marram

A physiological or biochemical adaptation is one that ensures the correct functioning of cell processes. For example, the yeast *Saccharomyces* can respire sugars aerobically or anaerobically depending on how much oxygen is available.

Marram grass shows the following physiological adaptations:

- The ability to roll its leaf is due to the action of specialised hinge cells in the lower epidermis. These cells lose water when water is scarce and lose their turgidity – this rolls the leaf more tightly. When water is available, the hinge cells become turgid, opening up the leaf to allow easier access for carbon dioxide for photosynthesis.
- The guard cells work in a similar way to open and close the stomata. Non-turgid cells close the stomata and turgid cells open the stomata.
- Marram is not very salt tolerant, but it maintains a cell water potential that is lower than other plants. This enables it to survive in the salty conditions found beside the sea.
- The leaves contain many lignified cells that provide support when turgidity is lost – this keeps the leaf upright when water is not available.

Moles

Since adaptations are selected by the environment, it is quite possible for two unrelated species living in similar habitats to evolve similar adaptations. Where those species adopt a similar lifestyle they may evolve to look very similar.

A mole is a burrowing mammal that feeds on small animals in the soil. Marsupial moles live in Australia and are part of a group of mammals that have been evolving separately from placental mammals for up to 100 million years. Despite this separate evolution and being unrelated, marsupial moles and placental moles (see Figure 1) share a number of characteristics and look remarkably similar. They have the following characteristics:

- cylindrical body
- small eyes
- strong front legs
- large claws on front legs
- short fur
- short tail
- nose with tough skin for protection.

This is known as **convergent evolution**.

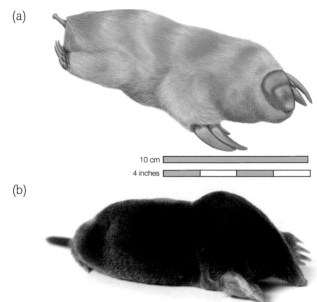

Figure 1 (a) Marsupial mole and (b) placental mole.

Questions

1. Write out three lists to categorise the following adaptations as anatomical, physiological or behavioural: flagella in bacteria, red pigments in seaweed, squinting in bright light, brightly coloured petals, ability to see colour, low pH in the stomach, possession of wings, use of salivary amylase to digest starch.

2. The Tasmanian wolf is only distantly related to the European wolf. Explain why the dentition (set of teeth) of these two animals is very similar.

3. Explain why whales can be mistakenly identified as fish.

4. Anatomical, physiological and behavioural adaptations are all interlinked. Explain why they should be considered together, rather than as isolated adaptations.

5. Select an animal such as a wolf. Describe three anatomical adaptations, three physiological adaptations and three behavioural adaptations of the animal.

4.3 (9) Natural selection and evolution

By the end of this topic, you should be able to demonstrate and apply your knowledge and understanding of:

* the mechanism by which natural selection can affect the characteristics of a population over time
* how evolution in some species has implications for human populations

Natural selection

An individual that has a characteristic which helps it survive in its environment is more likely to live long enough to reproduce. The process of evolution works by selecting individuals with particular adaptations to survive and reproduce. These adaptations are therefore passed from one generation to the next. Over a long period of time – possibly over very many generations – more and more individuals in the population will have that adaptive characteristic. We say that the adaptation has been selected.

How natural selection works

1. Mutation creates alternative versions of a gene (alleles).
2. This creates genetic variation between the individuals of a species (intraspecific variation).
3. Once variety exists, then the environment can 'select'. When resources are scarce, the environment will select those variations (characteristics) that give an advantage. There is a selection pressure.
4. Individuals with an advantageous characteristic will survive and reproduce.
5. Therefore they pass on their advantageous characteristics (inheritance).
6. The next generation will have a higher proportion of individuals with the successful characteristics. Over time, the group of organisms becomes well adapted to its environment (adaptation).

It is important to realise that variation must occur before evolution can take place. It is *genetic variation* that is important for evolution. Variation due to environmental factors will not be passed on to offspring.

> **LEARNING TIP**
>
> A species does not create mutations in order to adapt to a new environment. The correct sequence is that a mutation occurs, causing genetic variation. This variation may exist for many generations before selection occurs. When the environment changes, selection can occur.

Evolution today

Is evolution still going on? Most certainly, yes. Whenever a species or a group of organisms is placed under a new selection pressure, different characteristics will be selected. Evolution will occur. This is most obvious in organisms that have a short life cycle.

Pesticide resistance in insects

Some insects are pests. They eat our food crops, or cause damage to them. They can also act as vectors which transmit pathogens. Humans have devised ever more ingenious ways to kill insects. But some insects always survive.

Pesticides are chemicals designed to kill pests. Insecticides specifically kill insects. An insecticide applies a very strong selection pressure. If the individual insect is susceptible, then it will die. If it has some form of resistance, then the individual may survive. This will allow the individual with some resistance to reproduce and pass on the resistance characteristic. So the resistance quickly spreads through the whole population.

Resistance to pesticides was first documented in 1914, when scale insects were found to be resistant to inorganic insecticides. As we have introduced new classes of insecticide such as cyclodienes, carbamates, formamidines, organophosphates and pyrethroids, cases of resistance have been documented within 20 years. Resistance can arise within as little as two years.

DID YOU KNOW?

Resistance to insecticides has developed in different ways. Here are two examples:

- Pyrethroid insecticides are used to treat mosquito nets (for protection over beds in countries where malaria occurs). Mosquitoes, which carry the malaria parasite, have developed resistance to pyrethroids. They have evolved to produce an enzyme that can break down the pyrethroids.

- DDT is an insecticide that binds to a receptor on the plasma membrane of certain cells in insects. Insect populations have become resistant to DDT. Mutations in the genes coding for the cell-surface receptors have altered the shape of the receptor molecules. As a result, the shape is no longer complementary to the shape of the pesticide molecule. The DDT molecule does not bind to these modified receptor molecules.

When insects become resistant to pesticides, it can lead to another problem. It can cause the pesticide to accumulate in the food chain. If insects are resistant, they survive applications of these chemicals. The insects may then be eaten by their predators. The predators receive a larger dose of the insecticide, and it is quite possible for the insecticide to move all the way up the food chain. In this way, humans may receive quite large doses of insecticide. Because of increasing resistance and the fact that DDT accumulates in the food chain, it has been banned in many areas. However, DDT is still used in household-spraying programmes in some countries.

DID YOU KNOW?

The problem of accumulation in food chains is caused because insecticides like DDT are persistent. They do not break down in the ecosystem. DDT accumulates in fat tissue, and can now be found in almost all organisms, from the penguins of the Antarctic to the polar bears of the Arctic.

Microorganisms

The use of antibiotics (see topic 4.1.9) is a very powerful selection pressure on bacteria. When you take antibiotics, most of the bacteria are killed. But there may be one, or a few, that are resistant to the antibiotic. They are rarely completely unaffected by the antibiotic – but they are more resistant than most. Once most of the bacteria have been killed, you tend to feel better. So many people stop taking the antibiotics before they have finished the prescribed course. This allows the resistant bacteria to survive and reproduce to create a resistant strain of bacteria. Overuse and incorrect use of antibiotics has led to strains of bacteria that are resistant to virtually all the antibiotics in use. Some doctors now prescribe multiple antibiotics. This greatly reduces the chances that some bacteria will survive.

LEARNING TIP

Be careful not to say that microorganisms become immune to the antibiotic. Immunity implies activation of an immune system – microorganisms do not have an immune system.

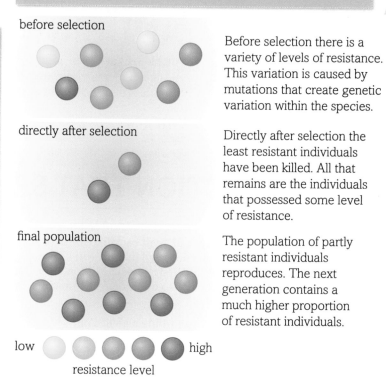

before selection

Before selection there is a variety of levels of resistance. This variation is caused by mutations that create genetic variation within the species.

directly after selection

Directly after selection the least resistant individuals have been killed. All that remains are the individuals that possessed some level of resistance.

final population

The population of partly resistant individuals reproduces. The next generation contains a much higher proportion of resistant individuals.

low ● ● ● ● ● high
resistance level

Figure 1 The selection of resistant microorganisms.

Some bacteria have gained a particularly wide range of resistance. The so-called 'superbug', MRSA, is one. MRSA stands for meticillin-resistant *Staphylococcus aureus*. But it might as well stand for multiple-resistant *Staphylococcus aureus*. This bacterium has developed resistance to an ever-increasing range of stronger and stronger drugs. This is an example of an 'evolutionary arms race'. Medical researchers are struggling to develop new and effective drugs, but the bacterial populations rapidly become resistant to them.

Questions

1. Explain how a particular colour of fur may be advantageous to (a) a predator, and (b) a prey species.

2. What factors may cause a struggle to survive amongst members of a population?

3. Explain why a resource will only become a selective force when it is limited.

4. Explain why evolution occurs in a shorter time in populations of microorganisms than in populations of mammals.

5. Explain why evolution tends to happen in short bursts.

THINKING BIGGER

ANTIBIOTICS AND RESISTANCE

Antibiotics have been used to combat bacterial infection since the discovery of penicillin by Alexander Fleming. However, overuse and misuse have led to the development of resistance in some bacteria. This article looks at some of the difficulties in developing new antibiotics.

A MALFUNCTION THAT SPAWNS FRANKENSTEIN BUGS

David Cameron pronounced this week that we are entering a post-antibiotics era which will see us "cast back into the dark ages of medicine where treatable infections and injuries will kill once again". The prime minister has ordered a review by economist Jim O'Neill into why industry has failed to deliver any new antibiotics for decades.

The battalions of bacteria have meanwhile marched on, relentlessly replicating and evolving and spewing out new generations of randomly mutated daughters. Some of these will have genetic mutations that allow them to survive medicinal onslaught, and replicate in ever greater numbers.

The common skin infection bacterium Staphylococcus aureus, or staph, for example, takes about 30 minutes to divide. If a bacterium entered your body this morning via a cut on your hand it would have spawned a colony of more than 1m bacteria, collectively containing about 300 mutations, by the time you go to bed. Your immune system will probably fight it off. But if it does not, within days you might need an antibiotic, such as meticillin, a modified form of penicillin.

If you really drew the short straw, the bacterium was not simple staph but its dangerous, meticillin-resistant cousin, or MRSA. Again, many healthy people can overcome it but some, especially the elderly or those with organ damage, cannot. Several strains have even outwitted vancomycin, a so-called treatment of last resort. Each year, 25 000 people across Europe die from drug-resistant bacterial infections.

The virtual halt in antibiotic development since the 1980s is a conspicuous market failure – but a dismally predictable one. If it takes a pharmaceutical company $1bn – as the industry often claims – to coax a new drug through the development pipeline, the last thing it wants is for its elixir to be used on as few people as possible, for the shortest possible time, as is absolutely necessary with antibiotics.

Even healthy pharmaceutical companies struggle to make the business case for developing them. AstraZeneca recently scaled back research into some promising antibiotics to focus on chronic conditions such as heart disease and respiratory disorders, which provide a steady supply of long-term consumers.

Incentives for pharmaceutical companies are likely to feature in Mr O'Neill's report, due out next spring.

As things stand, the World Health Organisation warns the bacteria are closing in: "This serious threat [of resistance] is no longer a prediction for the future; it is happening right now in every region of the world and has the potential to affect anyone, of any age, in any country."

The more antibiotics leach into the environment, the tougher our bacterial foes become. Some developing countries offer antibiotics without prescriptions, leading to overuse and misuse; nascent healthcare systems contain hospitals that flush antibiotic-laden sewage into the water supply.

In other states, up to 80% of antibiotic use is in animal husbandry. China's pig farms, where antibiotics are used mainly as growth promoters, have become crucibles for new, drug-resistant bacterial strains.

Until we restock the medicine cabinet, individual survival will become a macabre lottery.

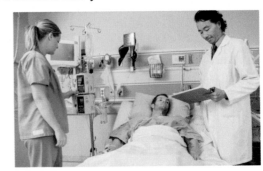

Figure 1 MRSA is a potentially lethal infection in hospitals.

Source
- Ahuja, A. A malfunction that spawns Frankenstein bugs. *Financial Times*, 4 July 2014.

Where else will I encounter these themes?

1.1   2.1   2.2   2.3   2.4   2.5

Let's start by considering the nature of the writing in the article.

1. This extract was adapted from the article *'A malfunction that spawns Frankenstein bugs'* by Anjana Anuja in the newspaper Financial Times. Read the article again and find two places in the article which suggest this newspaper might have a financial focus.
2. The name of 'staph' is given in full. What is scientifically incorrect about the way it has been written?

> You may want to consider what additional information might be required by an interested reader.

Now we will look at the biology in, or connected to, this article.

3. MRSA is a bacterium that causes a communicable disease. What is meant by a communicable disease?
4. To what genus does MRSA belong?
5. It is more accurate to state that these superbugs are resistant rather than immune to antibiotics. Explain why it is not correct to say that they are immune.
6. Suggest how the use of antibiotics in animal husbandry can promote the growth of pigs.
7. MRSA is a bacterium. Construct a table to compare bacterial cells and viruses.
8. Explain how new strains of MRSA can arise every few years.
9. It is often quoted that a possible source of new drugs to combat bacteria such as MRSA is plants in forest regions. Suggest why plants may have developed a molecule that can be effective against MRSA.
10. Saving habitats that may contain potentially useful medicinal plants is an ecological reason for conservation of species and habitats. State one economic reason and one aesthetic reason for conservation of biodiversity.

Activity

Draw a large diagram of a bacterial cell. Annotate the diagram with features of the cell that are different from a eukaryotic cell.

Draw a flow chart to explain how meticillin-resistant *Staphylococcus aureus* could arise from non-resistant strains of the bacterium.

The article suggests that China's pig farms 'have become a crucible for new, drug-resistant bacterial strains'. Write an article for a newspaper which could be published in China. The article should argue the case for reducing the level of antibiotic use in these pig farms. Remember that your arguments could include healthcare and animal welfare issues as well as productivity and economic issues.

Practice questions

1. What is the correct taxonomic hierarchy? [1]

   A. kingdom – class – phylum – order – family – genus – species

   B. kingdom – phylum – class – family – order – genus – species

   C. kingdom – class – phylum – family – order – genus – species

   D. kingdom – phylum – class – order – family – genus – species

2. Which row describes the main features of the protoctista? [1]

   A. single celled or multicellular, eukaryotic

   B. single celled or multicellular, prokaryotic

   C. single celled, prokaryotic

   D. multicellular, eukaryotic

3. Which type of graph should be used to display data about variation which is discontinuous? [1]

   A. a histogram

   B. a scatter gram

   C. a bar chart

   D. a line graph

4. Which row shows the correct way to write out the scientific name for humans? [1]

   A. *Homo Sapiens*

   B. *homo sapiens*

   C. *homo Sapiens*

   D. *Homo sapiens*

5. Which row describes the type of variation shown by hair colour in humans? [1]

   A. interspecific, continuous

   B. interspecific, discontinuous

   C. intraspecific continuous

   D. intraspecific discontinuous

   [Total: 5]

6. The following questions will test your knowledge of biological classification.

   (a) What is meant by the term natural classification? [2]

   (b) Explain how phylogeny is related to classification. [3]

   (c) Figure 1 shows a phylogenetic tree of species.

   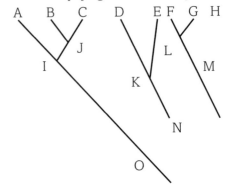

   Figure 1 A phylogenic tree of species.

   (i) What is the relationship of species M to species L, F, G and H? [1]

   (ii) Suggest which two species are most similar to each other? [1]

   (iii) Explain your choice above. [3]

   (iv) Explain why species A, B, C, J and I are known as monophyletic. [1]

   [Total: 11]

7. Figure 2 shows a number of world leaders.

   Figure 2

   (a) Identify two features that show continuous variation between these world leaders. [2]

   (b) What factor(s) cause this type of variation? [1]

   (c) Identify one feature that shows discontinuous variation between these world leaders. [1]

   (d) What factor(s) cause this type of variation? [1]

   (e) Using the theory of natural selection explain why unrelated organisms may show many of the same adaptive features. [7]

   [Total: 12]

8. Charles Darwin developed a theory of evolution by natural selection.

    (a) What four observations made by Darwin led to the theory of natural selection? [4]

    (b) How did the fossil record lend evidence to Darwin's theory? [3]

    (c) Explain how DNA has been used to provide more evidence for Darwin's theory. [8]

    [Total: 15]

9. The structure of biological molecules such as proteins has been used as evidence for evolution and for confirming our ideas about phylogeny.

    (a) Explain how the structure of proteins can reveal how closely related two species are. [4]

    (b) Not all proteins are suitable to establish how closely related species are. Suggest why a protein such as haemoglobin is not suitable to establish relatedness between many species. [3]

    (c) The similarity of proteins can be established by using antibodies. Agglutinins produced to combine with a protein from one species will combine with similar proteins in related species but will combine less well with those from more distantly related species. The amount of agglutination can be measured using a colorimeter. A scientist tested an agglutinin manufactured to combine with protein from species A on five other species. The table below shows the results of the tests.

Species	Relative level of agglutination (%)
A	100
B	6
C	75
D	98
E	23
F	5

    (i) The scientist concluded that species D is most closely related to species A. Explain this decision. [1]

    (ii) Explain why the agglutinins were able to produce 98% agglutination of the proteins in a different species. [3]

    (iii) Explain why proteins found in more distantly related species produce less agglutination. [3]

    [Total: 14]

10. *Clostridium difficile* is a microorganism that causes disease. *C. difficile* is prokaryotic. Small populations may live in the large intestines and are usually harmless. However, they can become harmful after treatment with antibiotics which reduces the levels of other gut flora.

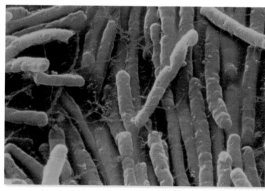

Figure 3

    (a) What type of organism is *C. difficile*? [1]

    (b) State two features of this group of organisms not shared with other groups. [2]

    (c) *C. difficile* has become resistant to many antibiotics. Describe how this resistance has arisen. [6]

    (d) Figure 4 shows the number of cases of *C. difficile* each year in the UK.

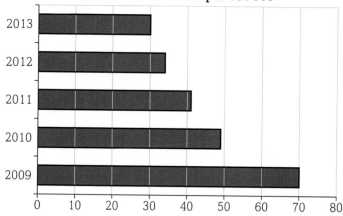

Number of cases per 100 000

Figure 4

    (i) Describe the trend shown by the data. [2]

    (ii) Calculate the percentage change in the number of cases from 2008–2013. [3]

    [Total: 14]

Maths skills

In order to be able to develop your skills, knowledge and understanding in Biology, you will need to have developed your mathematical skills in a number of key areas. This section gives more explanation and examples of some key mathematical concepts you need to understand. Further examples relevant to your AS/A Level Biology studies are given throughout the book.

Arithmetic and numerical computation

Using standard form

Dealing with very large or small numbers can be difficult. To make them easier to handle, you can write them in the format $a \times 10^b$. This is called standard form.

To change a number from decimal form to standard form:

- Count the number of positions you need to move the decimal point by until it is directly to the right of the first number which is not zero.
- This number is the index number that tells you how many multiples of 10 you need. If the original number was a decimal, your index number must be negative.

Here are some examples:

Decimal notation	Standard form notation
0.000 000 012	1.2×10^{-8}
15	1.5×10^1
1000	1×10^3
3 700 000	3.7×10^6

Using ratios, fractions and percentages

Ratios, fractions and percentages help you to express one quantity in relation to another with precision. Ratios compare like quantities using the same units. Fractions and percentages are important mathematical tools for calculating proportions.

Ratios

A ratio is used to compare quantities. You can simplify ratios by dividing each side by a common factor. For example $12:4$ can be simplified to $3:1$ by dividing each side by 4.

> **EXAMPLE**
>
> *Divide 180 into the ratio $3:2$*
> Our strategy is to work out the total number of parts then divide 180 by the number of parts to find the value of one part.
> total number of parts = $3 + 2 = 5$
> value of one part = $180 \div 5 = 36$
> answer = $3 \times 36 : 2 \times 36 = 108 : 72$

Check your answer by making sure the parts add up to 180: $72 + 108 = 180$

Fractions

When using fractions, make sure you know the key strategies for the four operators:

To add or subtract fractions, find the lowest common multiple (LCM) and then use the golden rule of fractions. The golden rule states that a fraction remains unchanged if the numerator and denominator are multiplied or divided by the same number.

> **EXAMPLE**
>
> $\frac{1}{2} + \frac{1}{5} = \frac{5}{10} + \frac{2}{10} = \frac{7}{10}$

To multiply fractions together, simply multiply the numerators together and multiply the denominators together.

> **EXAMPLE**
>
> $\frac{2}{7} \times \frac{4}{9} = \frac{8}{63}$

To divide fractions, simply invert or flip the second fraction and multiply.

> **EXAMPLE**
>
> $\frac{2}{3} \div \frac{7}{9} = \frac{2}{3} \times \frac{9}{7} = \frac{18}{21} = \frac{6}{7}$

Percentages

When using percentages, it is useful to recall the different types of percentage questions.

To increase a value by a given percentage, use a percentage multiplier.

> **EXAMPLE**
>
> *Increase 30 mg by 23%.*
> If we increase by 23%, our new value will be 123% of the original value. We therefore multiply by 1.23.
> answer = $30 \times 1.23 = 36.9$ mg

To decrease a value by a given percentage, you need to focus on the part that is left over after the decrease.

> **EXAMPLE**
>
> *Decrease 30 mg by 23%.*
> If we decrease by 23%, our new value will be $100 - 23 = 77\%$ of the original value. We therefore multiply by 0.77.
> answer = $30 \times 0.77 = 23.1$ mg

To calculate a percentage increase, use the following equation:

$$\text{Percentage change} = \frac{\text{difference between values}}{\text{original value}} \times 100$$

To calculate percentage decrease, use the same equation but remember that your answer should be negative.

The volume of a solution increased from 40 ml to 50 ml. Calculate the percentage increase.

change in volume = 10 ml

percentage increase = $\frac{10}{40} \times 100 = 25\%$

Here are some examples:

Exact number	To one s.f.	To two s.f.s.	To three s.f.s.
45 678	50 000	46 000	45 700
45 000	50 000	45 000	45 000
0.002 755	0.003	0.002 8	0.002 76

Algebra

Using algebraic equations

Using algebraic equations is a very important skill for finding the value of an unknown quantity. In the real world, letters are used to symbolise important variables such as the blood sugar level of a diabetic or the irregular heartbeat of a patient.

The key rule to remember when using equations is that any operation that you apply to one side of the equation must also be applied to the other side.

EXAMPLE

Find the value of x in the following equation: $7x - 6 = 36$
adding 6 to each side gives $7x = 42$
dividing each side by 7 gives $x = 6$

Changing the subject of an equation

It can be very helpful to rearrange an equation to express the variable in which you are interested in terms of other variables. Always remember that any operation that you apply to one side of the equation must also be applied to the other side.

EXAMPLE

The diameter of a cell measured under the light microscope at magnification ×100 is 2 mm. Calculate the actual size.
You may remember the equation: image size = actual size × magnification. However, this question is asking us to find the actual size given the image size and magnification. We can rearrange the equation to suit our needs:
image size = actual size × magnification
$\frac{\text{image size}}{\text{magnification}}$ = actual size
so actual size = $\frac{2}{100}$ = 0.02 mm

Handling data

Using significant figures

Often when you do a calculation, your answer will have many more figures than you need. Using an appropriate number of significant figures will help you to interpret results in a more meaningful way.

Remember the 'rules' for significant figures:

* The first significant figure is the first figure which is not zero.
* Digits 1–9 are always significant.
* Zeros which come after the first significant figure are significant unless the number has already been rounded.

Principles of sampling and using Simpson's diversity index

When a scientist studies a population, it is not possible to study each organism in detail. Scientists therefore use sampling to estimate characteristics of the whole population by looking at a subset of individuals in the population. It is important that the sample chosen is representative of the habitat.

Once a suitable sample has been selected, it can be analysed. For example, Simpson's diversity index is a measure of biodiversity which takes into account both the species richness and the species abundance of an area. It is calculated by the formula:

$$D = 1 - [\Sigma(n/N)^2]$$

where n is the number of individuals of a particular species (or the percentage cover for plants), and N is the total number of all individuals of all species (or the total percentage cover for plants).

Graphs

Understand that $y = mx + c$ represents a linear relationship

Two variables are in a linear relationship if they increase at a constant rate in relation to one another. If you plotted a graph with one variable on the x-axis and the other variable on the y-axis, you would get a straight line. Any linear relationship can be represented by the equation $y = mx + c$ where the gradient of the line is m and the value at which the line crosses the y-axis is c. An example of a linear relationship is the relationship between degrees Celsius and degrees Fahrenheit, which can be represented by the equation $F = 9/5 C + 32$ where C is temperature in degrees Celsius and F is temperature in degrees Fahrenheit.

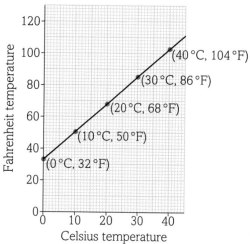

Conversion of Celsius temperatures to Fahrenheit

Figure 1

Calculate a rate of change from a graph showing a linear relationship

The rate of change from a graph showing a linear relationship is the gradient, or steepness, of the line. It is a measure of the rate of change of one variable, represented on the x-axis, in relation to the other variable, represented on the y-axis.

To calculate the rate of change:

1. Draw a right-angled triangle anywhere on the line.
2. Use the following equation to calculate the rate of change:

$$\text{gradient} = \frac{\text{difference on } y\text{-axis}}{\text{difference on } x\text{-axis}}$$

3. State the unit for your answer.

Draw and use the slope of a tangent to a curve as a measure of a rate of change

Sir Isaac Newton was fascinated by rates of change. He drew tangents to curves at various points to find the rates of change of graphs as part of his journey towards discovering calculus – an amazing branch of mathematics. He argued that the gradient of a curve at a given point is exactly equal to the gradient of the tangent of a curve at that point.

To calculate the rate of change:

1. Use a ruler to draw a tangent to the curve.
2. Calculate the gradient of the tangent using the technique given for a linear relationship. This is equal to the gradient of the curve at the point of the tangent.
3. State the unit for your answer.

Applying your skills

You will often find that you need to use more than one maths technique to answer a question. In this section, we will look at two example questions and consider which maths skills are required and how to apply them.

EXAMPLE

Hydrogen peroxide is a toxic by-product of respiration and is made in all living cells. Cells make the enzyme catalase in order to convert the toxin into water and oxygen. In order to study the effect of temperature on catalase activity, an experiment was set up using the equipment shown in Figure 2. The volume of oxygen released in 30 seconds was measured using the gas syringe. The results of the experiment are shown in the graph (Figure 3).

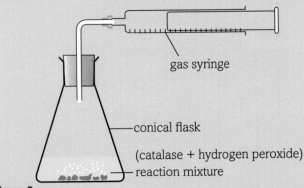

gas syringe

conical flask

(catalase + hydrogen peroxide)

reaction mixture

Figure 2

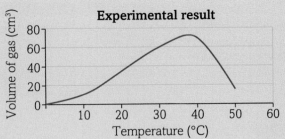

Figure 3

(a) Calculate the percentage increase in volume of oxygen produced from 10 °C to 40 °C.

(b) Calculate the rates of each temperature test at 20, 40 and 50 °C and interpret the results.

(c) A further experiment is carried out where the volume of oxygen is recorded over the entire time of the reaction at 10 °C and 40 °C. The results are shown below:

	Total volume of oxygen released (cm³)									
Temp.	0 s	10 s	20 s	30 s	40 s	50 s	60 s	70 s	80 s	90 s
10 °C	0	2	5	9	16	22	28	33	35	35
40 °C	0	7	15	27	33	35	35	35	35	35

Display both sets of results on an appropriately scaled graph.

(d) Calculate the difference in rate between both reactions at 15 s.

Answers

(a) The volume of gas at 10 °C = 10 cm³.
The volume of gas at 40 °C = 70 cm³.
The percentage increase = $(70 - 10)/10 \times 100 = 600\%$ increase

(b) $20\,°C = \dfrac{35}{30} = 1.17\ \text{cm}^3\,\text{s}^{-1}$

$40\,°C = \dfrac{70}{30} = 2.33\ \text{cm}^3\,\text{s}^{-1}$

$50\,°C = \dfrac{15}{30} = 0.5\ \text{cm}^3\,\text{s}^{-1}$

The rate has doubled between 20 °C and 40 °C, but at 50 °C, the rate has decreased.

(c)

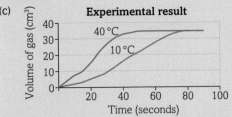

(d) We draw a tangent to each curve at 15 seconds so that we can use the gradient of the curve to calculate the rate.

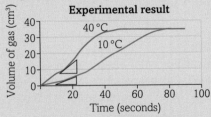

We can then use the following equation to calculate gradient:

$$\text{gradient} = \frac{\text{difference on } y\text{-axis}}{\text{difference on } x\text{-axis}}$$

rate of reaction at 10 °C at 15 seconds = $\frac{3}{10} = 0.3\ \text{cm}^3\,\text{s}^{-1}$

rate of reaction at 40 °C at 15 seconds = $\frac{8}{10} = 0.8\ \text{cm}^3\,\text{s}^{-1}$

difference in rate between reactions at 15 seconds
$= 0.8 - 0.3 = 0.5\ \text{cm}^3\,\text{s}^{-1}$

EXAMPLE

A photomicrograph of a T helper cell was taken using an electron microscope set at a magnification of ×50 000. In the image, several organelles were clearly identified and measured.

(a) In the table below, calculate the actual object length of each organelle.

Organelle	Image length (mm)	Object length (μm)
nucleus	240	
endoplasmic reticulum	360	
lysosome	10	
mitochondrion	120	

(b) A lysosome is a spherical organelle. Calculate the surface area and volume of a lysosome.

(c) Calculate the surface area to volume ratio of a lysosome.

Answers

(a) The question tells us that the magnification is ×50 000.

We know that image size = actual size × magnification

To make it easier to use, we can rearrange this equation as

$$actual\ size = \frac{image\ size}{magnification}$$

$$actual\ length\ of\ nucleus = \frac{240}{50\,000} = 0.004\,8\ mm$$

$$actual\ length\ of\ endoplasmic\ reticulum = \frac{360}{50\,000} = 0.007\,2\ mm$$

$$actual\ length\ of\ lysosome = \frac{10}{50\,000} = 0.000\,2\ mm$$

$$actual\ length\ of\ mitochondrion = \frac{120}{50\,000} = 0.002\,4\ mm$$

Before we can put these figures in the table, we need to convert to μm. 1 mm = 1000 μm so we need to multiply each figure by 1000.

Organelle	Image length (mm)	Object length (μm)
nucleus	240	4.8
endoplasmic reticulum	360	7.2
lysosome	10	0.2
mitochondrion	120	2.4

(b) You should be familiar with the following formulae, where r is radius:

surface area of sphere $= 4\pi r^2$

volume of sphere $= \frac{4}{3}\pi r^3$

From (a) you know that the diameter of the lysosome is 0.2 μm. This means that the radius must be 0.1 μm.

surface area of lysosome $= 4\pi(0.1)^2 = 4\pi \times 0.01 = 0.125\,7\ \mu m^2$ to 4 d.p.

volume of lysosome $= \frac{4}{3}\pi(0.1)^3 = \frac{4}{3}\pi \times 0.001 = 0.004\,2\ \mu m^3$ to 4 d.p.

(c) It is simplest and most accurate to use the exact expressions from (b) involving π, rather than the final answers which have been rounded.

surface area-to-volume ratio $= 4\pi \times 0.01 : \frac{4}{3}\pi \times 0.001$

We can simplify by multiplying each side by 1000 and dividing each side by π:

surface area-to-volume ratio $= 40 : \frac{4}{3}$

Now we can divide each side by 4 and multiply each side by 3 to get:

surface area-to-volume ratio $= 30 : 1$

Preparing for your exams

Introduction

The way that you are assessed will depend on whether you are studying for the AS or the A Level qualification. Here are some key differences:

- AS students will sit two exam papers, each covering all of the content of the AS specification.
- A Level students will sit three exam papers, each covering content from both years of A Level learning. The third paper can include questions from all six modules covered during the two years of study.
- A Level students will also have their competence in key practical skills assessed by their teacher in order to gain the Science Practical Endorsement. The endorsement will not alter the overall grade but if you pass the endorsement this will be recorded on your certificate and can be used by universities as part of their conditional offer of a place on a course.

The tables below give details of the exam papers for each qualification.

AS exam papers

Paper	Paper 1: Breadth in Biology	Paper 2: Depth in Biology
Topics covered	Modules 1–4	Modules 1–4
% of the AS qualification	50%	50%
Length of exam	1 hour 30 minutes	1 hour 30 minutes
Marks available	70 marks	70 marks
Question types	20 marks multiple-choice followed by 50 marks short structured response	Structured questions and extended writing including two level of response questions
Experimental methods?	Yes	Yes
Mathematics	A minimum of 10% of the marks across both papers will be awarded for mathematics at GCSE higher tier level or above	

A Level exam papers

Paper	Paper 1: Biological processes	Paper 2: Biological diversity	Paper 3: Unified biology	Science Practical Endorsement
Topics covered	Modules 1,2,3 and 5	Modules 1,2,4 and 6	Modules 1–6	Assessed by teacher throughout course. Does not count towards A Level grade but a 'pass' result will be reported on A Level certificate. It is likely that you will need to maintain a separate record of practical activities carried out during the course.
% of the A Level qualification	37%	37%	26%	
Length of exam	2 hours 15 minutes	2 hours 15 minutes	1 hour 30 minutes	
Marks available	100 marks	100 marks	70 marks	
Question types	15 marks multiple-choice followed by 85 marks structured response and extended writing including two level of response questions	15 marks multiple-choice followed by 85 marks structured response and extended writing including two level of response questions	Structured response and extended writing including two level of response questions	
Experimental methods?	Yes	Yes	Yes	
Mathematics	A minimum of 10% of the marks across all three papers will be awarded for mathematics at GCSE higher tier level or above			

Exam strategy

Arrive equipped

Make sure you have all of the correct equipment needed for your exam. As a minimum you should take:

- pen (black, ink or ball-point)
- pencil (HB)
- ruler (ideally 30 cm)
- rubber (make sure it's clean and doesn't smudge the pencil marks or rip the paper)
- calculator (scientific)

Ensure your answers can be read

Your handwriting does not have to be perfect but the examiner must be able to read it! When you're in a hurry it's easy to write key words that are difficult to decipher.

Plan your time

Note how many marks are available on the paper and how many minutes you have to complete it. This will give you an idea of how long to spend on each question. Be sure to leave some time at the end of the exam for checking answers. A rough guide of a minute a mark is a good start, but short answers and multiple-choice questions may be quicker. Longer answers might require more time.

Understand the question

Always read the question carefully and spend a few moments working out what you are being asked to do. The command word used will give you an indication of what is required in your answer.

Be scientific and accurate, even when writing longer answers. Use the technical terms you've been taught.

Always show your working for any calculations. Marks may be available for individual steps, not just for the final answer. Also, even if you make a calculation error, you may be awarded marks for applying the correct technique.

Plan your answer

Questions marked with a * are level of response questions. The examiners will be looking for a line of reasoning in your response. Here, marks will be awarded for your ability to logically structure your answer showing how the points that you make are related or follow on from each other where appropriate. Read the question fully and carefully (at least twice!) before beginning your answer.

Make the most of graphs and diagrams

Diagrams and sketch graphs can earn marks – often more easily and quickly than written explanations – but they will only earn marks if they are carefully drawn and fully annotated.

- If you are asked to read a graph, pay attention to the labels and numbers on the x- and y-axes. Remember that each axis is a number line.
- If asked to draw or sketch a graph, always ensure you use a sensible scale and label both axes with quantities and units. If plotting a graph, use a pencil and draw small crosses ($\times$) or dots with a circle around them ($\odot$) for the points.
- Diagrams must always be neat, clear and fully labelled or annotated.

Check your answers

For open-response and extended writing questions, check the number of marks that are available. If three marks are available, have you made three distinct points? It can be helpful to number or bullet-point your response.

For calculations, read through each stage of your working. Substituting your final answer into the original question can be a simple way of checking that the final answer is correct. Another simple strategy is to consider whether the answer seems sensible. Pay particular attention to using the correct units.

Sample AS questions – multiple choice

The following spirometer trace shows the results of an experiment. Soda lime was used to extract carbon dioxide from exhaled air.

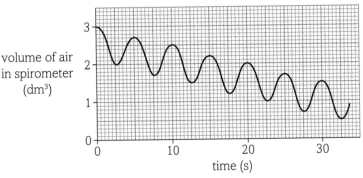

What is the rate of oxygen consumption in the experiment?

A $1.0\,dm^3$

B $3.0\,dm^3\,min^{-1}$

C $5.0\,dm^3\,min^{-1}$

D 12 breaths min^{-1}

Your answer ☐

Question analysis

Multiple-choice questions look easy until you try to answer them. Very often they require some working out and thinking.

In multiple choice questions you are given the correct answer along with three incorrect answers (called distractors). You need to select the correct answer and write the appropriate letter in the box provided.

If you change your mind, put a line through the box and write your new answer next to it.

Verdict

This is a weak answer because:

● This candidate has probably made the mistake of measuring down to the bottom of the trough in the trace just after 30 seconds. That gives a reading of $2.5\,dm^3$ in about 30 seconds which gives the answer $5\,dm^3$ when multiplied by 2.

● The candidate has measured to the wrong part of the trace and has also been careless in not converting to $dm^3\,min^{-1}$ properly. The trough is actually at about 32 seconds so it is not accurate to simply multiply by 2 to get a figure per minute.

If you have any time left at the end of the paper go back and check your answer to each part of a multiple-choice question so that a slip like this does not cost you a mark.

Sample AS questions – short structured

Fig 1.1 is a diagram of a plant cell.

(a) (i) Name the cell components labelled A and B.

A ....

B ....

(ii) State the **functions** of the components labelled **C** and **D**

C ....

....

....

D ....

....

....

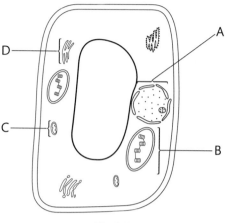

Fig 1.1

The command word in part (a) is 'name' and in part (b) it is 'state'. This indicates that all you need to do is write down the correct terms. You do not need to give further description or explanation. This would only waste your limited time in the examination.

Question analysis

- Generally one piece of information is required for each mark given in the question. There are two marks available for each part of this question so make sure you make two distinct points.
- The answer lines have been labelled to help you.
- Clarity and brevity are the keys to success on short structured questions. For one mark, it is often not necessary to write complete sentences.

Average student answer

In part (a)

A is the nuclear membrane

B is a chloroplast

In part (b)

Controls what enters and leaves the nucleus

Photosynthesis

Look closely at the label for A. It has a bracket on the end showing that the whole nucleus is being labelled. If a part of the nucleus were required as the answer then the label line would end on that part of the nucleus.

In part (b) it is important to read the question carefully and note that the question refers to different parts of the cell, not the same parts as used in part (a).

Verdict

- This is an average answer because:
- The student has recognised that the organelle is the nucleus but has given a name for only part of that organelle. The chloroplast has been correctly identified.
- The student has assumed that the functions required are for the components labelled in part (a). The functions given are accurate for the components named by the student in part (a). However, part (b) actually requires two steps: you need to identify organelles C and D and then write down their functions.

Sample AS questions – extended writing

Describe the arrangement and functions of two named components of a cell surface membrane. [4]

The command word in this question is 'describe'. This means that you need to provide a detailed account. The depth of the answer should be judged from the marks allocated for the question. You do not need to include any justifications or reasons. Four marks are available so four distinct points should be made. However, you are asked to describe something about the arrangement of two components and give the functions of the same two components of the membrane. Make sure that the arrangement and function do apply to the same component. This could be answered in the form of a small table in your examination paper.

Question analysis

- With any question worth four or more marks, think about your answer and the points that you need to make before you write anything down. It may be worth writing a few notes to help organise your thoughts – these could be written at the back of the examination paper.
- Keep your answer concise, and the information you write down relevant to the question. You will not gain marks for writing down biology that is not relevant to the question (even if correct) but it will cost you time.
- Remember that you can use bullet points or annotated diagrams in your answer.

Average student answer

A cell membrane is made up of a phospholipid bilayer. Within this bilayer there are some proteins that span the membrane and others that are free to move within the membrane. These form channels through which charged particles can pass across the membrane. Other features of the membrane include cholesterol, which sits within the bilayer and helps to stabilise the membrane structure. Glycoproteins and glycolipids provide binding sites for hormones.

At this level, your answers need technical terms and clarity in expression otherwise you will find yourself losing marks. It is also important to follow the rubric of the question.

Verdict

This is an average answer because:

- The student has made many good points.
- The question asks for the arrangement and function of two components. Therefore the examiner will mark the first two components named. That is phospholipid and protein in this response.
- The student has described an arrangement for phospholipids as a bilayer but has not given a function.
- The student has described both an arrangement and a function for proteins.
- The last two sentences go on to describe other components of the membrane. As these are the third and fourth components mentioned they will not be marked by the examiner – even though the candidate gave both an arrangement and a function for cholesterol. Writing these last two sentences was not a good use of time.

Sample AS questions – extended writing level of response

Haemoglobin is a protein that carries oxygen in the blood of all mammals. The structure of haemoglobin can vary slightly between species.

- Llamas live at high altitudes and camels live at low altitudes.
- At high altitudes the partial pressure of oxygen is low.
- Llama and camel haemoglobin consists of 2 α subunits and 2 β subunits.
- Each subunit contains a haem group and is able to bind to one molecule of oxygen.
- In the β subunits, one amino acid present in camel haemoglobin has been replaced by a different amino acid in llama haemoglobin.

Describe how the structure of llama haemoglobin is likely to be different from that of camel haemoglobin with reference to the four levels of protein structure. [6]

> Six marks are available so six points need to be made. The command word is 'describe' but you are also told to refer to the four levels of protein structure.
>
> You are given some hints in the stem of the question including a description of haemoglobin consisting of four polypeptides (two α and two β subunits) and that llama haemoglobin differs from camel haemoglobin by one amino acid.
>
> You need to organise your answer so that it follows a logical line of reasoning.

Question analysis

- There will be two questions in your second examination which will test your ability to organise your response with a clear and logical structure.
- These questions will be allocated either 6 or 9 marks.
- Your response will be assessed for the level of organisation and whether the information presented is relevant.
  - For level 1 (1–2 marks or 1–3 marks) the information is basic and presented in an unstructured way.
  - For level 2 (3–4 marks or 4–6 marks) there is a line of reasoning and some structure is evident. The information is for the most part relevant and is supported by some evidence.
  - For level 3 (5–6 marks or 7–9 marks) there is a well-developed line of reasoning and a logical structure. The information is relevant and substantiated.
- It is vital to plan out your answer before you write it down. There is always space given on an exam paper to do this so just jot down the points that you want to make before you answer the question in the space provided. This will help to ensure that your answer is coherent and logical and that you don't end up contradicting yourself.
- It helps with longer extended writing questions to think about the number of marks available and how they might be distributed.

Average student answer

The primary structure is the sequence of amino acids in the polypeptide chain. One change in the amino acid sequence means that the primary structure has been changed.

This could cause a change in the secondary structure affecting the shape of the α helix or the β pleated sheet. This is because the hydrogen bonding that holds the secondary structure may have been altered by the presence of the different amino acid.

The different R-group on the new amino acid could also affect the bonding between nearby R groups. This might alter the three dimensional shape of the whole polypeptide chain – this is the tertiary structure.

> You are given some hints in the stem of the question and you are reminded that proteins have four levels of structure – you need to write something about each level. To organise the answer it would be best to start at the primary structure and work through the levels of structure finishing with quaternary structure. Don't forget to substantiate your answer by explaining why each statement is true.

Verdict

This is an average to good answer because:

- The student has kept to relevant detail.
- The student has organised the information with a clear line of thought.
- Each statement about a change in structure has been explained by suggesting how the bonding within the molecule may be affected.
- However, the student has omitted to mention any possible effect on the quaternary structure. There may be no effect, but this should be clearly stated.

Sample AS questions – practical

Many enzymes are associated with non-protein molecules known as cofactors. Some cofactors are small inorganic ions.

Rennin is an enzyme that is involved in the digestion of milk. It converts soluble caseinogen in milk into insoluble casein. The cofactor Ca^{2+} is associated with this reaction.

A student wished to investigate the effect of Ca^{2+} on the action of rennin.

Describe how the student could carry out this investigation and produce valid results. [5]

> Notice that the question says 'describe how the student could carry out this investigation to produce valid results'. The command word 'describe' requires a suitable level of detail. It is also important that the results must be valid. This means that the variables have been controlled well enough to make the results comparable with each other.

Question analysis

- There will be questions in your exams which assess your understanding of practical skills and draw on your experience of the core practicals. For these questions, think about:
  - how apparatus is set up
  - the method of how the apparatus is to be used
  - how readings are to be taken
  - how to make the readings reliable
  - how to control any variables
  - possible limitations and improvements.

- It helps with extended writing questions to think about the number of marks available and how they might be distributed. For example, if the question asked you to give the arguments for and against a particular case, then assume that there would be equal numbers of marks available for each side of the argument and balance the viewpoints you give accordingly. However, you should also remember that marks will also be available for giving an overall conclusion so you should be careful not to omit that.

- It is vital to plan out your answer before you write it down. There is always space given on an exam paper to do this so just jot down the points that you want to make before you answer the question in the space provided. This will help to ensure that your answer is coherent and logical and that you don't end up contradicting yourself.

Average student answer

Mix 10 cm^3 milk with 1 cm^3 rennin solution and 1 cm^3 of 1 M calcium ion solution.

Time how long the milk takes to form solids.

Dilute the calcium ion solution to form 0.1, 0.25, 0.5 and 0.75 M solutions.

Repeat with the same volumes of milk and rennin but use 1 cm^3 of each of the diluted calcium ion solutions each time.

> A suitable level of detail means stating suitable volumes and concentrations. The results can be made valid by ensuring that the volumes and concentrations used are the same in all the readings. Some mention should be made of all the other possible variables.

Verdict

This is an average answer because:

- The student has quoted suitable volumes and concentrations to use.
- The student has kept the volumes the same in each test and has altered just one variable (the concentration of calcium ions).
- The student has used five different concentrations of calcium ions.
- However, there are some important details that have been missed. The student has not described carrying out the test with no calcium ions.
- The student does not mention repeating the experiment at each concentration of calcium ions enabling the reliability to be assessed, to check for anomalies and to calculate a mean.
- The student has not described controlling other variables, such as temperature and pH, to ensure the results are valid.
- The student has not detailed how the end point should be assessed – i.e. how to tell when solids have been formed.

Sample AS questions – calculation

On a biology field trip a pair of students collected some data about plant species in a area of ash woodland. Their results are shown in Table 4.1.

Species	Number of individuals (n)	n/N	$(n/N)^2$
Dog's mercury	40		
Wild strawberry	13	0.13	0.0169
Common avens	43		
Wood sorrel	4		
	$N =$		$\Sigma(n/N)^2 =$
			$1-(\Sigma(n/N)^2) =$

Table 4.1

(a)(i) Use the information in the table to work out the Simpson's index of diversity (D) for the area of woodland sampled using the formula:

$$D = 1-(\Sigma(n/N)^2)$$

Where: n = number of individuals of a particular species.
N = total number of individuals in all species
Σ = sum of

Complete Table 4.1.

You may use the space below for your working.

> The command here is 'calculate'. This means that you need to obtain a numerical answer to the question, showing relevant working. If the answer has a unit, this must be included.
>
> Finding the numbers requires you to use the appropriate numbers from the table given.

Question analysis

● The important thing with calculations is that you must show your working clearly and fully. The correct answer on the line will gain all the available marks; however, an incorrect answer can gain all but one of the available marks if your working is shown and is correct.

● Show the calculation that you are performing at each stage and not just the result. When you have finished, look at your result and see if it is sensible.

● At some point during your answer you will need to do some kind of sum, and the skills are to decide
  – which numbers you need
  – which operation you need.

Average student answer

Species	Number of individuals (n)	n/N	$(n/N)^2$
Dog's mercury	40	40/100 = 0.40	$(0.4/100)^2 = 0.1600$
Wild strawberry	13	0.13	0.0169
Common avens	43	0.43	0.1849
Wood sorrel	4	0.4	0.16
	$N = 100$		$\Sigma(n/N)^2 = 0.5218$
			$1-(\Sigma(n/N)^2) = 0.4782$

The student has not calculated n/N correctly for Wood sorrel. Therefore the value of $(n/N)^2$ is incorrect. This is an easy error to make if you use a calculator and do not think about the answer you have written. 4/100 is not 0.4, it is 0.04.

Verdict

This is an average answer because:

- The student has calculated most of the parts correctly.
- The working is shown in the top row of the table.
- The student has followed the calculation through appropriately.
- However, one part of the calculation is incorrect and this renders the final answer incorrect. AS students will sit two exam papers, each covering 50% of the content of the AS specification.

Glossary

α-glucose: glucose in which the hydrogen atom on carbon atom number one projects above the plane of the ring.

***ab initio* protein modelling:** a model is built based on the physical and electrical properties of the atoms in each amino acid in the sequence.

accuracy: how close a measured or calculated value is to the true value.

active immunity: where the immune system is activated and manufactures its own antibodies.

active site: an indented area on the surface of an enzyme molecule, with a shape that is complementary to the shape of the substrate molecule.

active transport: the movement of substances against their concentration gradient (from low to high concentration of that substance) across a cell membrane, using ATP and protein carriers.

adaptation: a characteristic that enhances survival in the habitat.

adhesion: the attraction between water molecules and the walls of the xylem vessel.

affinity: a strong attraction.

agglutination: the clumping of insoluble antigen molecules caused by crosslinking by antibodies that have a number of binding sites.

agglutinins: antibodies that cause pathogens to stick together.

allele: a version of a gene; also called gene variant.

alveoli: tiny folds of the lung epithelium to increase the surface area.

amino acids: monomers of all proteins, and all amino acids have the same basic structure.

amphiphilic: attracted to both water and fat – containing hydrophobic (lipophilic) and hydrophilic (lipophobic) parts.

amylopectin molecule: a molecule of polysaccharide with glycosidic bonds between carbon 1 and 4, and branches formed by glycosidic bonds between carbon 1 and 6. It is a constituent of starch.

amylose molecule: a molecule of polysaccharide with long straight chains of between 100 and 1000 α-glucose molecules. It is a constituent of starch. Like maltose, it has glycosidic bonds between carbon 1 and 4.

anatomical adaptations (anatomy): structural features.

anatomy: the branch of science concerned with studying the bodily structure of living organisms.

angina pectoris: a condition marked by severe pain in the chest, resulting from an inadequate blood supply, and therefore lack of oxygen, to the heart muscle that causes the coronary arteries to spasm (tighten).

anion: a negatively charged ion.

anomaly: result that does not fit the expected trend or pattern.

antibiotic: a chemical which prevents the growth of microorganisms. Antibiotics can be antibacterial or antifungal.

antibodies: specific proteins released by plasma cells that can attach to pathogenic antigens.

antigen-presenting cell: a cell that isolates the antigen from a pathogen and places it on the plasma membrane so that it can be recognised by other cells in the immune system.

antigen: a membrane-bound molecule used to recognise pathogens.

anti-toxins: antibodies that render toxins harmless.

aorta: the main artery of the body in mammals.

apoplast pathway: route by which water travels through the cell walls and in spaces between cells of plant tissue when travelling from roots to xylem and from xylem to leaves.

apoptosis: the death of cells which happens as a normal part of an organism's growth and development.

archaea: prokaryotic microorganisms of similar size to bacteria but having some differences of metabolism.

arithmetic mean: the average value of the numbers in a collection, found by dividing the sum of all the values by the number of values in the collection.

arteries: vessels that carry blood away from the heart.

arterioles: small blood vessels that distribute the blood from an artery to the capillaries.

artificial classification: a classification based on just one (or a few) characteristic(s).

artificial immunity: immunity that is achieved as a result of medical intervention.

artificial insemination: the medical or veterinary procedure of injecting semen, collected from a male animal, into the vagina or uterus of a female of the same species.

asexual reproduction: some multicellular organisms and single-celled protoctists such as *Amoeba* and *Paramecium* divide by mitosis to produce new individuals. They are genetically identical to the parent.

assimilates: substances that have become a part of the plant.

asymptomatic: not having any symptoms.

atria: thin-walled chambers of the heart that receive the blood from the veins, and then pass it to the ventricles.

atrio-ventricular node (AVN): a patch of tissue, in the heart, at the top of the septum that conducts the excitation wave from the atria to the ventricles.

atrio-ventricular valves: valves between the atria and the ventricles, which ensure that blood flows in the correct direction.

β-glucose: glucose in which the hydrogen atom on carbon atom number one projects below the plane of the ring.

B memory cells: cells that remain in the blood for a long time, providing long-term immunity.

bacteria: the plural of bacterium. Also see **prokaryotic microorganisms**.

bacterium: a member of a large group of unicellular microorganisms that have cell walls made of murein but lack membrane-bound organelles and a nucleus. Their DNA floats free in the cytoplasm.

behavioural adaptations: the ways that behaviour is modified for survival.

binary fission: a type of division found in prokaryotic cells and organelles such as chloroplasts and mitochondria.

binomial system: a system that uses the genus name and the species name to avoid confusion when naming organisms.

biodiversity: a measure of the variation found in the living world.

blood: the fluid used to transport materials around the body.

Bohr effect: the effect that extra carbon dioxide has on the haemoglobin, explaining the release of more oxygen.

Bohr shift: a change in the shape of the haemoglobin dissociation curve in the presence of carbon dioxide.

bordered pits: the part of plant cell walls which allow the exchange of fluids between tracheids or vessel elements.

bradycardia: a slow heart rhythm.

breathing rate: the number of breaths per minute.

bronchi and **bronchioles:** smaller airways leading into the lungs.

buccal cavity: the mouth.

buffer: a solution that resists changes in pH, so keeps the pH stable.

callose: a large polysaccharide deposit that blocks old or damaged phloem sieve tubes.

canker: a sunken lesion in tree bark caused by necrosis.

capillaries: very small vessels with very thin walls.

carbaminohaemoglobin: a compound of haemoglobin and carbon dioxide, and is one of the forms in which carbon dioxide exists in the blood, within red blood cells. 10% of carbon dioxide is carried in blood this way.

carbohydrates: a group of molecules containing C, H and O.

carbonic acid: a very weak acid formed when carbon dioxide reacts with water.

carbonic anhydrase: the enzyme that catalyses the combination of carbon dioxide and water.

cardiac cycle: the sequence of events in one full beat of the mammalian heart.

cardiac muscle: specialised muscle found in the walls of the heart chambers.

cartilage: a form of connective tissue.

Casparian strip: an impermeable, waterproof substance (suberin) in the walls of the endodermal cells of plant roots. It creates a water tight seal between the cells, preventing water entering the xylem via the apoplast pathway.

catalyst: chemical that speeds up the rate of a reaction and remains unchanged and reusable at the end of the reaction.

cation: a positively charged ion.

chloride shift: the movement of chloride ions into the erythrocytes to balance the charge as hydrogencarbonate ions leave the cell.

chromatids: replicates of chromosomes.

chromatography: a technique for the separation of a mixture by passing it in solution or suspension through a medium in which the components of the mixture move at different rates.

circulatory system (double): one in which the blood flows through the heart twice for each circuit of the body.

circulatory system (single): one in which the blood flows through the heart once for each circuit of the body.

ciliated epithelium: a layer of cells that have many hair-like extensions called cilia.

CITES: Convention on International Trade in Endangered Species.

class: a taxonomic group of organisms that all possess the same general traits, e.g. the same number of legs.

classification: the process of placing living things into groups.

climate change: significant, long-lasting changes in weather patterns.

clonal expansion: an increase in the number of cells by mitotic cell division.

clonal selection: selection of a specific B or T cell that is specific to the antigen.

closed circulatory system: one in which the blood is held in vessels.

coenzymes: small organic non-protein molecules that bind *temporarily* to the *active site* of enzyme molecules, either just before or at the same time that the substrate binds.

cofactor: a substance that has to be present to ensure that an enzyme-catalysed reaction takes place at the appropriate rate. Some cofactors (**prosthetic groups**) are part of the enzyme structure, and others (mineral ion cofactors and organic **coenzymes**) form temporary associations with the enzyme.

cohesion: the attraction between water molecules caused by hydrogen bonds.

collenchyma cells: cells that have thick cellulose walls and strengthen vascular bundles and outer parts of stems, whilst also allowing some flexibility in these regions.

colorimeter: an instrument for measuring the absorbance of different wavelengths of light in a solution.

common ancestor: the most recent individual from which a set of organisms in a group are directly descended.

companion cells: plant cells that help to load sucrose into the sieve tubes.

comparative protein modelling: one approach is protein threading, which scans the amino acid sequence against a database of solved structures and produces a set of possible models which would match that sequence.

competitive inhibition: inhibition of an enzyme, where the inhibitor molecule has a similar shape to that of the substrate molecule and competes with the substrate for the enzyme's active site. It blocks the active site and prevents formation of enzyme-substrate (ES) complexes.

computer modelling: a model of a process which is created on a computer, often used for processes that can need the increased calculation speed.

concentration: the abundance of molecules per unit volume.

concentration gradient: a measurement of how the concentration of a substance changes from one place to another, often across a membrane.

condensation: the conversion of a vapour or gas to a liquid.

condensation reaction: reaction that occurs when two molecules are joined together with the removal of water.

conformational change: a change in the shape of a macromolecule.

conjugated protein: a protein associated with a non-protein component.

connective tissue: a widely distributed animal/mammalian tissue consisting of cells in an extracellular matrix of protein and polysaccharide; includes bone, cartilage and blood; areolar and adipose tissue.

conservation *ex situ*: conservation outside the normal habitat of the species.

conservation *in situ*: carrying out active management to maintain the biodiversity in the natural environment.

continuous variation: variation where there are two extremes and a full range of values in between.

convergent evolution: the process whereby organisms not closely related independently evolve similar traits as a result of being adapted to similar environments or ecological niches.

coronary arteries: arteries supplying blood to the heart muscle.

correlation coefficient: a measure of how closely two sets of data are correlated. A value of 1 means perfect correlation.

cotransport: transport across a cell membrane, using a carrier or channel protein, of two substances, both moving in the same direction – for example, both moving into the cell.

countercurrent flow: where two fluids flow in opposite directions.

Countryside Stewardship Scheme: a scheme to encourage farmers and other land owners to manage parts of their land in a way that promotes conservation.

covalent bonds: formed when electrons are shared between atoms. These bonds are very strong.

crenated: a shrivelled animal cell that has lost water by osmosis.

cytochrome *c*: a type of cytochrome; an iron-containing protein found within inner mitochondrial membranes and that forms part of the electron transport chain.

cytokines: hormone-like molecules used in cell signalling to stimulate the immune response.

cytokinesis: cytoplasmic division following nuclear division, resulting in two new daughter cells.

cytolysis: the process in animal cells where, if a lot of water molecules enter, the cell will swell and burst as the plasma membrane breaks.

cytology: the study of cell structure and function.

cytoskeletal motor proteins: molecular motors such as myosins, kinesins and dyneins.

datalogger: an electronic device that records data over time or in relation to location either with a built-in instrument or sensor or via external instruments and sensors.

denaturation: a process in which proteins lose their tertiary structure and can no longer function. Their shape unravels due to extremes of pH or heat.

denatured: the irreversible change of shape/loss of tertiary structure of proteins; caused by high temperatures or extremes of pH.

deoxyribose: a five-carbon sugar derived from the five-carbon sugar ribose by replacement of a hydroxyl group by hydrogen, at carbon atom 2.

diaphragm: a layer of muscle beneath the lungs.

dicotyledonous plants: plants with two seed leaves and a branching pattern of veins in the leaf.

diastole: the relaxing phase of the cardiac (heartbeat) cycle.

differential staining: stains that bind to specific cell structures, staining each structure differently so the structures can be easily identified within a single preparation.

differentiation: process by which stem cells become specialised into different types of cell.

diffusion: movement of molecules from an area of high concentration of that molecule to an area of low concentration; it may or may not be across a membrane; it does not involve metabolic energy (ATP).

digestive system: the organs and glands in the body that are responsible for digestion beginning with the mouth and extending through the oesophagus, stomach, small intestine, and large intestine, ending with the rectum and anus.

dilate: to make or become wider, larger, or more open.

diploid: cell in which the nucleus has two complete sets of chromosomes.

direct transmission: passing a pathogen from host to new host, with no intermediary.

disaccharides: any of a class of sugars whose molecules contain two monosaccharide residues joined by a condensation reaction.

discontinuous variation: where there are distinct categories and nothing in between.

dissection: to cut apart tissues, organs or organisms for visual or microscopic study of their structure.

dissociation: releasing the oxygen from oxyhaemoglobin.

disulfide links: also called disulfide bridges or disulfide bonds; strong covalent bond (where electrons are shared) between two sulfur atoms, within a (protein) molecule. These bonds are not broken by heat but can be broken by reducing agents.

DNA polymerase: enzyme that catalyses formation of DNA from activated deoxyribose nucleotides, using single-stranded DNA as a template.

domain: the highest taxonomic rank. There are three domains: Archaea, Eubacteria and Eukaryotae.

double helix: shape of DNA molecule, due to coiling of the two sugar–phosphate backbone strands into a right-handed spiral configuration.

ecosystem a community of interacting organisms and their physical environment.

ectopic heartbeat: an extra or an early beat of the ventricles.

elastic fibres: protein fibres that can deform and then recoil to their original size.

elastin: a type of protein made by cross-linking a polypeptide called tropoelastin. Tropoelastin has a coiled structure. The cross-linking and coiling make elastin a strong and extensible protein. It is found in structures in living organisms, such as elastic cartilage and ligaments, where they need to stretch or adapt their shape as part of life processes.

electrocardiogram: a trace that records the electrical activity of the heart.

electron micrograph: a photograph of an image seen using an electron microscope.

elliptocytosis: cells being more elliptical in shape than they usually are.

electrophoresis: the movement of charged particles/molecules in a fluid or gel under the influence of an electric field.

embryo-transfer: a step in the process of assisted reproduction in which embryos are placed into the uterus of a female with the intent to establish a pregnancy.

endemicity: refers to degree of a condition being endemic – always present in an area/community.

endocytosis: bulk transport of molecules, too large to pass through a cell membrane even via channel or carrier proteins, into a cell.

endothelium: the inner layer or lining of a blood vessel, made of a single layer of cells.

environmental variation: variation caused by response to environmental factors such as light intensity.

endodermis: a layer of cells surrounding the vascular tissue in the root of a plant.

enzyme cascade: a sequence of successive activation reactions involving enzymes, which is characterised by a series of amplifications stemming from an initial stimulus. The product of each preceding reaction catalyses the next reaction.

enzyme-product complex: enzyme molecule with product molecule(s) in its active site. The two are joined temporarily by non-covalent forces.

enzyme-substrate complex: enzyme molecule with substrate molecule(s) in its active site. The two are joined temporarily by non-covalent forces.

epidemic: a rapid spread of disease through a high proportion of the population.

epidermal tissue: tissue consisting of epidermal cells – cells that form the outer layer of cells of a multicellular organism. Usually has a protective function.

epithelial cells: cells that constitute lining tissue in animals.

epithelial tissues: lining or covering tissue, consisting of epithelial cells.

erythrocyte: a red blood cell.

ester bond: a bond formed by condensation reaction between the –OH group of a carboxylic acid and the –OH group of an alcohol, to produce an ester

ES complex: see **enzyme-substrate complex**

Eubacteria: taxonomic domain, consisting of true bacteria – prokaryotic microorganisms not archaea.

Eukaryotae: taxonomic domain consisting of organisms that have eukaryotic cells.

eukaryotic: cells having a true nucleus/organisms with eukaryotic cells.

eukaryotic cell cycle: series of events in a eukaryotic cell leading to its replication to produce two daughter cells; consists of interphase, mitosis and cytokinesis.

evaporation: the change of state of a liquid into a vapour at a temperature below the liquid's boiling point. Evaporation occurs at the surface of a liquid where some molecules of liquid with high kinetic energy escape.

evolution: the gradual process by which the present diversity of living organisms has developed from earlier forms during the last 3000 million years of the history of the Earth.

exocytosis: the bulk transport of molecules, too large to pass through a cell membrane even via channel or carrier proteins, out of a cell.

extant still in existence; surviving.

extinction: when the last living member of a species dies and the species ceases to exist.

extracellular: outside the cell.

eyepiece graticule: a measuring device. It is placed in the eyepiece of a microscope and acts as a ruler when you view an object under the microscope.

facilitated diffusion: movement of molecules from an area of high concentration of that molecule to an area of low concentration, across a partially permeable membrane via protein channels or carriers; it does not involve metabolic energy (ATP).

family: a group of closely related genera, e.g. within the order Carnivora we might recognise the 'dog' family and the 'cat' family.

fatty acids: have a carboxyl group (–COOH) on one end, attached to a hydrocarbon tail, made of only carbon and hydrogen atoms. This may be anything from 2 to 20 carbons long.

fertilisation: the fusion of male and female gamete nuclei.

fetal haemoglobin: the type of haemoglobin usually found only in the fetus.

fibrillation: uncoordinated contraction of the atria and ventricles.

fibrous protein: has a relatively long, thin structure, is insoluble in water and metabolically inactive, often having a structural role within an organism.

filaments: slender branches of tissue that make up the gill. They are often called primary lamellae.

flaccid: plant tissue where there is no turgor – the tissue is soft.

fluid mosaic model: theory of cell membrane structure with proteins embedded in a sea of phospholipids.

gamete: sex cell, e.g. ovum/spermatozoon.

gene: a length of DNA that codes for a polypeptide or for a length of RNA that is involved in regulating gene expression.

genetic erosion: a process whereby an already limited gene pool of an endangered species of plant or animal diminishes even more when individuals from the surviving population die off without getting a chance to meet and breed with others in their endangered low population.

genetic variation: variation caused by possessing a different combination of alleles.

genome: the total DNA content of a cell or an individual.

genus: a group of closely related species.

gill filaments: slender branches of tissue that make up the gill of a fish. They are often called primary lamellae.

glucose: a 6-carbon monosaccharide sugar.

glycerol: has three carbon atoms. It is an alcohol, which means it has free –OH groups.

globular protein: has molecules of a relatively spherical shape, which are soluble in water, and often have metabolic roles within the organism.

glycocalyx: all the carbohydrate molecules on the exterior of a cell surface membrane.

glycogen molecule: the energy store in humans; large polysaccharide molecule made of many glucose residues joined by condensation reactions and like amylopectin, has glycosidic bonds between carbon atoms 1 and 4, and branches formed by glycosidic bonds between carbon atoms 1 and 6.

glycolipid: lipid/phospholipid with a chain of carbohydrate molecules attached.

glycoprotein: protein with a chain of carbohydrate molecules attached.

glycosidic bond: a bond formed between two monosaccharides by a condensation reaction.

growth factors: a substance, such as a vitamin, hormone or cytokinin, which is required for the stimulation of growth in living cells.

goblet cells: cells that secrete mucus.

guard cell: in leaf epidermis, two of these cells surround stomata.

habitat: where an organism lives.

haemoglobin: the red pigment used to transport oxygen in the blood.

haemoglobinic acid: the compound formed by the buffering action of haemoglobin as it combines with excess hydrogen ions.

haemolysis: lysis of animal cells, in this case it is referring to lysis of red blood cells.

haemolytic anaemia: anaemia with chronic premature destruction of red blood cells.

haploid: having only one set of chromosomes; represented by the symbol 'n'.

helicase: enzyme that catalyses the breaking of hydrogen bonds between the nitrogenous pairs of bases in a DNA molecule.

hazard: a factor that has the potential to cause harm.

herd vaccination: using a vaccine to provide immunity to all or almost all of the population at risk.

heterotropic ossification: overgrowth of bone, often in the wrong place, e.g. muscle tissue.

high-power drawing: a drawing showing detail of some individual cells.

histamine: a compound which is released by mast cells in response to injury and in allergic and inflammatory reactions, causing contraction of smooth muscle and dilation of capillaries.

homologous chromosomes: matching chromosomes, containing the same genes at the same places (loci). They may contain different alleles for some or all of the genes.

homozygosity: in a diploid cell or organism the state where both copies of a given gene are the same allele.

hormone: a chemical produced in glands and that travels to its target cells via the blood. Later broken down in the liver. Involved with communication and control.

hydathodes: Structures in plants that can release water droplets which may then evaporate from the leaf surface.

hydrocarbon: a compound consisting of only hydrogen and carbon.

hydrogen bond: a weak interaction that can occur wherever molecules contain a slightly negatively charged atom bonded to a slightly positively charged hydrogen.

hydrogencarbonate ion: HCO_3^-.

hydrolysis reaction: reaction that occurs when a molecule is split into two smaller molecules with the addition of water.

hydrophilic: attracted to water.

hydrophobic: repelled by water.

hydrophyte: a plant adapted to living in water or where the ground is very wet.

hydrostatic pressure: the pressure that a fluid exerts when pushing against the sides of a vessel or container.

hypertension: long term high blood pressure.

hyphae each of the branching filaments that make up the mycelium of a fungus.

immune response: the reaction of the cells and fluids of the body to the presence of a substance which is not recognised as a constituent of the body itself.

immune system: the organs and processes of the body that provide resistance to infection and toxins. Organs include the thymus, bone marrow, and lymph nodes.

***in vitro* fertilisation:** a process where an egg is surgically removed from the ovaries and fertilised with sperm in a laboratory.

indirect transmission: passing a pathogen from host to new host, via a vector.

inflammation: swelling and redness of tissue caused by infection.

inhibitor: a substance that reduces or stops a reaction.

inorganic ions: charged particles of inorganic (not carbon-based) substances, e.g. Mg^{2+}, Ca^{2+}.

integumentary system: the organ system that protects the body from various kinds of damage, such as loss of water or abrasion from outside. The system comprises the skin and its appendages (including hair, scales, feathers, hooves and nails).

intercalated discs: gap junctions between muscle cells in the heart muscle. They enable the heart muscle cells to fit tightly together and help to facilitate synchronised contraction of the heart muscle.

intercostal muscles: muscles between the ribs. Contraction of the external intercostal muscles raises the rib cage.

interleukins: signalling molecules that are used to communicate between different white blood cells.

interphase: phase of cell cycle where the cell is not dividing; it is subdivided into growth and synthesis phases.

interspecific variation: the differences between species.

intracellular: inside the cell.

intraspecific variation: the variation between members of the same species.

ionic bond: a type of chemical bond that involves the electrostatic attraction between oppositely charged ions.

karyotype: a photomicrograph of chromosomes in a cell.

keratin: a fibrous protein forming the main structural constituent of hair, feathers, hooves, nails, claws, horns.

keratinocytes: an epidermal cell that produces keratin.

keystone species: one that has a disproportionate effect upon its environment relative to its abundance.

kinetic energy: the energy of motion.

kingdom: taxonomic group; traditionally there are five main kingdoms: Plantae, Animalia, Fungi and Protoctista are all Eukarya whose cells possess a nucleus. All those single celled organisms that do not possess a nucleus are grouped into the kingdom Prokaryota.

lamellae: folds of the filament to increase surface area. They are also called secondary lamellae or gill plates.

leucocyte: a white blood cell.

lignification: the deposition of lignin in the walls of xylem vessels.

lignin: the waterproof substance that impregnates the walls of xylem vessels. When plant xylem vessels are lignified they are woody.

limiting factor: an environmental factor that limits the rate of a biological process. When a process is controlled by a number of factors, the factor that is in least supply will limit the process. If this factor is increased then the process will proceed at a faster rate. If it is decreased the process will proceed at a slower rate.

lipids: a group of substances that are soluble in alcohol rather than water. They include triglycerides, phospholipids, glycolipids and cholesterol.

lipophilic: attracted to fat (lipids).

lipophobic: repelled by fat (lipids).

locus: the position of a gene on a chromosome.

longitudinal section: a section cut lengthways.

low-power plan: a drawing showing distribution of cells but no individual cells shown.

lymph: the fluid held in the lymphatic system, which is a system of tubes that returns excess tissue fluid to the blood system.

lymphatic system: a network of vessels and organs that help maintain the internal fluid environment of the body; also transports fat and proteins and makes some blood cells. Receives tissue fluid that has passed out of blood capillaries and bathed cells. Lymph drains into blood vessels in neck region. Lymph organs include the tonsils, thymus, spleen and lymph nodes.

macromolecule: a very large, organic molecule.

macrophages: large phagocytic cells that ingest and digest pathogens and present the pathogen's antigens to other cells of the immune system.

magnification: the number of times larger an image appears, compared to the size of the object.

marine conservation zones: areas of the sea set aside to conserve the diversity of species and habitats.

median: the number that separates a data set into two halves; half the set is above the median value and the other half is below the median value. In a sample there may be no specimen that actually has the median value.

meiosis: type of nuclear division that results in the formation of cells containing half the number of chromosomes of the parent cell.

meristem: an area of unspecialised cells, in plants, that can divide and differentiate into other cell types.

mesenchyme: connective tissue.

mesoderm: the middle of the three layers in the early embryo; gives rise to connective tissue, muscles and part of the gonads (ovaries and testes).

mesophyll: a type of cell found in plant leaves.

metabolic/metabolism: the chemical reactions that take place inside living cells or organisms.

micrometer: a precise measuring device and not a unit of measurement. Note the difference in spelling. It is a small scale on a microscope slide that can be viewed under a microscope and used to calibrate the value of eyepiece divisions at different magnifications.

micrometre: equal to one millionth (10^{-6}) of a metre. It is the standard unit for measuring cell dimensions. Note the difference in spelling between micrometre and micrometer.

microscopy: the use of the microscope to study small organisms/objects.

mitosis: type of nuclear division that produces daughter cells genetically identical to each other and to the parent cell.

mode: the most common value amongst a group/data set.

monoculture: a crop consisting of one strain of one species.

monocytes: the largest white blood cells; usually have a large kidney-shaped nucleus.

monomer: a small molecule which binds to many other identical molecules to form a polymer.

monosaccharide: any of the class of sugars (e.g. glucose) that cannot be hydrolysed to give a simpler sugar.

mucous membrane: specialised epithelial tissue that is covered by mucus.

muscle tissue: highly cellular, well vascularised (has many blood vessels) tissues responsible for most types of body movement. Muscle cells are called fibres, contain the proteins actin and myosin, and can contract. Three types of muscle tissue: smooth, skeletal and cardiac.

musculo-skeletal system: the combination of the skeletal muscles and skeleton working together; includes the bones, muscles, tendons and ligaments of the body.

mutation: a change to the genetic material of an organism, either to a gene or to a chromosome. The changing of the structure of a gene, resulting in a variant allele that may be transmitted to subsequent generations; caused by the alteration of single base units in DNA, or the deletion, insertion, or rearrangement of larger sections of genes or chromosomes. May involve loss of a portion of a chromosome, or an abnormal chromosome number.

mycelium the vegetative part of a fungus, consisting of a network of fine white filaments (hyphae).

myocardial infarction: a heart attack.

myofibrils: microscopic fibres that make up the larger fibres of skeletal (striated) muscle.

myogenic muscle: muscle that can initiate its own contraction.

nanometre: one thousandth (10^{-3}) of a micrometre. It is therefore one thousand millionth (10^{-9}) of a metre. It is a useful unit for measuring the sizes of small organelles within cells and for measuring the size of large molecules.

natural classification: a classification that reflects the evolutionary relationships between organisms.

natural immunity: immunity achieved through normal life processes.

natural selection: the term used to explain how features of the environment apply a selective force on the reproduction of individuals in a population.

necrosis: cell death caused by disease or injury; it may subsequently limit the spread of a pathogen.

nervous system: the central nervous system (brain and spinal cord) and the peripheral nervous system – the network of nerve cells (neurones) that transmit nerve impulses between parts of the body and the central nervous system. Fast acting control system to detect stimuli and to bring about responses in muscles and glands.

nervous tissue: the main component of the nervous system. Consists of neurones and supporting cells.

neutrophil: a type of white blood cell that engulfs foreign matter and traps it in a large vacuole (phagosome), which fuses with lysosomes to digest the foreign matter.

non-competitive inhibition: the inhibition of an enzyme where the competitor molecule attaches to a part of the enzyme molecule but NOT the active site. It changes the shape of the active site and prevents ES complexes forming as the enzyme active site is no longer complementary in shape to the substrate molecule.

nucleotide: molecule consisting of a five-carbon sugar, a phosphate group and a nitrogenous base.

oncotic pressure: the pressure created by the osmotic effects of the solutes in a solution.

open circulatory system: one in which the blood is not held in vessels .

operculum: a bony flap that covers and protects the gills.

opsonins: proteins that bind to the antigen on a pathogen and then allow phagocytes to bind.

optimum: best.

optimum pH: the pH at which an enzyme works best, at its fastest rate.

optimum temperature: the temperature at which an enzyme works best, at its fastest rate.

order: taxonomic group; a subdivision of the class using additional information about the organisms, e.g. the class Mammalia is divided into meat-eating mammals (order Carnivora) and vegetation-eating mammals (order Herbivora).

organ: collection of tissues working together to perform a function/related functions.

organ system: a number of organs working together to carry out an overall life function.

organelles: small structures within cells, each of which carries out a specific function.

osmosis: passage of water molecules down their water potential gradient, across a partially permeable membrane.

ossification: process of changing cartilage to bone by depositing calcium phosphate.

ostia: pores in the heart of an insect that allow blood from the body to enter the heart.

ovalocytosis: cells being more oval in shape than they usually are.

oxygen tension: measured in units of pressure (kPa). See partial pressure.

oxygen uptake: the volume of oxygen absorbed by the lungs in one minute.

oxyhaemoglobin: a molecule of haemoglobin with oxygen molecules loosely bound to it. When haemoglobin takes up oxygen, it becomes oxyhaemoglobin.

palisade cells: closely-packed photosynthetic cells within leaves.

pandemic: an infectious disease which spreads rapidly across continents.

partial pressure: the concentration of oxygen is measured by the relative pressure that it contributes to a mixture of gases. This is called the partial pressure of oxygen or pO_2.

parenchyma: a packing tissue in plants which fills spaces between other tissues. In roots parenchyma cells may store starch. In leaves some (called chlorenchyma) have chloroplasts and can photosynthesise. In aquatic plants aerenchyma tissue is parenchyma with air spaces to keep the plant buoyant.

partially permeable: allows some, but not all, substances to pass through.

passive immunity: immunity achieved when antibodies are passed to the individual through breast feeding or injection.

pathogen: a microorganism that causes disease.

penicillin: an antibiotic or group of antibiotics produced naturally by certain mould fungi, now usually prepared synthetically.

pepsin: an enzyme that digests protein in the stomach of mammals.

peptide bond: a bond formed when two amino acids are joined by a condensation reaction.

pericycle: a thin layer of meristem tissue between the endodermis and the phloem in plant roots.

peristalsis: the involuntary contraction and relaxation of the muscle layers of the intestine or another canal within the body, creating wave-like movements which push the contents of the canal forward.

personalised medicine: the development of designer medicines for individuals.

phagocytosis: a type of endocytosis where large solid particles or small organisms are taken into a cell.

phloem: tissue that carries products of photosynthesis, in solution, within plants.

phospholipid: molecule consisting of glycerol, two fatty acids and one phosphate group.

photomicrograph: a photograph of an image seen using an optical microscope.

phylogeny: the study of the evolutionary relationships between organisms.

phylum: a major subdivision of the kingdom. A phylum contains all the groups of organisms that have the same body plan, e.g. possession of a backbone.

physiological adaptations (physiology): affect the way that processes work.

pinoendocytosis: if cells ingest liquids by endocytosis, this is called pino(endo)cytosis.

pinocytosis: a type of endocytosis where liquid is taken into a cell.

plasma: the fluid portion of the blood.

plasma cells: derived from the B lymphocytes, these are cells that manufacture antibodies.

plasma membrane: the cell surface membrane.

plasmodesmata: gaps in the cell wall containing cytoplasm that connects two cells.

plasmolysed: plant cell where the contents have shrunk due to loss of water by osmosis and the plasma membrane has separated from the cell wall.

plasmolysis: the process in which the protoplast of a plant cell shrinks as a result of water loss and the plasma membrane detaches from the cell wall.

platelets: small colourless disc-shaped cell fragments without a nucleus, found in large numbers in blood and involved in clotting.

pluripotent: able to differentiate into any type of cell.

polar: where the charge is not evenly distributed across the particle.

polymer: a large molecule made from many smaller molecules called monomers.

polymorphic gene locus: a locus that has more than two alleles.

polynucleotide: a large molecule consisting of many nucleotides.

polypeptide: polymer made of many amino acid units joined together by peptide bonds. Insulin is a polypeptide of 51 amino acids.

polysaccharides: polymers of monosaccharides that are made of hundreds or thousands of monosaccharide monomers bonded together.

potometer: a device that can measure the rate of water uptake as a leafy stem transpires.

precision: the closeness of agreement between measured values obtained by repeated measurements.

primary defences: those that prevent pathogens entering the body.

primary immune response: the initial response caused by a first infection.

primary structure: the sequence of amino acids found in a protein molecule.

prokaryotic microorganisms: unicellular microorganisms which have cell walls made of murein but lack membrane-bound organelles and an organised nucleus.

product: molecule produced from substrate molecules, by an enzyme-catalysed reaction.

prosthetic group: a non-protein component that forms a permanent part of a functioning protein molecule.

proteins: large polymers comprised of long chains of amino acids.

protoctista: phylum containing many eukaryotic organisms that cannot be classed as fungi, animals or plants.

pulmonary artery: the artery carrying blood from the right ventricle of the heart to the lungs for oxygenation.

pulmonary vein: the vein carrying oxygenated blood from the lungs to the left atrium of the heart.

Purkyne tissue: consists of specially adapted muscle fibres that conduct the wave of excitation from the AVN down the septum to the ventricles.

Q_{10}: temperature coefficient, calculated by dividing the rate of reaction at $(T + 10)$ °C by rate of the reaction at T °C.

quadrat: a simple square frame used to define the sample area.

qualitative testing: test that shows the presence or absence of a substance but does not indicate how much of the substance is present.

qualitative data: data that does not involve quantity (numbers).

quantitative data: data that does involve quantity (numbers).

quaternary structure: protein structure where a protein consists of more than one polypeptide chain. For example, insulin has a quaternary structure.

random error: statistical fluctuations in the measured data due to the precision limitations of the measurement device.

regenerative medicine: stem cells may be used to populate a bioscaffold of an organ, and then directed to develop and grow into specific organs for transplanting.

reproductive system: a collection of organs that work together for the purpose of producing a new organism, including gonads (ovaries and testes) and uterus.

residual volume: the volume of air that remains in the lungs after forced expiration.

resolution: the clarity of an image; the higher the resolution the clearer the image.

respiration: a process in living organisms involving the release of energy from food; aerobic respiration occurs in the presence of oxygen and involves the release of carbon dioxide; anaerobic respiration occurs in the absence of oxygen and may involve the production of carbon dioxide and ethanol, or the production of lactic acid; both aerobic and anaerobic respiration involve the oxidation of complex organic substances, such as glucose or fatty acids.

respiratory system: a biological system consisting of specific organs and structures used for the process of breathing, gaseous exchange and respiration in an organism.

ribose: a five-carbon sugar which occurs widely in nature as a constituent of nucleosides, RNA and several vitamins and coenzymes.

ribosome: a small organelle consisting of RNA and associated proteins found in large numbers in the cytoplasm of prokaryotic and eukaryotic cells, and on rough endoplasmic reticulum of eukaryotic cells. Each ribosome consists of two subunits. Amino acids are assembled into polypeptides at ribosomes, during a process called translation, where the ribosome moves along and reads instructions from a length of messenger RNA.

ring vaccination: used when a new case of a disease is reported. All people who have been in contact with/live close to, the patient are vaccinated.

risk: the level of exposure to a hazard.

risk assessment: a way of managing risks by reducing exposure to hazards.

root hair cells: epidermal cells of young roots with long hair-like projections.

sampling techniques: techniques used to collect samples, for example random sampling, systematic sampling and stratified sampling.

sarcomere: the smallest contractile unit of a muscle.

sclerenchyma cells: plant cells that have lignified walls and are used to strengthen stems and leaf midribs.

secondary defences: defences which attack pathogens that have entered the body.

secondary immune response: a more rapid and vigorous response caused by a second or subsequent infection by the same pathogen.

secondary structure: the coiling or folding of an amino acid chain, which arises often as a result of hydrogen bond formation between different parts of the chain. The main forms of secondary structure are the helix and the pleated sheet.

selective breeding: the selection of specific traits in plants or animals through breeding programmes.

semi-conservative replication: how DNA replicates, resulting in two new molecules, each of which contains one old strand and one new strand. One old strand is conserved in each new molecule.

semilunar valves: valves that prevent blood re-entering the heart from the arteries.

septum: the wall that separates two chambers.

sieve plates: the perforated end walls of the sieve tube elements in phloem.

sieve tube elements: make up the tubes in phloem tissue that carry sap up and down the plant. The sieve tube elements are separated by sieve plates.

significant figures: the digits of a number that have a meaning and contribute to the number's precision.

Simpson's index of diversity: a measure of the diversity of a species in a habitat.

sink: a part of the plant where materials are removed from the transport system; for example, the roots receive sugars and store them as starch. At another time of year, the starch may be converted back to sugars and transported to a growing stem – so the roots can also be a source!

sino-atrial node (SAN): the heart's pacemaker. It is a small patch of tissue that sends out waves of electrical excitation at regular intervals in order to initiate contractions.

sister chromatids: when a chromosome has replicated, the resulting two identical chromatids produced are sister chromatids.

smooth muscle: involuntary muscle that contracts without the need for conscious thought.

soil depletion: the loss of soil fertility caused by removal of minerals by continuous cropping.

source: a part of the plant that loads materials into the transport system; for example, the leaves photosynthesise and the sugars made are moved to other parts of the plant.

species: a group of organisms that can freely interbreed to produce fertile offspring.

species evenness: a measure of how evenly represented the species are.

species richness: a measure of how many different species are present.

spherocytosis: cells being more spherical in shape than they usually are.

spiracle: an external opening or pore that allows air in or out of the tracheae.

spirometer: a device that can measure the movement of air into and out of the lungs.

squamous epithelial cells: flattened epithelial cells arranged in a layer.

stage graticule: a precise measuring device. It is a small scale that is placed on a microscope stage and used to calibrate the value of eyepiece divisions at different magnifications.

standard deviation: a measure of the spread around a mean.

statins: chemicals that competitively inhibit an enzyme that catalyses the production of cholesterol in the liver.

stem cell: unspecialised cell able to express all of its genes and divide by mitosis.

stomata: minute pores in the epidermis of the leaf or stem of a plant, which allow movement of gases in and out of the intercellular spaces.

Student's *t*-test: a test used to compare two means.

substrate: molecule that is altered by an enzyme-catalysed reaction.

succulent: a plant that stores water in its stem which becomes fleshy and swollen.

surface area to volume ratio: the surface area of an organism divided by its volume, expressed as a ratio.

symbiosis: the relationship where two organisms coexist for their mutual benefit.

symplast pathway: the route taken by water as it moves from cell to cell via the cell cytoplasm.

synthesis: making large molecules from smaller ones.

synthetic biology: the re-engineering of biology. This could be the production of new molecules that mimic natural processes, or the use of natural molecules to produce new biological systems that do not exist in nature.

systematic error: errors that may be inherent in the equipment and are repeated at every replicate. However, if the percentage error is known, a calculation can be made to find the margin of error.

systemic circulation: the part of the cardiovascular system which carries oxygenated blood away from the heart to the body, and returns deoxygenated blood back to the heart.

systole: the phase of the heartbeat when the heart muscle contracts and pumps blood from the chambers e.g. from atria to ventricles or from ventricles into the arteries.

T helper cells: cells that release signalling molecules to stimulate the immune response.

T killer cells: cells that attack and destroy our own body cells that are infected by a pathogen.

T memory cells: cells that remain in the blood for a long time, providing long-term immunity.

T regulator cells: cells that are involved with inhibiting or ending the immune response.

tachycardia: a rapid heart rhythm.

temperature: the degree or intensity of heat present in a substance or object, especially as expressed according to a comparative scale.

tendinous cords: cord-like tendons in heart ventricles that connect the papillary muscles in the ventricle walls to the tricuspid valve or to the bicuspid valve in the heart, to prevent the valves turning inside out.

tertiary structure: the overall three-dimensional shape of a protein molecule. Its shape arises due to interactions including hydrogen bonding, disulfide bridges, ionic bonds and hydrophobic interactions.

tidal volume: the volume of air inhaled or exhaled in one breath, usually measured at rest.

tissue: group of cells that work together to perform a specific function/set of functions.

tissue fluid: the fluid surrounding the cells and tissues.

toxin: a substance that is poisonous to living cells/organisms.

trachea: the main airway leading from the back of the mouth to the lungs in mammals.

tracheal fluid: the fluid found at the ends of the tracheoles in the tracheal system in insects.

tracheal system: a system of air-filled tubes in insects.

transcription: the process of making messenger RNA from a DNA template.

transect: a line across a habitat.

translation: formation of a protein, at ribosomes, by assembling amino acids into a particular sequence according to the coded instructions carried from DNA to the ribosome by mRNA.

translocation: the transport of assimilates through a plant.

transmission: passing a pathogen from an infected individual to an uninfected individual.

transpiration: the loss of water vapour from the aerial parts of a plant, mostly through the stomata in the leaves.

transport: the movement of substances such as oxygen, nutrients, hormones, waste and heat around the body.

transverse section: a section cut crossways.

triglycerides: lipid molecules consisting of glycerol, and three fatty acids.

turgid: a swollen state of plant cells that have taken in water by osmosis and reached their maximum state of swelling. The cell wall now exerts a pressure and prevents any more water entering the cell.

tylose: an outgrowth from parenchyma cells into xylem vessels (vascular tissue used for water and mineral transport throughout a plant) which can block the vessel.

ultrastructure: structures within cells.

unsaturated fatty acid: fatty acid lacking the full complement of hydrogens. There are double bonds between some of the adjacent carbon atoms, giving a kink in the long hydrocarbon chain.

urinary system: also known as the renal system. Consists of the two kidneys, ureters, the bladder, and the urethra. Each kidney consists of millions of functional units called nephrons. This system deals with removal of nitrogenous waste and with osmoregulation.

vaccination: a way of stimulating an immune response so that immunity is achieved.

vacuolar pathway: the route taken by water as it passes from cell to cell via the cytoplasm and vacuoles.

validity: if an investigation provides the answer to the research question it is valid. It is measuring what it says it is measuring. For an investigation to be valid control variables have to be controlled so that only the independent variable is causing the change.

variation: the presence of variety – the differences between individuals.

vascular tissue: consists of cells specialised for transporting fluids by mass flow.

vasodilation: an increase in the diameter of the lumen of arterioles allowing increased blood flow to specific regions of the body.

vector: an organism that carries a pathogen from one host to another.

veins: vessels that carry blood back to the heart.

vena cava: a large vein carrying deoxygenated blood into the heart. There are two in humans, the inferior vena cava (carrying blood from the lower body) and the superior vena cava (carrying blood from the head, arms, and upper body).

venom: a harmful secretion injected into a victim by a venomous organism such as a snake, insect or spider.

ventilation: the refreshing of the air in the lungs, so that there is a higher oxygen concentration than in the blood, and a lower carbon dioxide concentration .

ventricles: the lower chambers of the heart, their walls create high pressure to pump the blood out of the heart and into arteries.

venules: small blood vessels that collect blood from capillaries and lead into the veins.

viable: capable of surviving or living successfully, especially under particular environmental conditions, relating to a plant, animal, or cell.

vital capacity: the greatest possible volume of air that can be expelled from the lungs after taking the deepest possible breath.

water potential: measure of the tendency of water molecules to diffuse from one region to another.

water vapour potential gradient: a difference in the concentration of water (vapour) molecules inside the leaf compared to outside.

waxy cuticles: these prevent water collecting on the cell surfaces. Since pathogens collect in water and need water to survive, the absence of water is a passive defence.

wildlife reserves: areas set aside for the conservation of species or habitats.

xerophyte: a plant adapted to living in dry conditions.

xylem: tissue that carries water and mineral ions from the roots to all parts of the plant.

xylem vessels: the tubes which carry water up the plant.

zygote: cell produced after fertilisation of two gametes during sexual reproduction.

Index